ASYMPTOTIC METHODS IN SINGULARLY PERTURBED SYSTEMS

MONOGRAPHS IN CONTEMPORARY MATHEMATICS
Formerly CONTEMPORARY SOVIET MATHEMATICS
Series Editor: Revaz Gamkrelidze, Steklov Institute, Moscow, Russia

ASYMPTOTIC METHODS IN SINGULARLY PERTURBED SYSTEMS
E. F. Mishchenko, Yu. S. Kolesov, A. Yu. Kolesov, and N. Kh. Rozov

ASYMPTOTICS OF OPERATOR AND PSEUDO-DIFFERENTIAL EQUATIONS
V. P. Maslov and V. E. Nazaikinskii

COHOMOLOGY OF INFINITE-DIMENSIONAL LIE ALGEBRAS
D. B. Fuks

DIFFERENTIAL GEOMETRY AND TOPOLOGY
A. T. Fomenko

HOMOLOGY OF ANALYTIC SHEAVES AND DUALITY THEOREMS
V. D. Golovin

LINEAR DIFFERENTIAL EQUATIONS OF PRINCIPAL TYPE
Yu. V. Egorov

THE OBLIQUE DERIVATIVE PROBLEM OF POTENTIAL THEORY
A. I. Yanushauskas

OPTIMAL CONTROL
V. M. Alekseev, V. M. Tikhomirov, and S. V. Fomin

THEORY OF OPERATORS
V. A. Sadovnichiĭ

THEORY OF SOLITONS: The Inverse Scattering Method
S. Novikov, S. V. Manakov, L. P. Pitaevskii, and V. E. Zakharov

TOPICS IN MODERN MATHEMATICS: Petrovskii Seminar No. 5
Edited by O. A. Oleinik

ASYMPTOTIC METHODS IN SINGULARLY PERTURBED SYSTEMS

E. F. Mishchenko
Steklov Mathematical Institute
Moscow, Russia

Yu. S. Kolesov
A. Yu. Kolesov
Yaroslavl' State University
Yaroslavl', Russia

N. Kh. Rozov
Moscow State University
Moscow, Russia

Translated from Russian by
Irene Aleksanova

CONSULTANTS BUREAU • NEW YORK AND LONDON

Library of Congress Cataloging-in-Publication Data

Asymptotic methods in singularly perturbed systems / E.F. Mishchenko
  ... [et al.].
      p.   cm. -- (Monographs in contemporary mathematics)
   Includes bibliographical references and index.
   ISBN 0-306-11034-2
   1. Differential equations--Asymptotic theory. I. Mishchenko, E.
F. (Evgeniĭ Froĭovich) II. Series.
QA372.A83 1994
515'.35--dc20   94-27791
                                                                     CIP

This translation is published under an agreement with the
Russian Copyright Agency (VAAP)

ISBN 0-306-11034-2

©1994 Consultants Bureau, New York
A Division of Plenum Publishing Corporation
233 Spring Street, New York, N.Y. 10013

All rights reserved

No part of this book may be reproduced, stored in a retrieval system, or transmitted in any form or
by any means, electronic, mechanical, photocopying, microfilming, recording, or otherwise, without
written permission from the Publisher

Printed in the United States of America

Preface

There are many books devoted to ordinary differential equations containing small parameters (small perturbations). The investigation of the dependence of solutions, in a finite time interval, on regular perturbations (the small parameter regularly appears on the right-hand sides of the equations) was carried out by Poincaré and was practically completed long ago. However, problems connected with singular perturbations still attract the attention of mathematicians.

This is what we understand by a singularly perturbed system: a system of differential equations dependent on a small parameter is said to be singularly perturbed if, as the parameter tends to zero, Cauchy's resolvent operator for the main range of time values and initial conditions from bounded sets (or the Poincaré operator) converges, in a suitable topology, to a limit object acting in a space of smaller dimension. In different cases this general idea of a singularly perturbed system becomes specific and leads to numerous important and interesting problems.

A certain class of these problems was only recently considered in monographic literature. This class includes problems connected with the so-called relaxation oscillations, a phenomenon well known to physicists, mechanicians, chemists, and ecologists. Van der Pol, Andronov, Haag, Dorodnitsyn, Stoker, Zheleztsov and others were the first to study relaxation oscillations. A comprehensive study of this phenomenon is hindered by considerable mathematical difficulties and requires the development of new asymptotic methods in the theory of differential equations. These methods, interesting in themselves, lead to the statement of new mathematical problems.

The first monograph devoted to these problems was the book by E. F. Mishchenko and N. Kh. Rozov entitled *Differential Equations with Small Parameters and Relaxation Oscillations* (Nauka, Moscow, 1975). (English translation: E. F. Mishchenko and N. Kh. Rozov, *Differential Equations with Small Parameters and Relaxation Oscillations*, Plenum Press, New York-

London, 1980). This monograph sums up the investigations begun in 1955 by Pontryagin and Mishchenko.

Another book later appeared in which relaxation oscillations were considered not as a specific illustrative example but as an object of study interesting in itself. It was the monograph by J. Grasman, *Asymptotic Methods for Relaxation Oscillations and Applications,* Applied Mathematical Sciences, **63**, Springer, New York, 1987. However, this book only contains a comprehensive review of special methods of solving separate problems published in scientific journals but does not suggest a systematic theory.

In particular, this monograph continues and completes, in a certain sense, the construction of a full asymptotic theory of relaxation oscillations begun in the works of Pontryagin, Mishchenko, and Rozov mentioned above.

In addition, it contains the results of investigation of a number of new problems, including problems connected with relaxation oscillations in systems of parabolic partial differential equations studied by Yu. S. Kolesov and A. Yu. Kolesov. The main results given in the book were obtained by us in recent years.

As far as possible we have taken into consideration many works by other authors (see References).

Below we give the contents of the book, chapter by chapter.

Introduction. Relaxation oscillations

Here we give, on an intuitive level, a general description of the phase space of a singularly perturbed multidimensional system of ordinary differential equations, and, in particular, of the mechanism of origination of relaxation oscillations in systems of this kind. This mechanism is illustrated by way of specific examples both in second-order systems (van der Pol's equation) and in multidimensional systems. The concepts of *fast* and *slow* motions are introduced and the notions of *junction points* and some others are introduced and widely used. The reader will need the material given here in order to grasp the content of Chapter 1.

Chapter 1. Theorem on the C^1-proximity of the solutions of a relaxation and a relay system and the asymptotics of relaxation oscillations

In this chapter, a singularly perturbed system is considered as a special small perturbation of the relay system which we construct in accordance with a certain law proceeding from a singularly perturbed system. This approach makes it possible to prove the theorem on the C^1-proximity of the solutions of a singularly perturbed and a relay system, and this, in turn, allows us to complete the solution, in the general form, of the problems concerned with the stability and local uniqueness of a relaxation cycle and with the computation of the full asymptotics of its trajectory and period that remained unsolved.

As an application of the theorem on the C^1-proximity, we have proved the existence (under definite conditions) of a countable number of dichotomous cycles for a relaxation system of this kind, in which x is of dimension 1 and y is of dimension 2.

The chapter is concluded with the solution of a problem on the tori of relaxation systems posed as long ago as the 1960's by Pontryagin and a new problem on relaxation tori.

Chapter 2. Relaxation oscillations in a medium with diffusion

Here we consider a problem on the oscillations of a singularly perturbed system of parabolic equations. Problems of the "reaction–diffusion" type may be of interest to those who carry out investigations in some fields of biophysics, chemistry, and ecology. In these fields, there is usually a spatially homogeneous cycle in mathematical models (i.e., a cycle independent of a space variable), and it is important to know its stability properties. The chapter begins with the theorem on the C^1-proximity of the solutions of the original system and the solutions of a certain parabolic relay system. This theorem allows us to obtain results for the original system similar to the results presented in Chapter 1 and, in particular, to establish criteria of stability and instability of a spatially homogeneous relaxation cycle.

In this chapter we analyze a number of mathematical models arising in applications.

Chapter 3. Structure of the neighborhood of a relaxation cycle

In this chapter, we show how a suitable modification of the Bogolyubov–Mitropol'skii asymptotic method can be used to construct a normal form of mappings in the neighborhood of a relaxation cycle.

The results obtained are used, for instance, to solve the problem on the bifurcation of spatially nonhomogeneous periodic solutions from a spatially homogeneous relaxation cycle. This can be regarded as an extension of the well-known Turing–Prigogine bifurcation theorem to the oscillation case for a new class of evolutionary equations.

Chapter 4. Duck-trajectories of relaxation systems

The so-called duck-trajectories were discovered and investigated by the methods of nonstandard analysis by a group of French mathematicians (F. Diener and M. Diener). In this chapter, these and even more general results are obtained by means of the classical methods of asymptotic analysis that we have developed. In particular, we have studied some phenomena connected with duck-trajectories that occur in the van der Pol system with harmonic external action on slow variables.

Here we also study the mechanism of the origination of a relaxation cycle from the equilibrium state. In particular, we investigate the relationship between this phenomenon and the phenomenon of the delay of the loss of stability discovered by Pontryagin.

Chapter 5. Nonclassical relaxation auto-oscillations

This chapter is devoted to singularly perturbed systems of ordinary differential equations known in applications as Lotka–Volterra systems. It turns out that under definite conditions nonstandard relaxation oscillations occur in these systems, namely, the fast component is δ-like and the slow component is close to a discontinuous periodic function.

A similar situation was found in systems that describe relaxation oscillations in a medium with diffusion, encountered when the well-known Belousov–Zhabotinskii reaction is simulated.

Chapter 6. Autowave processes in singularly perturbed systems of reaction–diffusion type

This chapter is based on the so-called method of quasinormal forms of constructing autowave processes in parabolic systems with small diffusion. It is substantiated with an essential use of a certain technique of investigating the problem of stability of solutions of linear parabolic equations which smoothly depend on a small parameter and a special technique of constructing multifrequency oscillations. The combination of these techniques makes it possible to prove the theorems on the bifurcation of invariant tori. As one of the applications of the obtained results, we want to mention the turbulence problem, more precisely, the consideration of the well-known hypothesis of L. D. Landau concerning the mechanism of the origination of chaos in dynamic systems.

Note that the numeration of the formulas begins anew in each section (the first digit is the number of the section and the second digit is the number of the formula within the section). The same refers to the theorems and other propositions and to the figures.

The Authors

Contents

Introduction. Relaxation oscillations 1

    I. Regular and singular dependence of solutions on
       a parameter 1

    II. Second-order relaxation systems 7

    III. Arbitrary-order relaxation systems 15

Chapter 1. Theorem on the C^1-proximity of the solutions of a relaxation and a relay system and the asymptotics of relaxation oscillations 28

    1. Statement of the problem, preliminary information,
       and heuristic arguments 29

    2. Proof of the theorem on the C^1-proximity.................... 40

    3. Principle of reduction in the neighborhood of a junction
       point 54

    4. Asymptotics of multidimensional relaxation oscillations
       in the neighborhood of a junction point 57

    5. Uniqueness, asymptotics, and stability of a relaxation cycle ... 77

    6. Stochastic character of the behavior of trajectories
       in three-dimensional relaxation systems 80

    7. Pontryagin's problem on the tori of singularly perturbed
       systems 82

    8. Stable relaxation tori in three-dimensional systems 87

Chapter 2. Relaxation oscillations in a medium with diffusion 91

    9. On a certain class of parabolic relay systems 92

    10. Theorem on the C^1-proximity for relaxation parabolic systems 95

    11. Criterion of stability of a homogeneous relaxation cycle 99

    12. Analysis of some mathematical models of biophysics 105

Chapter 3. Structure of the neighborhood of a relaxation cycle 108

    13. Normal form of mapping 108

    14. Intrinsic resonance upon the loss of stability by a homogeneous cycle of a relay system 114

    15. Bifurcation of a relaxation cycle 117

Chapter 4. Duck-trajectories of relaxation systems 120

    16. Origination of duck-trajectories upon the violation of the normal switching condition 121

    17. Destruction of an invariant torus of van der Pol's system with harmonic input 136

    18. Pontryagin delay phenomenon and stable duck-cycles 143

    19. Instability of duck-cycles of multidimensional relaxation systems 156

Chapter 5. Nonclassical relaxation auto-oscillations. 162

    20. Systems of Lotka–Volterra type 162

    21. Relaxation cycles of systems of Lotka–Volterra type 172

    22. Construction of complete asymptotics of trajectories 174

    23. Multidimensional systems of Lotka–Volterra type 191

    24. A new type of relaxation oscillations 199

Contents

Chapter 6. Autowave processes in singularly perturbed systems of reaction–diffusion type 206

    25. Exponential dichotomy of solutions of linear parabolic equations 207

    26. Estimates of bounded solutions of linear differential equations 210

    27. Stability of solutions of linear parabolic equations 216

    28. Bifurcation of invariant tori of parabolic systems with small diffusion.. 228

    29. Applications of bifurcation theorems........................ 250

    30. Diffusion chaos 259

    31. Dynamics of systems with a terminal turn point 263

References 273

Introduction.
Relaxation Oscillations

When we describe the operation of a real object or the course of a real process by a differential equation, we always pass from the object (the process) itself to its idealized model. When constructing an idealized mathematical model, we inevitably neglect various small quantities, and then the question as to the degree to which this neglect distorts the real picture becomes the principal one.

This gives rise to a mathematical problem of the dependence of the solutions of differential equations on small parameters.

In this Introduction, we describe, on an intuitive level, the phase space of a singularly perturbed system of ordinary differential equations and explain the mechanism of the origination of the so-called relaxation oscillations. We give two examples to illustrate this mechanism. The material given here is necessary for the understanding of the next chapter.

I. Regular and Singular Dependence of Solutions on a Parameter

I.1. Regular Dependence on a Parameter

Let us consider an autonomous system of ordinary differential equations

$$\dot{x}_\nu = \mathcal{F}_\nu(x_1, \ldots, x_n, \varepsilon), \qquad \nu = 1, \ldots, n, \tag{I.1}$$

or, in vector form,

$$\dot{x} = \mathcal{F}(x, \varepsilon). \tag{I.2}$$

Here $x = (x_1, \ldots, x_n)$ is an n-dimensional vector of the Euclidean space R^n;

$$\mathcal{F}(x, \varepsilon) = (\mathcal{F}_1(x, \varepsilon), \ldots, \mathcal{F}_n(x, \varepsilon))$$

is an n-dimensioanl vector function of the arguments x and ε. For simplicity, we consider the parameter ε to be scalar and assume that it is nonnegative and *small*, i.e., ε satisfies the condition

$$0 \leq \varepsilon \leq \varepsilon_0, \qquad (I.3)$$

where ε_0 is a sufficiently small number.

Suppose that the right-hand sides of system (I.1), i.e., the functions $\mathcal{F}_\nu(x_1,\ldots,x_n,\varepsilon)$, $\nu = 1,\ldots,n$, are defined and continuous at all points of a certain domain $G \subset R^{n+1}$ of variation of the variables $x_1,\ldots,x_n,\varepsilon$, where ε belongs to the interval (I.3). If $\varepsilon \neq 0$ and the fixed point $(x_0,\varepsilon) \in G$, then we denote by

$$x = \varphi(t,\varepsilon) \qquad (I.4)$$

the solution of system (I.2) with the initial condition $x_0 = \varphi(t_0,\varepsilon)$. In addition to system (I.2), we consider the system

$$\dot{x} = \mathcal{F}(x,0), \qquad (I.5)$$

resulting from (I.2) when $\varepsilon = 0$. Let

$$x = \varphi_0(t) \qquad (I.6)$$

be its solution with the same initial condition $x_0 = \varphi_0(t_0)$. We suppose that this solution is defined on some *finite* time interval

$$t_0 \leq t \leq T. \qquad (I.7)$$

For a small $\varepsilon \neq 0$ the right-hand sides of system (I.2) and system (I.5) are almost similar. Then the following question naturally arises: what is the difference between their solutions (I.4) and (I.6) respectively? In many cases interesting for practical applications it is answered by the following classical theorem (see [87], for instance).

THEOREM. *If the right-hand side of system (I.2) is continuously differentiable with respect to x and continuous in ε in the domain G, then, for a sufficiently small ε, solution (I.4) is defined on the same interval (I.7) as solution (I.6) and can be represented as*

$$\varphi(t,\varepsilon) = \varphi_0(t) + \mathcal{R}_o(t,\varepsilon),$$

where $\mathcal{R}_0(t,\varepsilon) \to 0$ as $\varepsilon \to 0$ uniformly with respect to t on the interval (I.7).

If, in the domain G, the right-hand side of system (I.2) is continuously differentiable $m \geq 1$ times with respect to all the variables x, ε, then for a sufficiently small ε, solution (I.4) can be represented as

$$\varphi(t,\varepsilon) = \varphi_0(t) + \varepsilon\varphi_1(t) + \ldots + \varepsilon^{m-1}\varphi_{m-1}(t) + \mathcal{R}_m(t,\varepsilon),$$

Introduction

where, as $\varepsilon \to 0$, $\mathcal{R}_m(t,\varepsilon) \to 0$ as a quantity of order ε^m uniformly with respect to t on interval (I.7).

If, in the domain G, the right-hand side of system (I.2) is an analytic function of its arguments, then, for a sufficiently small ε, solution (I.4) can be represented as a series

$$\varphi(t,\varepsilon) = \varphi_0(t) + \sum_{m=1}^{\infty} \varepsilon^m \varphi_m(t),$$

converging uniformly on interval (I.7).

Note that this theorem makes it possible to estimate the deviation of solution (I.4) from solution (I.6) only on the *finite* interval (I.7). Generally speaking, the deviation of these solutions on an *infinite* interval is not small.

I.2. Singular Dependence on a Parameter

The theorem formulated in I.1 for the estimation of the deviation of solutions is inapplicable in the case where the dependence of the right-hand sides of system (I.1) on the parameter ε is *discontinuous* (or nonsmooth).

This is, in particular, the case when normal systems of ordinary differential equations include a small positive parameter ε as a multiplier of a part of the derivatives, i.e., when we have systems of the form

$$\begin{cases} \varepsilon \dot{x}_i = f_i(x_1, \ldots, x_k, y_1, \ldots, y_m), & i = 1, \ldots, k, \\ \dot{y}_j = g_j(x_1, \ldots, x_k, y_1, \ldots, y_m), & j = 1, \ldots, m, \end{cases} \quad (I.8)$$

where f_i and g_j are smooth functions of all of their $k + m = n$ arguments. Obviously, if we rewrite system (I.8) in the form (I.1), then the right-hand sides of the first k equations will contain functions $\frac{1}{\varepsilon} f_i$, which, in general, increase indefinitely as $\varepsilon \to 0$. A system of form (I.8) is said to be *singularly perturbed*.

We set $x = (x_1, \ldots, x_k) \in R^k$, $y = (y_1, \ldots, y_m) \in R^m$ and write system (I.8) in vector form:

$$\begin{cases} \varepsilon \dot{x} = f(x,y), \\ \dot{y} = g(x,y). \end{cases} \quad (I.9)$$

Here $f(x,y)$, $g(x,y)$ are vector functions of dimensions k and m respectively and ε is a scalar small positive parameter, i.e., $0 < \varepsilon \leq \varepsilon_0$, where ε_0 is a sufficiently small number. Acting as we did in I.1, we introduce a solution

$$x = \varphi(t,\varepsilon), \quad y = \psi(t,\varepsilon) \quad (I.10)$$

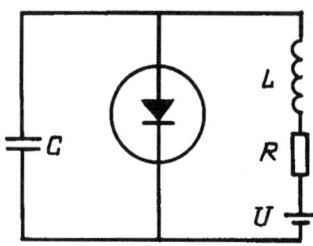

Fig. I.1

of system (I.9) satisfying the initial condition $x_0 = \varphi(t_0, \varepsilon)$, $y_0 = \psi(t_0, \varepsilon)$. However, if, alongside system (I.9), we consider a system formally obtained from (I.9) for $\varepsilon = 0$,

$$\begin{cases} 0 = f(x, y), \\ \dot{y} = g(x, y), \end{cases} \tag{I.11}$$

then we can easily see that, in general, it has no solution that would assume the initial value $x|_{t=t_0} = x_0$, $y|_{t=t_0} = y_0$.

In other words, in contrast to the situation we had in I.1, we do not now know for sure with what solution of system (I.11) the solution of (I.10) should be compared.

It should be especially pointed out that system (I.9) can be reduced to the form (I.2) in a way different from that indicated above, so that the parameter ε will smoothly enter into the right-hand side. For this to occur, it is sufficient to make a change of time $t = \varepsilon\theta$, and then we obtain

$$\frac{dx}{d\theta} = f(x, y), \qquad \frac{dy}{d\theta} = \varepsilon g(x, y). \tag{I.12}$$

The theorem from I.1 can certainly be applied to system (I.12), but this application is hardly of any use since it guarantees the proximity of solutions with the same initial values of system (I.12) and the system resulting from (I.12) for $\varepsilon = 0$ on a finite interval of the new time θ, i.e., only at an *infinitely small*, as $\varepsilon \to 0$, time interval t.

Let us consider two specific examples from physics that lead to normal systems of ordinary differential equations with a small parameter in some derivatives.

EXAMPLE I.1. Figure I.1 shows the principal scheme of a well-known physical instrument, a multivibrator on a tunnel diode, which serves as a

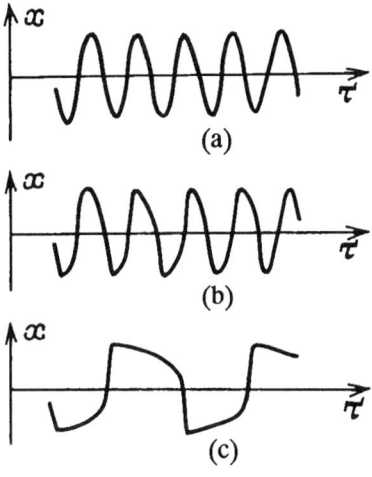

Fig. I.2

source of undamped periodic electric oscillations. With a certain idealization, *van der Pol's equation* can be used as a mathematical model of this oscillator (see [2]):

$$\frac{d^2x}{d\tau^2} + \lambda(x^2-1)\frac{dx}{d\tau} + x = 0, \tag{I.13}$$

where x is a quantity connected with the current intensity, τ is time, and the characteristics of the oscillator itself are described by one parameter λ.

We know that for any $\lambda > 0$, on the phase plane $(x, dx/d\tau)$, van der Pol's equation (I.13) has a unique limit cycle which is stable. This mathematical fact is adequate to the following experimentally observed physical phenomenon: if the characteristics of the oscillator are such that the corresponding value of the parameter λ is positive, then the oscillator generates auto-oscillations of a definite frequency and amplitude.

For sufficiently small values of the parameter $\lambda > 0$, Eq. (I.13) is close to the equation of a linear oscillator and the auto-oscillations in the oscillator are close to simple harmonic vibrations (Fig. I.2a). With the growth of λ, the difference between the auto-oscillations and harmonic vibrations increases (Fig. I.2b) and for sufficiently large values of λ the character of the auto-oscillations differs essentially (Fig. I.2c). As we used to say, the oscillations turn into *relaxation oscillations*.

For a *large* value of the parameter $\lambda > 0$, Eq. (I.13) can be easily

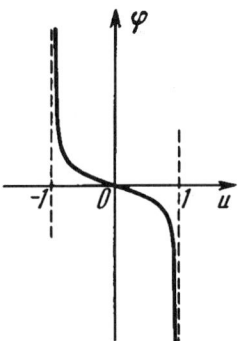

Fig. I.3

reduced to a system of equations of the form (I.9). Indeed, let us introduce new variables y, t and a parameter ε:

$$y = \int_0^x (x^2 - 1)\, dx + \frac{1}{\lambda}\frac{dx}{d\tau}, \qquad t = \frac{\tau}{\lambda}, \qquad \varepsilon = \frac{1}{\lambda^2}.$$

After obvious transformations, (I.13) yields the system (I.8) for $k = m = 1$:

$$\begin{cases} \varepsilon \dfrac{dx}{dt} = y - \tfrac{1}{3}x^3 + x, \\ \dfrac{dy}{dt} = -x. \end{cases} \qquad (\text{I}.14)$$

Here $\varepsilon > 0$ is a *small* parameter. In what follows we shall call (I.14) a *van der Pol system* by analogy with van der Pol's equation (I.13).

It follows from what we have said that a van der Pol system has a unique periodic solution (with an accuracy to within a time shift) which, for a sufficiently small $\varepsilon > 0$, describes a relaxation oscillation process.

EXAMPLE I.2. When small parasitic capacitances, inductances, etc., are taken into account, the operation of a number of electronic devices (such as the Frugauer generator, a symmetric multivibrator, and others, see [2]) can be described by systems of fourth-order differential equations of the form

$$\begin{cases} \varepsilon \dot{x}_1 = -\alpha(y_1 - y_2) + \varphi(x_1) - x_2, \\ \varepsilon \dot{x}_2 = \alpha(y_1 - y_2) + \varphi(x_2) - x_1, \\ \dot{y}_1 = x_1, \\ \dot{y}_2 = x_2 \end{cases} \qquad (\text{I}.15)$$

Introduction

where $\alpha > 0$ is a constant, and $\varphi(u)$, $-1 < u < 1$, is a smooth function whose graph is shown in Fig. I.3. Now if the small parasitic capacitances, inductances, etc., are not taken into account when a mathematical model is constructed, then the system

$$\begin{cases} -\alpha(y_1 - y_2) + \varphi(x_1) - x_2 = 0, \\ \alpha(y_1 - y_2) + \varphi(x_2) - x_1 = 0, \\ \dot{y}_1 = x_1, \\ \dot{y}_2 = x_2, \end{cases} \quad (I.16)$$

results from (I.15) for $\varepsilon = 0$.

It has long been known by physicists that these devices can excite periodic oscillations of an unusual nature, namely, at some time moments the currents (or voltages) vary jumpwise and in the intervals between these time moments they vary smoothly. Oscillations of this kind are known as *relaxation oscillations*.

However, all attempts to substantiate this phenomenon with the aid of system (I.16) were unsuccessful, and it was necessary to introduce additional physical hypotheses (for instance, a "jump hypothesis," see [2]). The phenomenon of relaxation oscillations was explained by purely mathematical means, without any additional physical hypotheses, only on the basis of system (I.15), when the necessity was discovered of taking into account small parasitic parameters in order to construct adequate mathematical models of this kind of physical devices. We give this explanation in III.2.

II. Second-Order Relaxation Systems

II.1. The van der Pol System

Let us consider a system of second-order equations

$$\varepsilon \dot{x} = f(x, y), \qquad \dot{y} = g(x, y), \quad (II.1)$$

where x and y are unknown scalar functions of the independent (time) variable t and ε is a small positive parameter. It is customary to write this as $0 < \varepsilon \ll 1$. We consider the right-hand sides of (II.1) to be defined and sufficiently smooth throughout the phase plane (x, y).

Formally setting $\varepsilon = 0$ in the *nondegenerate system* (II.1), we define a *degenerate system*

$$f(x, y) = 0, \qquad \dot{y} = g(x, y). \quad (II.2)$$

It is not a normal system of differential equations since the first equation is not differential (it is customary to call systems of this kind *differential-algebraic* or *hybrid* systems). It is clear that the degenerate system (II.2) has no solutions with an arbitrary initial point (x_0, y_0): since *all* its trajectories lie on the curve $\Gamma = \{(x,y) : f(x,y) = 0\}$, the condition $(x_0, y_0) \in \Gamma$ must be satisfied.

This is why we can pose the problem of the proximity of solutions of systems (II.1) and (II.2) only for those solutions of the nondegenerate system (II.1) whose initial points lie in a small (together with ε) neighborhood of the curve Γ. It turns out, however, that as $\varepsilon \to 0$, even these solutions do not always tend to the solutions of the degenerate system (II.2). It is also necessary to find the conditions under which the trajectory of system (II.1) can get from its initial point (x_0, y_0), lying at a finite distance from the curve Γ, to a small neighborhood of this curve, how much time this will take, and other things.

We shall carry out here an heuristic analysis of the behavior of the phase trajectories of system (II.1). In particular, we shall show that such a system may have periodic solutions which have the character of relaxation oscillations. See [77] for a full substantiation of the facts being considered.

For simplicity and visuality, we begin with an analysis of the van der Pol system, introduced in Example I.1 (see I.2):

$$\varepsilon \dot{x} = y - \frac{1}{3}x^3 + x, \qquad \dot{y} = -x. \tag{II.3}$$

The corresponding degenerate system is

$$y - \frac{1}{3}x^3 + x = 0, \qquad \dot{y} = -x. \tag{II.4}$$

Clearly, the trajectories of all its solutions lie on the curve Γ which is a cubic parabola $y = \frac{1}{3}x^3 - x$. If we speak of whole trajectories of the degenerate system (II.4), they are evidently five in number (Fig. II.1):

$$(-\infty, S_1), \qquad (+\infty, S_2), \qquad (0, S_1), \qquad (0, S_2), \qquad \text{point } 0,$$

the point 0 being the only equilibrium position. It is easy to understand that beginning its movement, say, from P_0 on the branch $(-\infty, S_1)$ of the curve Γ, the phase point of this system will reach S_1 in a *finite* time. However, none of the trajectories (II.4) begins at the point S_1, and therefore we cannot make any inference concerning the further movement of the phase point when we consider a degenerate system.

Introduction

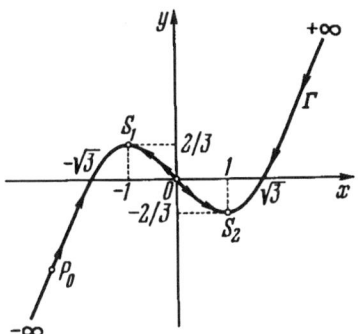

Fig. II.1

We take a nondegenerate system (II.3) and construct a vector field of the phase velocities in its phase plane (x, y) (Fig. II.2). Analyzing it, we can see that the trajectory of this system which begins at the arbitrary point $Q_0 \notin \Gamma$ first enters a small (together with ε) neighborhood of the branch $(-\infty, S_1)$ (or the branch $(+\infty, S_2)$ if the point Q_0 is in differential position), and then, during all of the time that follows, passes in the vicinity of the contour $Z_0 = P_2 S_1 P_1 S_2$ consisting of the horizontal line segments $S_1 P_1$, $S_2 P_2$ and the arcs $P_2 S_1$, $P_1 S_2$ of the curve Γ (Fig. II.3).

It is already clear that in the vicinity of the contour Z_0 there is a closed trajectory Z_ε of the van der Pol system (II.3) (Fig. II.4). Indeed, let us take (see Fig. II.3) a segment $L_1 L_2$ of a small finite length which is parallel to the x-axis and which cuts the arc $P_2 S_1$ at the interior point L. By virtue of what was said above, by means of the traversal of the trajectories of system (II.3), this segment is mapped into its part lying at a small (together with ε) distance from the point L. And this mapping has a fixed point through which the closed trajectory Z_ε passes.

We can easily infer from the equations of system (II.3) that the motion of its phase point along the segments of the cycle Z_ε lying in the vicinity of the arcs $P_2 S_1$ and $P_1 S_2$ of the contour Z_0 occurs with a *finite* velocity and, consequently, takes finite time. As to the segments of the cycle Z_ε lying in the vicinity of the segments $S_1 P_1$ and $S_2 P_2$ of the contour Z_0, they are traversed almost *instantaneously* since along these segments the horizontal component of the phase velocity vector has a value of order $1/\varepsilon$.

In other words, when the motion occurs along the closed trajectory Z_ε,

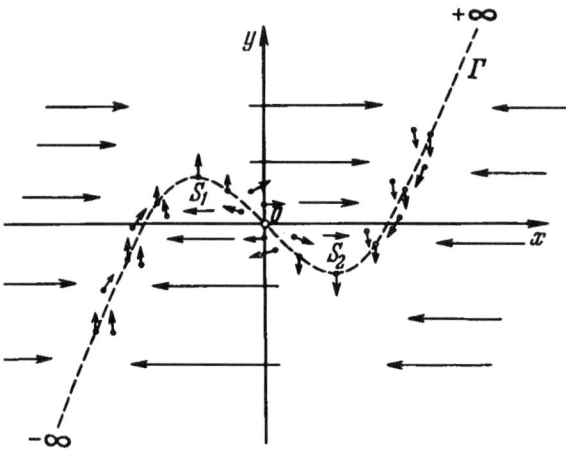

Fig. II.2

comparatively slow, smooth variations of the phase state of the van der Pol system alternate with fast jumpwise motions (Fig. II.5). A periodic motion of this kind is called a *relaxation oscillation*.

These are the general features of the phase pattern of the van der Pol system (II.3). A peculiarity of its phase trajectories is that they have *sections of fast motion* and *sections of slow motion* (in the vicinity of the segments S_1P_1, S_2P_2 and the arcs P_2S_1, P_1S_2 respectively) and sections of transition from one kind of section to the other, i.e., *junction* (or *breakoff*) *sections* (in the vicinity of the points S_1 and S_2) and *drop sections* (in the vicinity of the points P_1 and P_2).

II.2. Analysis of the Phase Plane

The peculiarities of the behavior of the trajectories of the van der Pol system are typical, in full measure, of the phase pattern of the second-order arbitrary system (II.1). The reason why the trajectories of this system pass in the vicinity of some sections of the curve Γ can be explained with the aid of kinematics, in terms of the stability (instability) of the equilibrium position of a special auxiliary first-order equation, and the phenomenon of breakoff, i.e., the transition from a slow motion to a fast one, can be explained by the bifurcation of this equilibrium position.

A connected set of the points of Γ at which the inequality $f'_x(x,y) <$

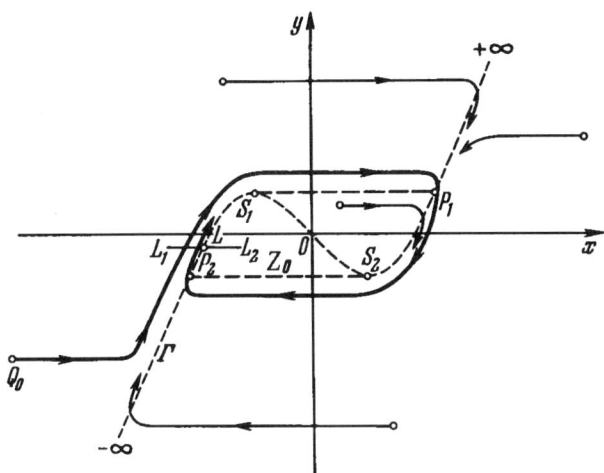

Fig. II.3

0 ($f'_x(x,y) > 0$ respectively) is satisfied is said to be a *stable* (*unstable*, respectively) section of the curve. Stable and unstable sections are divided by the points of Γ at which $f'_x(x,y) = 0$. We shall assume for simplicity that these *division points* are isolated on the curve Γ and are *nondegenerate* in the sense that at each of them $f''_x(x,y) \neq 0$.

For instance, in the case of the van der Pol system (II.3), the curve Γ (a cubic parabola, see Fig. II.1) consists of two stable sections $(-\infty, S_1)$ and $(+\infty, S_2)$ and one unstable section (S_1, S_2), and there are two division points S_1 and S_2.

Let us now take the first of the equations of system (II.1),

$$\varepsilon \dot{x} = f(x,y), \qquad (II.5)$$

and consider y in it as a parameter. For a fixed value of this parameter, say, for $y = y_1$, the autonomous equation (II.5) can have positions of equilibrium. Let $x = x_1$ be one of them. According to the definition of an equilibrium position, the equality $f(x_1, y_1) = 0$ holds true and, consequently, $(x_1, y_1) \in \Gamma$. The converse is also true, i.e., if $(x_1, y_1) \in \Gamma$, then it is clear that $x = x_1$ is an equilibrium position of Eq. (II.5) for $y = y_1$.

Thus, there is a one-to-one correspondence between the points of the curve Γ and the equilibrium positions of Eq. (II.5) for different values of y. To put it otherwise, we can identify Γ with the set of equilibrium positions

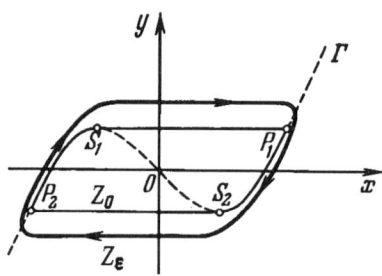

Fig. II.4

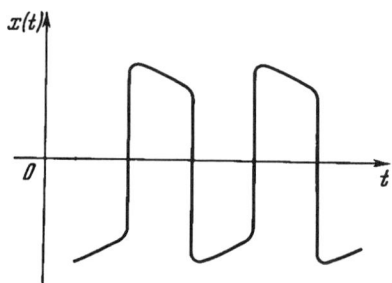

Fig. II.5

of the family of equations (II.5). We can easily see that in this case the stable sections of Γ correspond to the stable positions of equilibrium and the unstable sections correspond to the unstable equilibrium positions. The division points of the curve Γ correspond to the merging of the stable and the unstable equilibrium position, i.e., to the bifurcation values of the parameter y for Eq. (II.5).

For instance, in the case shown in Fig. II.6, the second coordinate y_2 of the division point S is a bifurcation value of the parameter y for the corresponding equation (II.5): for $y < y_2$ in the neighborhood of the point S it has two equilibrium positions (one stable and one unstable), for $y = y_2$ these equilibrium positions merge into one, and then, for $y > y_2$, there are no longer any equilibrium positions in the vicinity of the point S.

Using the arguments given above, we can interpret the motion of the

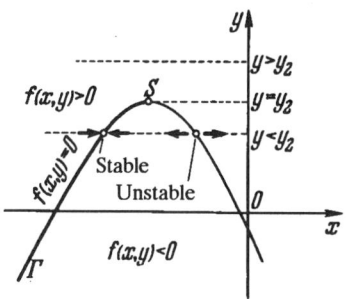

Fig. II.6

phase point along the arbitrary trajectory of system (II.1) (Fig. II.7) as follows. Recall that the vector of the phase velocity of this system (cf. Fig. II.2)

$$v(x, y) = \left(\frac{1}{\varepsilon}f(x, y), g(x, y)\right)$$

is defined at every point of the phase plane (x, y).

Let $Q_0(x_0, y_1)$ be the initial point of the trajectory. If it is at a finite distance from Γ, then, at this point, for a finite second component, the vector of the phase velocity has an infinitely large (as $\varepsilon \to 0$) first component. Consequently, a very rapid variation of the coordinate x begins for an almost invariable value of the coordinate y. This character of the motion is preserved until both components of the vector of the phase velocity become finite, i.e., until the phase point is at a distance of order ε from the curve Γ.

In other words, by virtue of Eq. (II.5) (for the indicated value of the parameter y), this stage of motion along the trajectory of system (II.1) is close to a fast motion along the horizontal straight line $y = y_1$, from the initial point $x = x_0$ into the neighborhood of the stable equilibrium position $x = x_1$ which is associated with $P_0(x_1, y_1) \in \Gamma$. (In the case where x_0 does not lie in the domain of attraction of some stable equilibrium position of Eq. (II.5) for $y = y_1$, the phase point of system (II.1) recedes to infinity at a high velocity almost along the straight line $y = y_1$.)

After the phase point of system (II.1) approaches the point P_0 as close as ε, it continues its motion with a finite velocity along the stable section of the curve Γ that contains the point P_0, remaining in a small (together with ε) neighborhood of this section. It seems to follow smoothly the stable equilibrium position of Eq. (II.5), which moves along Γ for the varying y. The value of y varies slowly because of the degenerate system (II.2).

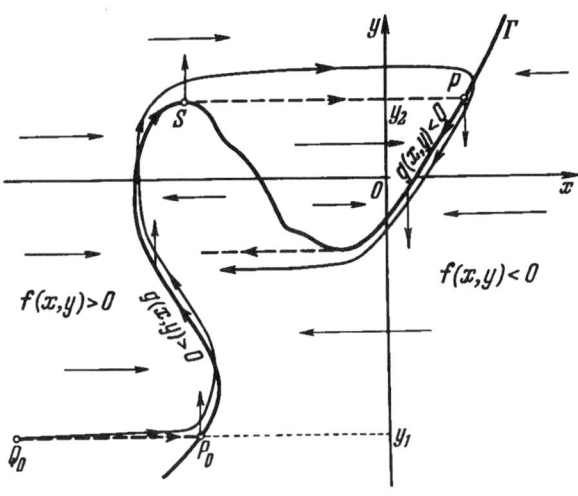

Fig. II.7

Suppose that there are no equilibrium positions of system (II.2) on the stable section of Γ (its endpoints inclusive), i.e., $g(x,y) \neq 0$ along it. The direction of variation of y, upwards or downwards from the initial value of y_1, is defined by the sign of the function $g(x,y)$ on this stable section.

In the process of its monotonic variation, the quantity y can reach, in a finite time, some value which is a bifurcation value for (II.5), say, $y = y_2$, that is associated with a division point $S(x_2, y_2)$ which is the endpoint of the stable section being considered. (If there is no bifurcation value, then the phase point of system (II.1) recedes to infinity along Γ.) By virtue of the nondegeneracy assumption, the point S is extremal for Γ, and therefore (cf. Fig. II.6), for $y = y_2$, the stable equilibrium position of (II.5) followed by the phase point of system (II.1) disappears.

After the disappearance of this equilibrium position, the phase point of system (II.1) leaves the small neighborhood of the point S and rapidly moves, almost along the straight line $y = y_2$, to a small neighborhood of another stable equilibrium position of Eq. (II.5) for $y = y_2$, which is associated with $P(x_3, y_2) \in \Gamma$. (If there is no new equilibrium position of this kind, the phase point of system (II.1) recedes to infinity along the straight line $y = y_2$.)

Then the part of the motion described above is repeated, with the substitution of P for P_0, and so on.

Thus, the trajectory of system (II.1) is a sequence of *fast motion* sections which lie along the segments that are parallel to the x-axis, and *slow motion* sections that lie along the arcs of the curve Γ. It is just the existence of alternating sections traversed in the times differing by their order (with respect to ε) that is a peculiarity of singularly perturbed systems.

According to the kinematic interpretation given above, we call the variable x in system (II.1) a *fast variable* and the variable y a *slow variable*. The division points of the curve Γ that are reached by the phase point of system (II.2) (of the type of the point S in Fig. II.7) during its motion are called *junction (breakoff) points*. The point of the curve Γ to whose neighborhood the phase point of (II.1) moves from the neighborhood of a certain junction point, of the type of the point P in Fig. II.7, is called a *drop point* following this junction point. Equation (II.5), in which y is regarded as a parameter, is called an *equation of fast motions* corresponding to system (II.1).

It may so happen that as a result of the successive alternation of sections of slow and fast motion a closed trajectory of the nondegenerate system (II.1) is formed (cf. Fig. II.4). Then the corresponding periodic solution of this system is a *relaxation oscillation*. A singularly perturbed system that admits of such a periodic solution is known as a *relaxation system*.

III. Arbitrary-Order Relaxation Systems

III.1. Analysis of a Phase Space

Let us consider an arbitrary-order system:

$$\varepsilon \dot{x} = f(x, y), \qquad \dot{y} = g(x, y), \qquad \text{(III.1)}$$
$$x \in R^k, \qquad y \in R^m, \qquad 0 < \varepsilon \ll 1.$$

We consider its right-hand sides to be defined and sufficiently smooth throughout the space R^n, $n = k + m$, which can naturally be represented as a direct sum of the k-dimensional subspace X^k and the m-dimensional subspace Y^m.

We put the *nondegenerate* system (III.1) into correspondence with the *degenerate system*

$$f(x, y) = 0, \qquad \dot{y} = g(x, y), \qquad \text{(III.2)}$$
$$x \in R^k, y \in R^m.$$

Its first equation

$$f(x, y) = 0, \qquad x \in R^k, \qquad y \in R^m, \qquad f: R^n \to R^k, \qquad \text{(III.3)}$$

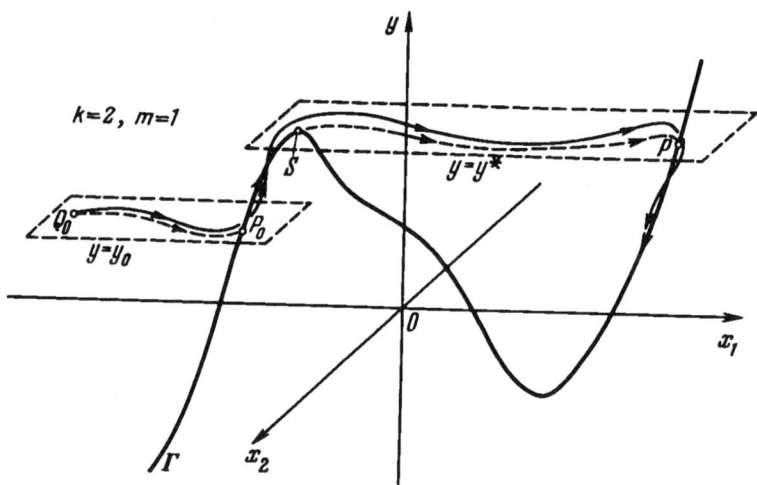

Fig. III.1

defines in R^n a smooth m-dimensional surface Γ. It is clear that all the trajectories of the degenerate system (III.2) lie on Γ, and therefore it has no solutions with the initial point $(x_0, y_0) \notin \Gamma$.

We shall present a schematic description of the qualitative peculiarities of the motion of system (III.1) along an arbitrary trajectory of system (III.1) and give a specific example showing the mechanism of origination of relaxation oscillations in a system of this kind.

We immediately see that the second component of the phase velocity vector of system (III.1),

$$v(x, y) = \left(\frac{1}{\varepsilon} f(x, y), g(x, y)\right),$$

has a finite value at any point (x, y) of the phase space R^n, and the first component is, in general, infinitely large (as $\varepsilon \to 0$). Therefore, the nondegenerate system (III.1) usually has *fast* and *slow motions*, fast motions occurring *far* from the surface Γ almost in parallel with the subspace X^k and slow motions occurring *close* to this surface, in its small (together with ε) neighborhood.

Suppose (see Fig. III.1, III.2, cf. Fig. II.7) that the initial point $Q_0(x_0, y_0)$ of the trajectory is at a finite distance from Γ. Then the vector x begins varying very quickly, whereas the value of the vector y remains almost the

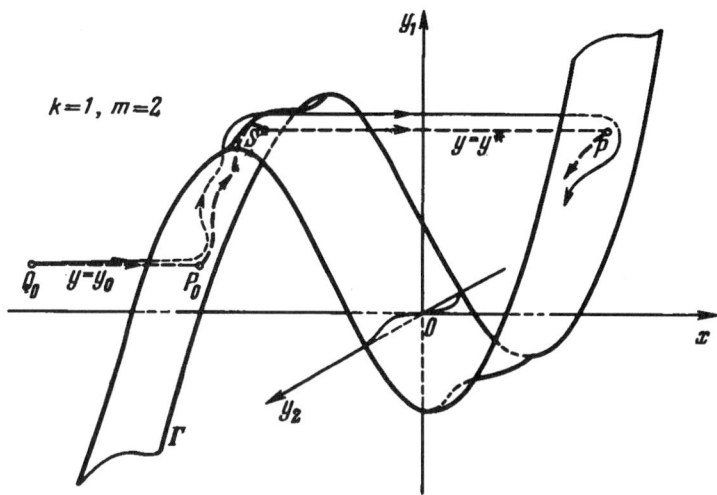

Fig. III.2

same, i.e., at this stage the motion along the trajectory of system (III.1) is close to a fast motion over the plane $y = y_0$ by virtue of the equation

$$\varepsilon \dot{x} = f(x, y), \qquad x \in R^k, \tag{III.4}$$

for the value $y = y_0$. Therefore, we call (III.4), in which $y \in R^m$ is regarded as a parameter, an *equation of fast motions* corresponding to the nondegenerate system (III.1).

Suppose that for every fixed $y \in R^m$ Eq. (III.4) can have only equilibrium positions as its stationary solutions. (Another interesting and important case in which the equation of fast motions has an exponentially stable periodic solution among its stationary solutions was considered in [80, 81] and is investigated below, in Sec. 7.) If $x = x^*$ is a stable equilibrium position of (III.4) for $y = y_0$, whose domain of attraction contains the initial point $x = x_0$, then its phase point quickly approaches this equilibrium position and, consequently, the phase point of the nondegenerate system (III.1) gets into the neighborhood of the point $P_0(x^*, y_0) \in \Gamma$.

Thereafter the variables x and y in the nondegenerate system (III.1) vary with similar velocities, and the motion of this system along the trajectory is slow in a small neighborhood of the surface Γ, as if following the stable equilibrium position of (III.4), which moves along Γ with the variation of y. The vector y varies with a finite velocity obeying the degenerate system (III.2).

The character of the motion of the phase point of system (III.1) in the vicinity of Γ remains the same until, for some bifurcation value of y, say, $y = y^*$, the traced stable equilibrium position disappears as a result of merging with some unstable equilibrium position of Eq. (III.4). Then, by analogy with the two-dimensional case, it is natural to expect that the phase point of system (III.1) will rapidly move (almost over the plane $y = y^*$) to the neighborhood of the point $P \in \Gamma$ corresponding to another stable equilibrium position of (III.4) for $y = y^*$.

Thus, in the neighborhood of the point S, corresponding to the bifurcation value of the parameter y, the phase motion of system (III.1) changes from a *slow* to a *fast* one. It is rather difficult to give a complete mathematical description of this change. We give this description in Chapter 1.

When the phase point of the nondegenerate system (III.1) is in a small neighborhood of the point P, a new section of slow motion begins. Thus, in the neighborhood of the point P the phase motion of system (III.1) also changes from a *fast* to a *slow* one.

It may so happen that the successive alternation of slow and fast motions (and the alternation of changes, respectively) results in the appearance of a closed trajectory of the nondegenerate system (III.1). The corresponding periodic solution of this system is known as a *relaxation oscillation*.

III.2. Analysis of Example I.2

We shall consider Example I.2 (see I.2) in order to illustrate the scheme of origination of relaxation oscillations.

Let us consider the system

$$\begin{cases} \varepsilon \dot{x}_1 = -\alpha(y_1 - y_2) + \varphi(x_1) - x_2, \\ \varepsilon \dot{x}_2 = \alpha(y_1 - y_2) + \varphi(x_2) - x_1, \\ \dot{y}_1 = x_1, \qquad \dot{y}_2 = x_2. \end{cases} \quad \text{(III.5)}$$

Note that the right-hand sides of this system are defined in the domain

$$\{(x_1, x_2, y_1, y_2) : -1 < x_1 < 1, -1 < x_2 < 1, y_1 \in R, y_2 \in R\}.$$

We shall use vector notations, $x = (x_1, x_2)$, $y = (y_1, y_2)$, for the sake of convenience.

The system of equations for fast motions, corresponding to (III.5), has the form

$$\begin{cases} \varepsilon \dot{x}_1 = -\alpha(y_1 - y_2) + \varphi(x_1) - x_2, \\ \varepsilon \dot{x}_2 = \alpha(y_1 - y_2) + \varphi(x_2) - x_1 \end{cases} \quad \text{(III.6)}$$

Introduction

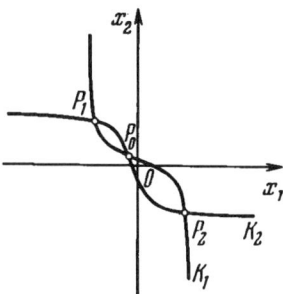

Fig. III.3

and is defined in the square $\Pi = \{x_1, x_2 : |x_1| < 1, |x_2| < 1\}$ of the plane (x_1, x_2). Here y_1 and y_2 are parameters. The equilibrium positions of this system are the points of intersection of the curves

$$K_1(x_1, x_2) \equiv -\alpha(y_1 - y_2) + \varphi(x_1) - x_2 = 0,$$
$$K_2(x_1, x_2) \equiv \alpha(y_1 - y_2) + \varphi(x_2) - x_1 = 0,$$

i.e., the curves that result from the translation of the graphs of the functions $x_2 = \varphi(x_1)$ and $x_1 = \varphi(x_2)$ (see Fig. I.3) along the coordinate axes x_2 and x_1 respectively.

Note that the set of intersection points of these curves for different values of the parameters y_1 and y_2 is precisely the surface Γ in the four-dimensional phase space R^4 which contains all the trajectories of the degenerate system

$$\begin{cases} -\alpha(y_1 - y_2) + \varphi(x_1) - x_2 = 0, \\ \alpha(y_1 - y_2) + \varphi(x_2) - x_1 = 0, \\ \dot{y}_1 = x_1, \quad \dot{y}_2 = x_2. \end{cases} \quad \text{(III.7)}$$

It is easy to verify that only the following three cases are possible, depending on the values of the parameters y_1 and y_2:

(a) system (III.6) has *three* equilibrium positions P_1, P_2, P_0, the positions P_1 and P_2 being stable nodes and P_0 being a saddle point (Fig. III.3),

(b) system (III.6) has *two* equilibrium positions $P_1 = P_0$ and P_2 or $P_2 = P_0$ and P_1, the first being an unstable saddle–node and the second being a stable node (Fig. III.4),

(c) system (III.6) has *one* equilibrium position P_2 or P_1, which is a stable node (Fig. III.5).

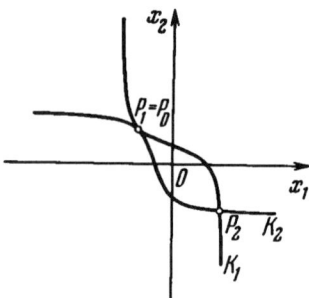

Fig. III.4

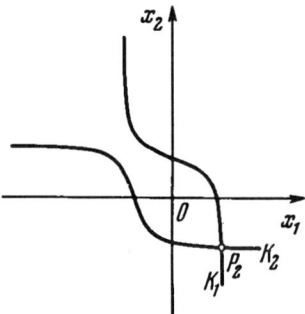

Fig. III.5

The phase patterns of system (III.6) corresponding to these three cases are schematically shown in Figs. III.6–III.8. It is easy to verify (for instance, with the use of Bendixson's criterion) that this system has no closed trajectories for any values of the parameters y_1 and y_2.

Let $Q_0(x_1^0, x_2^0, y_1^0, y_2^0)$ be the initial point of the trajectory of system (III.5). We suppose, for definiteness, that case (a) is realized for the values of the parameters $y_1 = y_1^0$, $y_2 = y_2^0$ for system (III.6) in the plane (x_1, x_2) (see Figs. III.3 and III.6). (Other possibilities will be considered below.) In addition, let us suppose that the point $x^0 = (x_1^0, x_2^0) \in \Pi$ is at a finite distance from the equilibrium positions P_1, P_2, P_0 (this means that the point Q_0 is at a finite distance from the surface Γ). Let us finally suppose that the initial point x^0 of system (III.6) for $y = y^0 = (y_1^0, y_2^0)$ belongs to the

Introduction

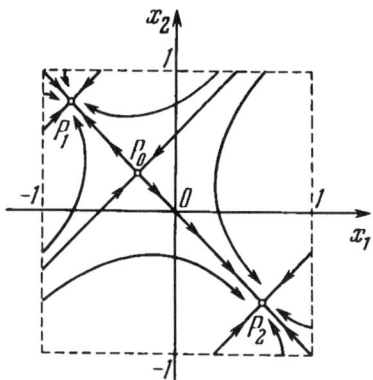

Fig. III.6

domain of attraction of, say, the stable node $P_1(-a, a)$, where $a > 0$.

Since, in accordance with system (III.5), the vector x rapidly varies when it is far from the surface Γ, whereas the vector y remains almost unchanged, we can suppose that the trajectory of this system, starting at the point $Q_0(x^0, y^0)$ and remaining close to the plane $y_1 = y_1^0$, $y_2 = y_2^0$, will quickly move to the neighborhood of the point $(-a, a, y_1^0, y_2^0) \in \Gamma$. The first two coordinates of this point are the coordinates of the stable equilibrium position P_1 to which the trajectory of system (III.6), which starts at the point x^0, tends for $y = y^0$ (cf. Figs. III.1, III.2).

Then the vector x and y begin varying at similar rates and the phase point of system (III.5) follows the stable equilibrium position P_1 (cf. Figs. III.1, III.2), which is slowly displaced along Γ with the variation of y_1 and y_2 by virtue of the degenerate system (III.7). It is easy to follow the qualitative character of this variation. Using the last two equations of system (III.7) and noting that the inequalities $x_1 < 0$, $x_2 > 0$ hold true in the vicinity of the point P_1 (see Fig. III.6), we can infer that the difference $y_1 - y_2$ decreases with an increase in t. This means that the curve $K_1(x_1, x_2) = 0$ moves upward along the x_2-axis and the curve $K_2(x_1, x_2) = 0$ moves to the left along the x_1-axis. As a result, after definite time intervals we have the situations shown in Figs. III.3, III.4, III.5 or in Figs. III.6, III.7, III.8 respectively. In other words, during its displacement the stable equilibrium position P_1 comes close to the saddle point P_0 and, at a certain moment, for the bifurcation value $y = y^* = (y_1^*, y_2^*)$, merges with it, and then disappears.

After this merging and disappearance of the stable equilibrium position

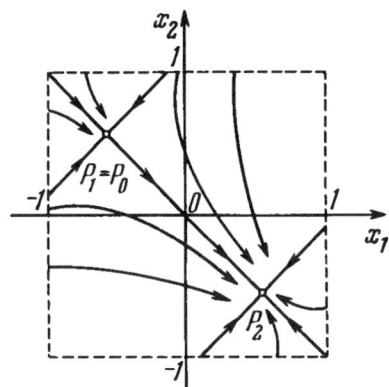

Fig. III.7

we are tracing, the vector x begins again varying rapidly, whereas the vector y remains almost unchanged. Therefore, remaining close to the plane $y = y_1^*$, $y_2 = y_2^*$, the phase point of system (III.5) will tend very rapidly to the stable equilibrium position of system (III.6) lying in this plane for $y = y^*$, i.e., to the stable node $P_2(b, -b)$, where $b > 0$. To put it otherwise, the phase point of system (III.5) will be quickly displaced to the neighborhood of the point $(b, -b, y_1^*, y_2^*) \in \Gamma$ (cf. Figs. III.1, III.2).

Now the vectors x and y again begin varying at similar rates, and the phase point of system (III.5) follows the stable equilibrium position P_2. However, the inequalities $x_1 > 0$, $x_2 < 0$ are valid in the vicinity of the point P_2 (see Fig. III.8), and therefore, as we can see from the last two equations of system (III.7), this time the difference $y_1 - y_2$ increases with an increase in t. Thus, the curve $K_1(x_1, x_2) = 0$ moves downward along the x_2-axis and the curve $K_2(x_1, x_2) = 0$ moves to the right along the x_1-axis. As a result, we have, in succession, the situations shown in Figs. III.5, III.4, III.3, III.9, III.10 (or in Figs. III.8, III.7, III.6, and so on, respectively). In other words, in the process of its displacement the stable equilibrium position P_2 comes close to the saddle point P_0 and then merges with it.

As soon as they merge, the phase point of system (III.5) moves very fast to the remaining stable equilibrium position P_1, and then the whole process is repeated.

Now we can feel, by intuition, that system (III.5) has a periodic solution that describes relaxation oscillation.

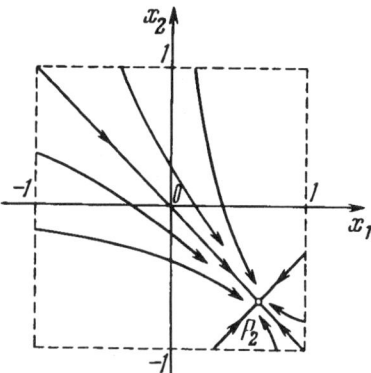

Fig. III.8

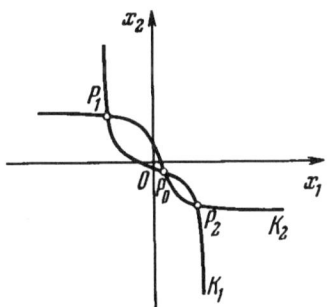

Fig. III.9

III.3. Solutions of a Degenerate System of Equations

Let us consider in greater detail the degenerate system (III.2) corresponding to the nondegenerate system (III.1).

We have noted that Eq. (III.3) defines, in the space R^n, a smooth m-dimensional surface Γ on which all the trajectories of the degenerate system (III.2) lie. We call the set of points of Γ at which *all* the eigenvalues of the matrix

$$A(x,y) = f'_x(x,y) =$$
$$= \left\| \frac{\partial}{\partial x_j} f_i(x_1, \ldots, x_k, y_1, \ldots, y_m) \right\|, \qquad i,j = 1, \ldots, k,$$

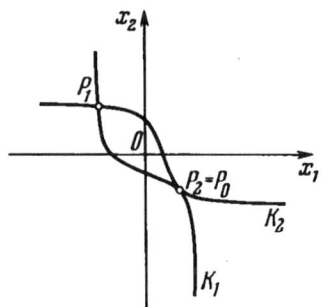

Fig. III.10

have negative real parts a *stable domain* of the surface Γ and denote this domain by Γ_-. We call the set of points of Γ at which the relation

$$\det A(x, y) = \det f'_x(x, y) = 0$$

is satisfied, a *breakoff section* and denote it by Γ_0. The breakoff section Γ_0 is an $(m-1)$-dimensional subset of the surface Γ and, in general, divides Γ into several parts.

Let us consider the fast-motion equation (III.4) corresponding to the nondegenerate system (III.1). For a fixed value of the parameter $y \in R^m$, say, for $y = y_0$, this equation can have an equilibrium position among its solutions. Let $x = x_0$ be one of them. Then, in accordance with the definition of an equilibrium position, the point $P_0(x_0, y_0)$ belongs to the surface Γ. The converse is obviously also true, i.e., if $(x_0, y_0) \in \Gamma$, then the point $x = x_0$ is an equilibrium position of the fast-motion equation (III.4) for the value $y = y_0$.

Thus, there is one-to-one correspondence between the points of the surface Γ and the equilibrium positions of the family of equations (III.4) for different values of the parameter $y \in R^m$. It is clear that in this case the stable domain Γ_- of the surface Γ corresponds to the stable equilibrium positions of Eq. (III.4).

In order to find the solutions of the degenerate system (III.2), it is natural to do the following: first to solve Eq. (III.3) for $x \in R^k$, i.e., to find the vector function $x = \Phi(y)$, and then substitute it into the second equation of the degenerate system (III.2) and solve the resulting normal system of differential equations

$$\dot{y} = g(\Phi(y), y), \qquad y \in R^m. \tag{III.8}$$

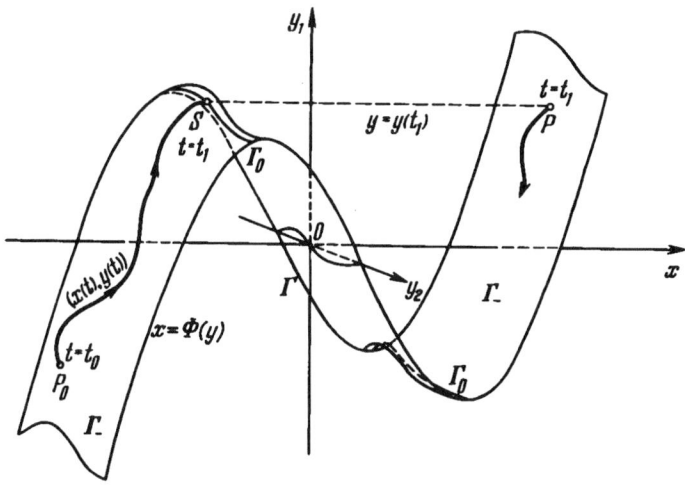

Fig. III.11

Let $P_0(x_0, y_0)$ be a point of the surface Γ, with $x_0 = \Phi(y_0)$ (Fig. III.11). If $y = y(t)$ is a solution of system (III.8) satisfying the initial condition $y(t_0) = y_0$, then it is easy to see that the set of functions

$$x = x(t) \equiv \Phi(y(t)), \qquad y = y(t) \tag{III.9}$$

is the solution of the degenerate system (III.2) with the initial point P_0 for $t = t_0$. At the moment $t = t_0$ the trajectory of solution (III.9) begins at the point $P_0 \in \Gamma$ and lies on the surface Γ.

It is easy to realize that this procedure of constructing solution (III.9) of the degenerate system (III.2) is only applicable for the values $t \geq t_0$ for which the relation $f(x, y(t)) = 0$ admits of the solution $x = \Phi(y)$ for x chosen above. By virtue of the theorem on an implicit function, we can state that solution (III.9) is certainly defined from the initial moment t_0 till the moment $t = t_1$ when $\det A(x(t), y(t))$ vanishes *for the first time* or, what is the same, when the trajectory of solution (III.9) gets to the breakoff section Γ_0 for the first time, falling on some point $S(x(t_1), y(t_1))$ (see Fig. III.11).

In the general case we cannot make any inference concerning the subsequent behavior of solution (III.9) if we consider only the degenerate system (III.2) itself, since no trajectories of this system emanate from the point S with an increase in time. However, if we interpret the trajectories of the degenerate system (III.2) as the limits, as $\varepsilon \to 0$, of the trajectories of the nondegenerate system (III.1), then, taking into account the arguments presented in III.1, we can extend solution (III.9), in a number of cases, to the case where $t > t_1$.

Suppose that for $t_0 \leq t < t_1$ the trajectory of solution (III.9) belongs to the stable domain Γ_- of the surface Γ and for $t = t_1$ this trajectory arrives at the point $S(x(t_1), y(t_1))$ on the breakoff section Γ_0. Since $\det A(S) = \det A(x(t_1), y(t_1)) = 0$, at least one of the eigenvalues of the matrix $A(S)$ is zero.

We make the following assumptions:

(a) the matrix $A(S)$ has *only one* zero eigenvalue, and the real parts of all eigenvalues of order $k - 1$ are negative;

(b) in the k-dimensional plane defined by the equation $y = y(t_1)$ in R^n there is a unique trajectory of the fast-motion equation (III.4) (for the indicated value of the parameter y) that adjoins, as $t \to -\infty$, the point $x(t_1)$ and tends, as $t \to +\infty$, to some stable equilibrium position $x = x^*$.

If these assumption are fulfilled, then we suppose that the phase point of the degenerate system (III.2) makes an *instantaneous jump*, for $t = t_1$, from the *junction point* $S(x(t_1), y(t_1)) \in \Gamma_0$ to the point $P(x^*, y(t_1)) \in \Gamma_-$, which is called a *drop point* and which comes next to the junction point S (see Fig. III.11). The subsequent motion of the phase point of the degenerate system (III.2), for $t > t_1$, follows the trajectory of this system emanating, at the moment $t = t_1$, from the point P. The procedure described above can be used to construct such a trajectory. If in the process of the motion of the trajectory constructed in this way the phase point of the degenerate system (III.2) again turns out to be, at some moment $t_2 > t_1$, on the breakoff section Γ_0, then one more instantaneous jump occurs, and so on.

The vector function $(x(t), y(t))$, $t \geq t_0$, obtained in accordance with the law described, is discontinuous, i.e., in general, at the moments t_1, t_2, etc., all the components of the vector x make jumps, but satisfies the degenerate system (III.2) everywhere outside of these points of discontinuity. It is natural to call such a function a *discontinuous solution* of a degenerate system. It should be emphasized that the *trajectory of a discontinuous solution* is a continuous curve in the phase space R^n and consists of alternating sections of the following two types:

(a) *slow-motion sections* lying in the stable domain Γ_- of the surface Γ that are traversed in a finite time,

(b) *fast-motion sections* between a junction point and a drop point that comes next to it, both of them lying in planes parallel to the subspace X^k and traversed instantaneously.

If the alternating sections of these types form a closed trajectory, then we say that a degenerate system has a *discontinuous periodic solution*.

The definition of a discontinuous solution is especially clear when x and y are scalar variables, i.e., for a degenerate system of the form (III.2) for $k = m = 1$. In this case the trajectory of the discontinuous solution consists of the arcs of the stable sections of the curve Γ that are traversed in a finite time and the horizontal sections from the junction point to the drop point traversed instantaneously (see Fig. II.7). The closed curve Z_0 shown in Figs. II.3, II.4 is a trajectory of the discontinuous periodic solution of such a system.

Thus we see that the *definition* of the trajectory of a discontinuous solution of a degenerate system given here (and, in particular, the definition of the trajectory of a discontinuous periodic solution) is based on a certain "hypothesis of a jump." However, if we *define* these trajectories as the limits, as $\varepsilon \to 0$, of the trajectories of the corresponding nondegenerate system (such a limiting process is substantiated in [77], for instance), then we no longer need the "hypothesis of a jump."

In the next chapter we shall give another promising approach based on the relationship between the singularly perturbed system (III.1) and the so-called relay system which is constructed, in a special way, by means of the functions $f(x, y)$ and $g(x, y)$.

Chapter 1

Theorem on the C^1-Proximity of the Solutions of a Relaxation and a Relay System and the Asymptotics of Relaxation Oscillations

At the beginning of the chapter we formulate and prove the theorem on the C^1-proximity which makes it possible to solve the problem of the uniqueness and stability of a multidimensional relaxation cycle. Then we describe the procedure of reducing an original relaxation system in the neighborhood of a junction point to a system whose right-hand sides regularly depend on a small parameter. This reduction principle plays a significant part in the construction of a complete asymptotics of solutions of a relaxation system in the neighborhood of a junction point. It is important that it allows us to use qualitative methods in order to solve the problem of "sewing together" different asymptotic formulas and that of the orders of the remainders.

As a direct application of the theorem on the C^1-proximity, we describe a situation in which a three-dimensional relaxation system with two slow derivatives has a countable number of dichotomous cycles (the corresponding conditions are formulated in terms of a limiting relay system).

At the end of the chapter we give the solution of Pontryagin's problem which is related to the tori of singularly perturbed systems and which generalizes the well-known theorem of Pontryagin–Rodygin. Then we generalize the Pontryagin–Rodygin theorem in another direction, namely, we formulate the conditions for the existence and stability of relaxation tori, i.e., tori such that when a trajectory moves along them, it jumps from one stable slow-motion surface to another an infinite number of times.

Theorem on the C^1-Proximity

The results given in this chapter were obtained in [36–38, 40–42].

1. Statement of the Problem, Preliminary Information, and Heuristic Arguments

1.1. Relay Systems

In what follows we suppose that the equation

$$f(x,y) = 0, \qquad x \in R^k, \qquad y \in R^m,$$

where $f(x,y)$ is a smooth vector function with values in R^k, defines in R^n, $n = k + m$, a smooth m-dimensional surface Γ that disintegrates into nonintersecting parts: $\Gamma = \Gamma_{1-} + \Gamma_{10} + \Gamma_+ + \Gamma_{2-} + \Gamma_{20}$. We set $A(x,y) = f'_x$. By definition, $(x,y) \in \Gamma_{1-} + \Gamma_{2-}$ if all the eigenvalues of the matrix A have negative real parts, $(x,y) \in \Gamma_{10} + \Gamma_{20}$ if it only has a prime zero eigenvalue and all the other eigenvalues remain in the left-hand complex half-plane, and, finally, $(x,y) \in \Gamma_+$ if at least one eigenvalue of the matrix A lies in the right-hand half-plane. We suppose, in addition, that $x = \Phi_1(y)$ or $x = \Phi_2(y)$ (Φ_1, Φ_2 are smooth vector functions) for (x,y) from Γ_{1-} or Γ_{2-} respectively.

Suppose we are given a smooth vector function $g(x,y)$ of the variables $x \in R^k$, $y \in R^m$ with values in R^m. Let us consider in R^m a differential equation with relay nonlinearity

$$\dot{y} = g(\Phi(y), y), \qquad \Phi(y) = \begin{cases} \Phi_1(y), \\ \Phi_2(y). \end{cases} \tag{1.1}$$

As usual, it is easier to define constructively the concept of the solution of this type of equation. For instance, suppose that the vector y_0 is such that $(\Phi_1(y_0), y_0) \in \Gamma_{1-}$. Under this condition the solution of Eq. (1.1) first coincides with that of the equation

$$\dot{y} = g(\Phi_1(y), y) \tag{1.2}$$

with the initial condition $y|_{t=0} = y_0$. Suppose that with an increase in t the point $(\Phi_1(y(t)), y(t))$ gets into the set Γ_{10} at some first moment t_0. Then a switching occurs and in the subsequent moments of time $y(t)$ is the solution of the equation

$$\dot{y} = g(\Phi_2(y), y) \tag{1.3}$$

with the initial condition $y|_{t=t_0} = y(t_0)$. We think that upon a subsequent increase in t, at some first moment of time $t_1 > t_0$, the point $(\Phi_2(y(t)), y(t))$ gets into the set Γ_{20}. Then another switching occurs and so on.

Suppose that the relay system (1.1) has a periodic solution $y_0(t)$ with two or more switchings on time intervals of period length T_0 and assume, for definiteness, that $(\Phi_1(y_0(0)), y_0(0)) \in \Gamma_{1-}$. The technique developed in [46, 59] makes it easy to investigate the problem of its stability under the following additional restriction.

CONDITION 1.1. Suppose that the increasing positive numbers $t_0, t_1, \ldots$ are successive moments of switching of the periodic solution $y_0(t)$ and, hence,

$$(\Phi_1(y_0(t_{2s})), y_0(t_{2s})) \in \Gamma_{10}, \qquad (\Phi_2(y_0(t_{2s+1})), y_0(t_{2s+1})) \in \Gamma_{20},$$

where $s = 0, 1, \ldots$. We introduce tangent planes to Γ at these points and then shift them to the origin. Then we consider their intersection with R^m. We also displace the vectors $g(\Phi_1(y_0(t_{2s})), y_0(t_{2s})), g(\Phi_2(y_0(t_{2s+1})), y_0(t_{2s+1}))$ to the origin. We say that the switching is normal if none of these vectors belongs to the corresponding plane in R^m introduced above.

In the plane which passes through the point $y_0(0)$ and is orthogonal to the vector $\dot{y}_0(0)$ we consider the Poincaré operator Π_0 along the trajectories of the relay system (1.1). By virtue of condition 1.1, it is defined in the neighborhood of the point $y_0(0)$ and is continuous (when we remove condition 1.1 in the general case, it does not possess this property any longer). The investigation of its smoothness is somewhat complicated by the fact that the derivatives of the functions Φ_1 and Φ_2 increase indefinitely in the norm when the vectors $(\Phi_1(y), y)$ and $(\Phi_2(y), y)$ approach Γ_{10} and Γ_{20} respectively. However, under the constraints imposed on A, the "infinity" of the derivatives of Φ_1 and Φ_2 is one-dimensional in nature, and this allows us to establish the continuous differentiability of the operator Π_0. (An accurate substantiation is connected with the introduction of special coordinates in the neighborhood of the switching points which reduces the problem to the consideration of a certain scalar differential equation whose analysis is trivial. We do not consider this problem in detail here since this coordinate system is described below.) Thus the following statement is valid.

LEMMA 1.1 *Suppose that condition* 1.1 *is satisfied and all eigenvalues of the matrix Π_0', calculated at the point $y_0(0)$, are less than unity in absolute value. Then the periodic solution $y_0(t)$ of the relay system* (1.1) *is orbitally exponentially stable.*

In what follows we shall need an algorithm for the investigation of the stability properties of the periodic solution $y_0(t)$ of the relay system (1.1). Obviously, upon its formal linearization we arrive at an equation in R^m with a δ-like coefficient, the concept of whose solution can be introduced in many ways. In our case its solutions between the neighboring switching moments are solutions of the equation

$$\dot{h} = [D(t) - C(t)A^{-1}(t)B(t)]h. \quad (1.4)$$

Here $D(t) = g'_y$, $C(t) = g'_x$, $A(t) = f'_x$, $B(t) = f'_y$ and the derivatives are taken for $y = y_0(t)$ and $x = \Phi_1(y_0(t))$ or $\Phi_2(y_0(t))$. When it passes through the switching moments t_p, the solution $h(t)$ of (1.4) suffers discontinuities of the first kind with the satisfaction of the relations

$$h(t_p + 0) = h(t_p - 0) + a(b, h(t_p - 0)), \qquad p = 0, 1, \ldots, \quad (1.5)$$

where $a = \dot{y}_0(t_p - 0) - \dot{y}_0(t_p + 0)$, $b = \operatorname{grad} T(y)|_{y=y_0(t_p)}$, and $T(y)$ is the time during which, moving from the point y along the trajectory of the relay system (1.1), we get into Γ_{10} or into Γ_{20} respectively. It follows from the properties of the derivatives of Φ_1 and Φ_2 that the matrix appearing on the right-hand side of (1.4) is not limited in the norm as $t \to t_{2s} - 0$ and $t \to t_{2s+1} - 0$ respectively. However, the singularities appearing here are summable because the singularity of the derivatives of Φ_1 and Φ_2 is one-dimensional (see the remark before Lemma 1.1).

Let us consider relations (1.5) in greater detail. We denote the semigroups generated by Eqs. (1.2) and (1.3), respectively, by $\Pi_1(t, y)$ and $\Pi_2(t, y)$. It is obvious that the phase flow defined by system (1.1) coincides with $\Pi_1(t, y)$ if $0 \leq t < T(y)$ and with $\Pi_2(t - T(y), \Pi_1(T(y), y))$ if t slightly exceeds $T(y)$. Using these formulas, we linearize the phase flow under consideration on the periodic solution $y_0(t)$ in the neighborhood of the switching point t_0. We obtain

$$h(t_0 + 0) - h(t_0 - 0) = (-\dot{y}_0(t_0 + 0) + \dot{y}_0(t_0 - 0))(\operatorname{grad} T|_{y=y_0(0)}, h(0)).$$

Here $h(t)$ is the trajectory of the linearized phase flow. Then we differentiate the obvious representation $T(y) = t_0 - \delta + T(y(t_0 - \delta))$ and make δ tend to zero. As a result we obtain

$$(\operatorname{grad} T|_{y=y_0(0)}, h(0)) = (\operatorname{grad} T|_{y=y_0(t_0)}, h(t_0 - 0)).$$

By virtue of the preceding formula we arrive at condition (1.5) for the switching moment t_0 being considered.

In the sequel we shall need an explicit formula for the vector b. Let us consider matrices $A = A(t_p - 0)$ and $B = B(t_p - 0)$. We denote by α and β the eigenvectors of the matrices A and A^*, respectively, corresponding to the zero eigenvalue, assuming that $(\alpha, \beta) = 1$. It turns out that

$$b = -B^*\beta/(\dot{y}_0(t_p - 0), B^*\beta), \quad (1.6)$$

where the denominator is nonzero by virtue of condition 1.1. To prove this equality, it is sufficient to introduce special coordinates in the neighborhood of the switching points, which are described below.

We have thus given a technique of linearizing the relay system (1.1) which makes it possible to construct a monodromy matrix of the linear periodic problem (1.4), (1.5). Note that by virtue of (1.6) it has a discontinuous periodic solution $\dot{y}_0(t)$, i.e., the monodromy matrix possesses an identity eigenvalue. Its other eigenvalues coincide with those of the matrix Π'_0. Now it follows from Lemma 1.1 that the well-known theorem of Andronov–Witt is extended to the case of relay systems.

In conclusion, it should be pointed out that system (1.4) with relations (1.5) results from the linearization of the relay system (1.1) on any one of its trajectories $y_0(t)$ (not necessarily periodic).

1.2. Properties of a Hybrid Dynamic System

Let us analyze the hybrid dynamic system

$$x = \Phi(y), \qquad \dot{y} = g(x, y). \tag{1.7}$$

Suppose for definiteness that the initial condition $(x(0), y(0))$ of its solution $(x(t), y(t))$, which is in the vicinity of Γ_{1-}, is such that as a result of projection in parallel with R^k we get into Γ_{1-}. First of all, we consider the curve $(x(t), y(0))$ to be a part of the trajectory of this solution, where $0 \le t < \infty$, $x(\infty) = \Phi_1(y(0))$, and $x(t)$ is the solution of the differential equation

$$\dot{x} = f(x, y), \qquad x \in R^k, \tag{1.8}$$

for $y = y(0)$. We assume that the "jump" from the point $(x(0), y(0))$ to the point $(\Phi_1(y(0)), y(0))$ is instantaneous. Then we find $y(t)$ from (1.2) and $x(t) = \Phi_1(y(t))$. The next "jump" occurs when the point $(x(t_0), y(t_0))$ gets into the set Γ_{10}. Suppose that for $y = y(t_0)$ the differential equation (1.8) has a solution $x(t)$ such that $x(-\infty) = x(t_0)$, $x(\infty) = \Phi_2(y(t_0))$. We also consider the curve $(x(t), y(t_0))$, where $-\infty < t < \infty$, to be a part of the trajectory of the solution of system (1.7) assuming that here, too, the motion along it is instantaneous. In the same way, we define the trajectories of the solutions of system (1.7) at the subsequent moments of time. Just as in 1.1, we fix some trajectory $(x_0(t), y_0(t))$ of the hybrid system (in accordance with what we said above $y_0(t)$ is the solution of system (1.1) with the initial condition $y|_{t=0} = y_0(0)$).

CONDITION 1.2. For $y = y_0(t_0)$, Eq. (1.8) has an exponentially stable one-dimensional integral manifold in the neighborhood of the equilibrium position $x_0 = \Phi_1(y_0(t_0))$. We think that the right-hand side of the equation on this manifold begins with a quadratic term. We suppose that the same is true for $y = y_0(t_1)$, etc.

It is significant that the "outgoing whiskers" from these equilibrium positions of (1.8) are unique. It is interesting that in this case we can have a continuum of the "ingoing whiskers." The system $\dot{x}_1 = x_1^2$, $\dot{x}_2 = -x_2$ can serve as a model example, its above-mentioned "whiskers" are defined by the relations $x_2 = c \cdot \exp(1/x_1)$, where $x_1 < 0$ and c is an arbitrary constant.

CONDITION 1.3. Let us consider in R^n balls of a sufficiently small radius whose centers lie on the unstable "whiskers" of Eq. (1.8) introduced in Condition 1.2. We think that the values $x(0)$ and y belonging to these balls, the solutions of (1.8) asymptotically approach Γ_{1-} or Γ_{2-} as $t \to \infty$.

The geometrical meaning of Condition 1.3 is trivial and the consequences of Condition 1.2 are relatively complicated. In their aggregate they ensure the existence and uniqueness of the sections of the trajectory of solutions of the hybrid system (1.7) related to Eq. (1.8). Moreover, these conditions imply the existence of a cycle of the hybrid system (1.7) which is a "rise" of the cycle of the relay system (1.1) from R^m to R^n if the trajectory $y_0(t)$ of the relay system is closed. We have thus arrived at the following statement.

LEMMA 1.2. *Suppose that, in addition to Lemma 1.1, Conditions 1.2 and 1.3 are fulfilled on the closed trajectory $y_0(t)$. Then the cycle Z_0 of the hybrid system (1.7) is orbitally exponentially stable.*

The proof of Lemma 1.2 follows from Lemma 1.1 since the sections of the trajectory of the hybrid system (1.7) related to Eq. (1.8) are traversed instantaneously, and this reduces the problem of the stability of the cycle Z_0 to the geometrical proximity of some curves ensured by Conditions 1.2 and 1.3.

Note that Conditions 1.1–1.3 are encountered in [77] in other terms. In the formulation given above, they imply the stability of the cycle of the hybrid system, a fact that plays a significant part in the proof of the existence of a relaxation cycle. It is also important that, owing to the relationship between the relay system and the hybrid system, there is an algorithm for the investigation of the stability of the cycle Z_0.

1.3. The Relationship Between the Hybrid System and the Relaxation System

Let us consider the relaxation system

$$\varepsilon \dot{x} = f(x,y), \qquad \dot{y} = g(x,y), \qquad 0 < \varepsilon \ll 1 \tag{1.9}$$

corresponding to the hybrid system (1.7). Suppose that $(x_0(t,\varepsilon), y_0(t,\varepsilon))$ is its trajectory with initial conditions $x_0(0,\varepsilon) = x_0$, $y_0(0,\varepsilon) = y_0$ such that the point x_0 lies in the domain of attraction of the equilibrium position $\Phi_1(y_0)$ of system (1.8) for $y = y_0$ and $(\Phi_1(y_0), y_0) \in \Gamma_{1-}$. Next we consider the linear system

$$\varepsilon \dot{h}_1 = A_{11}(t,\varepsilon)h_1 + A_{12}(t,\varepsilon)h_2, \qquad \dot{h}_2 = A_{21}(t,\varepsilon)h_1 + A_{22}(t,\varepsilon)h_2, \tag{1.10}$$

resulting from the linearization of system (1.9) on the indicated trajectory and denote by $h_1(t,\varepsilon)$, $h_2(t,\varepsilon)$ its solutions with initial conditions $h_1(0,\varepsilon) = h_1^0$, $h_2(0,\varepsilon) = h_2^0$ independent of ε. We denote by $(x_0(t), y_0(t))$ the solution of the hybrid system (1.7) with the same initial conditions $x_0(0) = x_0$, $y_0(0) = y_0$ and switching moments $t_0 < t_1 < \ldots$. Suppose that Conditions 1.1–1.3 are fulfilled on the trajectory $y_0(t)$ of the relay system (1.1).

THEOREM 1.1 *The limiting relations*

$$\lim_{\varepsilon \to 0}(x_0(t,\varepsilon), y_0(t,\varepsilon)) = (x_0(t), y_0(t)), \tag{1.11}$$

$$\lim_{\varepsilon \to 0}(h_1(t,\varepsilon), h_2(t,\varepsilon)) = (h_{10}(t), h_{20}(t)), \tag{1.12}$$

where $h_{20}(t)$ is the solution of system (1.4), (1.5) with the initial condition $h_{20}(0) = h_2^0$ and

$$h_{10}(t) = (\Phi_1'|_{y=y_0(t)})h_{20}(t) \quad or \quad (\Phi_2'|_{y=y_0(t)})h_{20}(t)$$

hold uniformly with respect to the initial conditions (x_0, y_0) belonging to any closed bounded set and are valid uniformly with respect to the vectors h_1^0, h_2^0 lying in a unit ball with center at zero and also with respect to t from a finite interval with deleted arbitrarily small fixed neighborhoods of the points $t = 0, t_0, t_1, \ldots$.

It should be emphasized that in the C-metric the proximity of solutions of a hybrid and a relaxation system follows from the asymptotics of the trajectories in the neighborhood of the junction points and the general properties of singularly perturbed systems that we give below. Therefore, we have to substantiate the limiting relation (1.12). Also make note of the fact that Theorem 1.1 makes it possible to formulate theorems on the correspondence between the structurally stable stationary behavior of the relay system (1.1) and that of the relaxation system (1.9). For instance, we can make the following statement, which generalizes the Pontryagin–Mishchenko theorem [74].

THEOREM 1.2. *Under the conditions of Lemma 1.2, in some neighborhood of the cycle Z_0 independent of ε the relaxation system (1.9) has a unique exponentially orbitally stable cycle $Z(\varepsilon)$ with period $T(\varepsilon)$, with $Z(\varepsilon) \to Z_0$, $T(\varepsilon) \to T_0$ as $\varepsilon \to 0$.*

To prove this theorem, we construct a generic plane at the point $(\Phi_1(y_0(0)), y_0(0))$. Note that, by virtue of the C-proximity of solutions of the hybrid and the relaxation system, the Poincaré operator along the trajectories of system (1.9) maps a certain convex set of this plane into itself. In addition, it follows from [77] (we shall speak of this later) that all cycles $Z(\varepsilon)$ of system (1.9) lying in the vicinity of Z_0 have the same asymptotics with an accuracy to within quantities of order ε. This is a decisive circumstance. Indeed, performing linearization on the cycles indicated above, we arrive at linear $T(\varepsilon)$-periodic systems of the form (1.10) whose monodromy matrices we denote by $W(\varepsilon)$. The limiting relation (1.12) allows us to infer that

$$\lim_{\varepsilon \to 0} W(\varepsilon) = \begin{pmatrix} 0 & -A^{-1}(T_0)B(T_0)V \\ 0 & V \end{pmatrix},$$

where $A(t)$, $B(t)$ are the same as in Eq. (1.4) and V is a monodromy matrix of the periodic problem (1.4), (1.5). We have proved Theorem 1.2.

Note that a statement similar to Theorem 1.2 can be found in [91], where the convergence of the derivative of the Poincaré operator of system (1.9), as $\varepsilon \to 0$, is substantiated. In order to solve the problem, which was also considered in [77], the author of [91] reduces system (1.9) to a certain canonical form in which only one fast variable remains, which is taken as the new time. In addition, in order to make the presentation more visual, he considers only two slow variables. The technique used in [91] is more specialized and essentially differs from ours. No attention is given in [91] to the algorithmic part, there is nothing similar there to the periodic problem (1.4), (1.5) which we use to formulate an effective criterion of the uniqueness and stability of the cycle.

Let us return to Theorem 1.1. The proof of the limiting relation (1.12) is rather cumbersome. To make it easier to grasp what we present below, we describe the general character of the variation of the functions $h_1(t,\varepsilon)$, $h_2(t,\varepsilon)$ with time which vary in the least trivial fashion when we pass through the neighborhoods of the junction points. We shall first consider the values of t that appear when the motion along the trajectory $(x_0(t,\varepsilon), y_0(t,\varepsilon))$ of the relaxation system (1.9) is slow. For these t the eigenvalues of the matrix A_{11} from (1.10) lie on the left of the imaginary axis, and this allows us to pass from system (1.10) to Eq. (1.4). It follows from the whole complex of the constructions that, when passing through the neighborhood of the junction point, the norm $h_1(t,\varepsilon)$ grows, assuming the order ε^{-1}. Moving along the trajectory $(x_0(t,\varepsilon), y_0(t,\varepsilon))$, we recede far from the junction point during the time of order ε. On this entire interval, the norm $h_1(t,\varepsilon)$ remains a quantity of order ε^{-1}. Now, when moving along the trajectory, we come close to the drop point, the norm $h_1(t,\varepsilon)$ falls to values of the order of unity, and the norm $h_2(t,\varepsilon)$ remains bounded because of the boundedness of the L-norm $h_1(t,\varepsilon)$ uniform with respect to ε. As concerns the purely mathematical techniques, the main difficulty lies in the proof of the boundedness of $h_2(t,\varepsilon)$, say, for $t_0 - \delta \leq t \leq t_0 + \delta$ and the calculation of the quantities

$$h_{20}(t_0 \pm 0) = \lim_{\delta \to 0} \lim_{\varepsilon \to 0} h_2(t_0 \pm \delta, \varepsilon). \qquad (1.13)$$

When all this is done, the validity of (1.5) can be easily proved.

1.4. Some Auxiliary Information

First of all, we shall describe the general properties of the solutions of system (1.9) and of the linearized system (1.10) in the neighborhood of the junction point. For our purpose, we shift the zero of the space R^n to the point $(\Phi_1(y_0(t_0)), y_0(t_0))$ and introduce a special coordinate system in its neighborhood. In R^k we consider a plane orthogonal to the vector β appearing on the right-hand side of formula (1.6), in which we choose some $k-1$ linearly independent base directions. We supplement them by a curvilinear one-dimensional direction which is an invariant integral manifold of Eq. (1.8) for $y = y_0(t_0)$. In R^m we take as the basis a vector $\dot{y}_0(t_0 - 0)$ and $m-1$ linearly independent directions lying in the plane orthogonal to vector (1.6). We denote the new coordinates by $\xi = (\xi_1, \ldots, \xi_k)$ and $\eta = (\eta_1, \ldots, \eta_m)$ respectively, supposing that ξ_1 measures the distances along the one-dimensional "whiskers" of (1.8) and η_1 corresponds to the direction of $\dot{y}_0(t_0 - 0)$.

After identity transformations and normalization of time, system (1.9) in the new variables assumes the form

$$\varepsilon\dot{\xi} = F(\xi,\eta), \qquad \dot{\eta} = G(\xi,\eta), \qquad (1.14)$$

where the first component $F_1 = \xi_1^2 + \eta_1 + \ldots$ of the vector function F does not include terms linear with respect to $\eta_2, \ldots, \eta_m$ and $F_j(\xi_1, 0, \ldots, 0) \equiv 0$ for $j = 2, \ldots, k$.

These properties of F are obvious consequences of the technique of transition to new coordinates described above, in which the zero approximation of the trajectory of system (1.9) can be analytically written in a more simple form. Indeed, as follows from [77], the relations

$$\eta_1(\xi_1,\varepsilon) = \eta_1^0(\xi_1) + O(\varepsilon^{2/3}), \quad \eta_j(\xi_1,\varepsilon) = \eta_j^0(\xi_1) + O(\varepsilon \ln\frac{1}{\varepsilon}), \qquad (1.15)$$

$$\xi_l(\xi_1,\varepsilon) = \xi_l^0(\xi_1) + O(\varepsilon^{2/3}), \quad j = 2,\ldots,m, \quad l = 2,\ldots,k, \qquad (1.16)$$

(an independent way of substantiating them will be explained later) hold true for $-q \leq \xi_1 \leq q$, where $q > 0$ is sufficiently small. In these relations $\eta_1^0(\xi_1) = -\xi_1^2 + \ldots$ for $\xi_1 \leq 0$ and $\eta_1^0(\xi_1) \equiv 0$ for $\xi_1 \geq 0$, the functions $\eta_j^0(\xi_1)$, $\xi_l^0(\xi_1)$ are also identically zero for positive ξ_1, and for $\xi_1 \leq 0$ the expansions of η_j^0 begin with cubic terms and those of ξ_l^0 begin with quadratic terms (in [77] the functions $\xi_l^0(\xi_1)$ do not vanish for $\xi_1 \geq 0$).

Everywhere in what follows we assume that system (1.10) is the result of the linearization of system (1.14) on the corresponding time intervals, the moment of time $t = 0$ on the trajectory being associated with $\xi_1 = -q$.

The coordinates that we have introduced make it possible to investigate the various integrals that appear when we analyze the properties of $h_1(t,\varepsilon)$, $h_2(t,\varepsilon)$, with the use of the substitution

$$dt = \varepsilon d\xi_1/F_1(\xi,\eta), \qquad (1.17)$$

which is connected in a natural way with the trajectory of system (1.9) and whose invertibility follows from the fact that the variable ξ_1 along the trajectory increases monotonically. It is essential that the structure of the function F_1 and formulas (1.15), (1.16) give the estimate

$$1/F_1(\xi,\eta) \leq N/\xi_1^2, \qquad \Delta_1 \varepsilon^{1/3} \leq \xi_1 \leq q, \qquad (1.18)$$

where $\Delta_1 > 0$ is sufficiently large, $N > 0$ (here and in what follows the letter N denotes positive constants whose exact meaning is inessential).

Below $\xi_1^{(t)}$ is the coordinate ξ_1 of a point on the trajectory of system (1.14), in particular, $\xi_1^{(0)} = -q$. We denote by τ_2 a moment of time such

that $\xi_1^{(\tau_2)} = q$. It is significant that the matrix $A_{11}(t,\varepsilon)$ from (1.10) possesses, for $0 \le t \le \tau_2$, a unique eigenvalue $\lambda(t,\varepsilon)$ that changes sign, and its other eigenvalues, for these t, lie on the left of the imaginary axis. The asymptotic relation
$$\lambda(t,\varepsilon) = \lambda_0(\xi_1^{(t)}) + O(\varepsilon^{2/3}), \qquad -q \le \xi_1^{(t)} \le q, \qquad (1.19)$$
where $\lambda_0(\xi_1) = 2\xi_1 + O(\xi_1^2)$, is valid for the leading eigenvalue of the matrix $A_{11}(t,\varepsilon)$. Below we explain the way of defining the function $\lambda_0(\xi_1)$. We denote by $\lambda_0(\xi,\eta)$ the leading eigenvalue of the matrix F'_ξ. Substituting the zero approximations from (1.15), (1.16) into $\lambda_0(\xi,\eta)$, we obtain $\lambda_0(\xi_1)$. Now the validity of the asymptotic relation (1.19) immediately follows from (1.15), (1.16) and from the structure of the vector function F.

Using (1.19), we infer that $\lambda(t,\varepsilon)$ has only one zero τ_1 when $0 \le t \le \le \tau_2$ and that $\xi_1^{(\tau_1)} = O(\varepsilon^{2/3})$. The calculation of the breakoff time of the trajectory of system (1.14) carried out in [77] leads to relations $\tau_1 = t_0 + +O(\varepsilon^{2/3})$, $\tau_2 - \tau_1 = O(\varepsilon^{2/3})$. The latter relation means that, as compared to ε, the interval of positivity of $\lambda(t,\varepsilon)$ is large. Proceeding from (1.17), (1.19) and calculating F_1 with due account of the terms of order $\varepsilon^{2/3}$ (which are essential here), we can additionally verify that
$$\int_{\tau_1}^{\tau_2} \lambda(t,\varepsilon)\, dt = \frac{2}{3}\varepsilon \ln \frac{1}{\varepsilon} + O(\varepsilon). \qquad (1.20)$$

(We omit the detailed derivation of (1.20) since in our construction it has only an heuristic value.) We shall prove in the next section that the norm $h_1(\tau_1,\varepsilon)$ is of order $\varepsilon^{-1/3}$. From this fact and from (1.20) it is now clear why the norm $h_1(t,\varepsilon)$ assumes the order ε^{-1} when passing the interval of positivity of $\lambda(t,\varepsilon)$.

The substantiation of the following proposition explains, to a certain degree, the arguments used in the preceding paragraph. (Here and henceforth $|\cdot|$ is the norm or the modulus.)

LEMMA 1.3. *For $0 \le \tau < t \le \tau_2$, the Cauchy matrix $K(t,\tau)$ of the system*
$$\varepsilon \dot{h} = A_{11}(t,\varepsilon) h \qquad (1.21)$$
satisfies the inequality
$$|K(t,\tau)| \le N K_0(t,\tau), \qquad K_0(t,\tau) = \exp\left\{\frac{1}{\varepsilon}\int_\tau^t \lambda(s,\varepsilon)\, ds\right\}. \qquad (1.22)$$

Theorem on the C^1-Proximity

PROOF. We introduce into consideration the matrix

$$U(t,\varepsilon) = U(\xi,\eta)|_{\xi=\xi(t,\varepsilon),\eta=\eta(t,\varepsilon)},$$

where $\xi(t,\varepsilon)$, $\eta(t,\varepsilon)$ are the components of the trajectory of system (1.14) for $0 \leq t \leq \tau_2$ and $U(\xi,\eta)$ gives a block-diagonal form to the matrix F'_ξ. Making the substitution $h = U(t,\varepsilon)v$ in (1.21), we obtain

$$\varepsilon \dot{v} = [C_1(t,\varepsilon) + \varepsilon C_2(t,\varepsilon)]v, \tag{1.23}$$

where $C_2(t,\varepsilon) = -U^{-1}\dot{U}$ and $C_1(t,\varepsilon) = \text{diag}\{\lambda(t,\varepsilon), \Lambda(t,\varepsilon)\}$. In this case, all eigenvalues of the matrix $\Lambda(t,\varepsilon)$ of order $k-1$ lie on the left of the imaginary axis. In addition, since q is small, the elements of the matrix $\Lambda(t,\varepsilon)$ are close to constants, and therefore

$$|K_1(t,\tau)| \leq N_1 \exp\left\{-\frac{N_2}{\varepsilon}(t-\tau)\right\},$$

where $K_1(t,\tau)$ is the Cauchy matrix of the equation $\varepsilon \dot{w} = \Lambda(t,\varepsilon)w$. From this equation and from the Gronwall–Bellman lemma it follows that, in order to prove this assertion, it is sufficient to make sure of the uniform (with respect to ε) boundedness of $C_2(t,\varepsilon)$ in the L_1-norm.

It follows from the choice of the matrix $U(t,\varepsilon)$ that

$$|C_2(t,\varepsilon)| \leq N|\dot{\xi}(t,\varepsilon)|. \tag{1.24}$$

Note that the integral of $|\dot{\xi}(t,\varepsilon)|$ taken along the interval $[0,\tau_2]$ is equal to the variation of the vector function $\xi(t,\varepsilon)$ or to the length of the curve. As to the latter, it is geometrically obvious that it is bounded uniformly with respect to ε. Indeed, choosing ξ_1 as a parameter on the curve and differentiating the asymptotic relations (1.16) for the trajectory of system (1.14), we make sure that

$$\frac{d\xi_l(\xi_1,\varepsilon)}{d\xi_1} = \frac{d\xi_l^0(\xi_1)}{d\xi_1} + O(\varepsilon^{1/3}), \quad -q \leq \xi_1 \leq q, \tag{1.25}$$

where $l = 2,\ldots,k$. From this the required fact follows trivially.

Let us prove the inequality

$$K_0(t,\tau) \leq N(\xi_1^{(t)}/\xi_1^{(\tau)})^2, \quad \Delta_1 \varepsilon^{1/3} \leq \xi_1^{(\tau)} < \xi_1^{(t)} \leq q, \tag{1.26}$$

often used at different stages of the substantiation of Theorem 1.1. For the ξ_1 being considered (by virtue of the properties of F_1 and formulas (1.15), (1.16)), $1/F_1 \leq 1/(\xi_1^2 - N\varepsilon^{2/3})$. Now we use the asymptotic representation

(1.19) and substitution (1.17) for the formula for K_0. As a result we arrive at the following inequality, equivalent to (1.26):

$$K_0(t,\tau) \leq N_1 \exp\left\{ \int_{\xi_1^{(\tau)}}^{\xi_1^{(t)}} \frac{2\xi_1}{\xi_1^2 - N_2 \varepsilon^{2/3}} d\xi_1 \right\}.$$

The facts presented above, which are basic facts, make it possible to substantiate the properties of the functions $h_1(t,\varepsilon)$, $h_2(t,\varepsilon)$, described in 1.3, which immediately yield the limiting relation (1.12). It is clear that the analysis of the asymptotic properties of the given functions requires the calculation and estimation of a large number of special integrals. Therefore, we devote a special section to this problem.

2. Proof of the Theorem on the C^1-Proximity

2.1. Asymptotic Properties of the Functions $h_1(t,\varepsilon)$, $h_2(t,\varepsilon)$ at the Beginning of the Breakoff Section

It follows from the asymptotic properties of the relaxation system trajectory that outside of a certain neighborhood of order $\varepsilon^{2/3}$ of the switching moments t_p of the relay system (1.1), to which the relaxation system in question converges uniformly with respect to t from any finite interval, as $\varepsilon \to 0$, we have the relations

$$\begin{aligned} A_{11}(t,\varepsilon) &= A(t) + O(\varepsilon^{2/3}), & A_{12}(t,\varepsilon) &= B(t) + O(\varepsilon^{2/3}), \\ A_{21}(t,\varepsilon) &= C(t) + O(\varepsilon^{2/3}), & A_{22}(t,\varepsilon) &= D(t) + O(\varepsilon^{2/3}), \end{aligned} \quad (2.1)$$

where the matrices A, B, C, D are the same as in (1.4), and matrices (2.1) are uniformly bounded on the whole interval of variation of t (the neighborhoods of the points t_p inclusive). Using (2.1) and applying the technique of asymptotic integration described in [7] to system (1.10), we make sure that on any interval $[\delta, t_0 - \delta]$, where $\delta > 0$ is fixed, the functions $h_1(t,\varepsilon)$, $h_2(t,\varepsilon)$ uniformly converge, as $\varepsilon \to 0$, to the functions

$$h_{10}(t) = -A^{-1}(t)B(t)h_{20}(t), \qquad (2.2)$$

$$h_{20}(t) = h_2^0 + \int_0^t [D(\tau) - C(\tau)A^{-1}(\tau)B(\tau)]h_{20}(\tau)\, d\tau. \qquad (2.3)$$

Note that as $t \to t_0 - 0$, function (2.2) has a singularity of the form $1/\sqrt{t_0 - t}$, i.e., according to (1.19) the norm $A^{-1}(t)$ increases as $1/\xi_1^{(t)}$, as $t \to t_0 - 0$, and, by virtue of the zero approximation formulas for the relaxation system trajectories, $\xi_1 \sim -\sqrt{-\eta_1}$, $\eta_1 \sim t - t_0$.

THEOREM 2.1. *Uniformly with respect to $t \in [0, \tau_1]$,*

$$\lim_{\varepsilon \to 0} h_2(t, \varepsilon) = h_{20}(t), \qquad (2.4)$$

where $h_{20}(t)$ is defined by (2.3). A similar limiting relation for $h_1(t, \varepsilon)$ holds true on any interval of the form $[\delta, t_0 - \delta]$, and the estimate

$$|h_1(\tau_1, \varepsilon)| \leq N\varepsilon^{-1/3} \qquad (2.5)$$

is true at the point τ_1.

PROOF. The first equation of (1.10) gives

$$h_1(t, \varepsilon) = K(t, 0)h_1^0 + \frac{1}{\varepsilon} \int_0^t K(t, \tau) A_{12}(\tau, \varepsilon) h_2(\tau, \varepsilon) \, d\tau. \qquad (2.6)$$

Substituting (2.6) into the second equation of (1.10), we arrive at the equality

$$h_2(t, \varepsilon) - \int_0^t [A_{22}(\tau, \varepsilon) h_2(\tau, \varepsilon) +$$

$$+ \frac{1}{\varepsilon} A_{21}(\tau, \varepsilon) \int_0^\tau K(\tau, \theta) A_{12}(\theta, \varepsilon) h_2(\theta, \varepsilon) \, d\theta] \, d\tau =$$

$$= h_2^0 + \int_0^t A_{21}(\tau, \varepsilon) K(\tau, 0) h_1^0 \, d\tau. \qquad (2.7)$$

Setting $h_2(t, \varepsilon) = h_{20}(t) + h_{21}(t, \varepsilon)$ in (2.7), we infer that $h_{21}(t, \varepsilon)$ is the solution of the integral equation

$$Lh_{21} = \Psi_1(t) + \Psi_2(t) + \Psi_3(t), \qquad (2.8)$$

where the Volterra operator L is defined by the left-hand side of (2.7) and

$$\Psi_1(t) = \int_0^t [A_{22}(\tau, \varepsilon) - D(\tau)] h_{20}(\tau) \, d\tau,$$

$$\Psi_2(t) = \int_0^t A_{21}(\tau, \varepsilon) K(\tau, 0) h_1^0 \, d\tau,$$

$$\Psi_3(t) = \int_0^t \left[\frac{1}{\varepsilon} A_{21}(\tau, \varepsilon) \int_0^\tau K(\tau, \theta) A_{12}(\theta, \varepsilon) h_{20}(\theta) \, d\theta + \right.$$

$$\left. + C(\tau) A^{-1}(\tau) B(\tau) h_{20}(\tau) \right] d\tau.$$

Let us prove that the norm of the operator L^{-1} in the space C of functions continuous on $[0, \tau_1]$ is uniformly bounded with respect to ε. With this aim in view, we consider a nonhomogeneous equation $Lh = f(t)$. Let $N_1 = \max_{0 \le t \le \tau_1} |f(t)|$. Then, by virtue of Lemma 1.3, the function $v(t) = |h(t)|$ satisfies the inequality

$$v(t) \le N_1 + N_2 \int_0^t v(\tau) \, d\tau + N_3 \frac{1}{\varepsilon} \int_0^t \left(\int_0^\tau K_0(\tau, \theta) v(\theta) \, d\theta \right) d\tau. \qquad (2.9)$$

Changing the integration order in the last term on the right-hand side of (2.9), we arrive at the inequality

$$\int_0^t \left(\int_\theta^t K_0(\tau, \theta) d\tau \right) v(\theta) \, d\theta \le \int_0^t \left(\int_\theta^{\tau_1} K_0(\tau, \theta) d\tau \right) v(\theta) \, d\theta. \qquad (2.10)$$

Taking into account (2.10) and (2.9) and using the Gronwall–Bellman lemma, we see that in order to prove the required fact, we have to verify the boundedness of the integral

$$\frac{1}{\varepsilon} \int_0^{\tau_1} \left(\int_\theta^{\tau_1} K_0(\tau, \theta) d\tau \right) d\theta. \qquad (2.11)$$

Let us prove that integral (2.11) (we denote it by J) has a limit, as $\varepsilon \to 0$, equal to $-\int_0^{t_0} \lambda_0^{-1}(\theta) \, d\theta$, where $\lambda_0(\theta)$ is the zero approximation of $\lambda(t, \varepsilon)$. With this aim in view, we divide (2.11) into two integrals along the intervals $[0, \tau_1 - \tau_{10}\varepsilon^{2/3}]$ and $[\tau_1 - \tau_{10}\varepsilon^{2/3}, \tau_1]$, where $\tau_{10} > 0$ is arbitrarily fixed. In the second integral the integration interval is asymptotically small, and this allows us to estimate it by the value $N\varepsilon^{1/3}$. We write the first integral in the form $J_1 + J_2$, dividing the inner integral in it by the point $\tau_1 - \tau_{10}\varepsilon^{2/3}$. For J_2 we have the estimate

$$J_2 = \frac{1}{\varepsilon} \int_{\tau_1 - \tau_{10}\varepsilon^{2/3}}^{\tau_1} K_0(\tau, \tau_1 - \tau_{10}\varepsilon^{2/3}) \, d\tau \int_0^{\tau_1 - \tau_{10}\varepsilon^{2/3}} K_0(\tau_1 - \tau_{10}\varepsilon^{2/3}, \theta) \, d\theta \le N\varepsilon^{1/3},$$

which becomes obvious if we put K_0 in the second factor under the differentiation sign, i.e., write it as the Stieltjes integral, and take into consideration that, by virtue of (1.19), $\lambda(\tau_1 - \tau_{10}\varepsilon^{2/3}, \varepsilon) = O(\varepsilon^{1/3})$, and use the inequality $K_0 \le 1$ in the first factor. We write J_1 in a similar way, in the form of the Stieltjes integral, and then pass to the limit in it. To substantiate the limiting process, we deviate, as usual, from the "singularity" of $\tau_1 - \tau_{10}\varepsilon^{2/3}$

Theorem on the C^1-Proximity 43

by a small value $\delta > 0$ and on the corresponding intervals use the inequality $\lambda(t,\varepsilon) \leq -N\sqrt{\tau_1 - \tau_{10}\varepsilon^{2/3} - t}$, $0 \leq t \leq \tau_1 - \tau_{10}\varepsilon^{2/3}$, following from (1.19) (see the remark after (2.3)). The last inequality leads to the estimates

$$\frac{1}{\varepsilon} \int_{\tau_1-\tau_{10}\varepsilon^{2/3}-\delta}^{\tau_1-\tau_{10}\varepsilon^{2/3}} \left(\int_{\theta}^{\tau_1-\tau_{10}\varepsilon^{2/3}} K_0(\tau,\theta)\,d\tau \right) d\theta \leq$$

$$\leq N_1 \varepsilon^{1/3} \int_{-\delta N_2/\varepsilon^{2/3}}^{0} \exp(-|u|^{3/2}) \int_{u}^{0} \exp(|v|^{3/2})\,dv\,du \leq N_3\sqrt{\delta},$$

$$\frac{1}{\varepsilon} \int_{0}^{\tau_1-\tau_{10}\varepsilon^{2/3}-\delta} \left(\int_{\tau_1-\tau_{10}\varepsilon^{2/3}-\delta}^{\tau_1-\tau_{10}\varepsilon^{2/3}} K_0(\tau,\theta)\,d\tau \right) d\theta \leq$$

$$\leq N_1 \varepsilon^{1/3} \int_{-\infty}^{-\delta N_2/\varepsilon^{2/3}} \exp(-|u|^{3/2})\,du \int_{-\delta N_2/\varepsilon^{2/3}}^{0} \exp(|v|^{3/2})\,dv \leq N_3 \varepsilon^{1/3},$$

that substantiate the possibility of the passage to the limit.

Thus the quantity (2.11) is bounded. To prove the limiting relation (2.4), we must show that as $\varepsilon \to 0$, the right-hand side of (2.8) tends to zero uniformly with respect to t, $0 \leq t \leq \tau_1$. For $\Psi_1(t)$ we have this from (2.1). Writing the integral $\Psi_2(t)$ in the Stieltjes form, we make sure that it is of order ε. Finally, acting as in the case of integral (2.11), we infer that as $\varepsilon \to 0$, $\Psi_3(t) \to 0$ uniformly with respect to t, where $0 \leq t \leq \tau_1$. We have thus proved (2.4). Taking (2.4) into account in (2.6), we make sure that a similar equality holds true for $h_1(t,\varepsilon)$ on any interval of the form $[\delta, t_0 - \delta]$.

Let us now prove estimate (2.5). Using Lemma 1.3 and taking into account the boundedness of $h_2(t,\varepsilon)$ on $[0,\tau_1]$ established above, we derive

$$|h_1(\tau_1,\varepsilon)| \leq N_1 + N_2 \frac{1}{\varepsilon} \int_0^{\tau_1} K_0(\tau_1,\tau)\,d\tau \qquad (2.12)$$

from (2.6).

We divide the integration interval in the integral on the right-hand side of (2.12) by the point $\tau_1 - \tau_{10}\varepsilon^{2/3}$ and get the required estimate if we rewrite the first integral in the Stieltjes form and recall that $\lambda(\tau_1 - \tau_{10}\varepsilon^{2/3},\varepsilon) = O(\varepsilon^{1/3})$ and use the inequality $K_0 \leq 1$ in the second integral. We have proved Theorem 2.1.

It follows from the asymptotic properties of the trajectory $(x_0(t,\varepsilon), y_0(t,\varepsilon))$ of system (1.9) that $(x_0(\tau_1,\varepsilon), y_0(\tau_1,\varepsilon)) \to (\Phi_1(y_0(t_0)), y_0(t_0))$ as $\varepsilon \to 0$, i.e., Theorem 2.1 is really connected with the beginning of the fast-motion section. We can also prove a similar proposition for the end of the fast-motion section. Here is what we mean. We set $\tau_3 = \tau_2 + \Delta_2\varepsilon \ln(1/\varepsilon)$, where $\Delta_2 > 0$ is sufficiently large. By virtue of the results given in [77], $(x_0(\tau_3,\varepsilon), y_0(\tau_3,\varepsilon)) \to (\Phi_2(y_0(t_0)), y_0(t_0))$ as $\varepsilon \to 0$, i.e., the analog of Theorem 2.1 again holds true for suitable values of $t \geq \tau_3$ if, of course, we first establish the boundedness of the quantities $h_1(\tau_3,\varepsilon)$, $h_2(\tau_3,\varepsilon)$ as $\varepsilon \to 0$. Therefore, it follows from Theorem 2.1 and from the corresponding statement for $t \geq \tau_3$ that quantities (1.13) can be obtained from the limiting relations

$$h_{20}(t_0 - 0) = \lim_{\varepsilon \to 0} h_2(\tau_1,\varepsilon), \qquad h_{20}(t_0 + 0) = \lim_{\varepsilon \to 0} h_2(\tau_3,\varepsilon).$$

2.2. The Boundedness of $h_2(t,\varepsilon)$ for $\tau_1 \leq t \leq \tau_3$

Let us prove the boundedness of $h_2(t,\varepsilon)$ for $\tau_1 \leq t \leq \tau_3$ which, as was pointed out in Sec. 1, is the central feature in the substantiation of the limiting relation (1.12). The importance of the boundedness of $h_2(t,\varepsilon)$ is also clear from the relation

$$h_2(\tau_3,\varepsilon) - h_2(\tau_1,\varepsilon) = \frac{1}{\varepsilon} \int_{\tau_1}^{\tau_3} A_{21}(\tau,\varepsilon) \int_0^\tau K(\tau,\theta) A_{12}(\theta,\varepsilon) h_2(\theta,\varepsilon) \, d\theta \, d\tau +$$

$$+ \int_{\tau_1}^{\tau_3} [A_{22}(\tau,\varepsilon) h_2(\tau,\varepsilon) + A_{21}(\tau,\varepsilon) K(\tau,0) h_1^0] \, d\tau, \qquad (2.13)$$

which follows from (2.7) and which we shall use later to calculate the limit of $h_2(\tau_3,\varepsilon)$ as $\varepsilon \to 0$. Indeed, under the condition of boundedness of $h_2(t,\varepsilon)$ on the interval $[\tau_1,\tau_3]$, the second term on the right-hand side of (2.13) can be estimated in the norm by a value of order $\varepsilon^{2/3}$, i.e., we shall have to analyze only the first term.

Let us first prove the boundedness of $h_2(t,\varepsilon)$ on the interval $[\tau_1,\tau_2]$. Rewriting the integral equation (2.7) for the initial moment of time τ_1 and reasoning as we did when we proved Theorem 2.1, we infer that the required fact follows from the boundedness of the quantities

$$|h_1(\tau_1,\varepsilon)| \int_{\tau_1}^{\tau_2} K_0(\tau,\tau_1) \, d\tau, \qquad \frac{1}{\varepsilon} \int_{\tau_1}^{\tau_2} \left(\int_\theta^{\tau_2} K_0(\tau,\theta) \, d\tau \right) d\theta. \qquad (2.14)$$

Theorem on the C^1-Proximity

We divide these integrals into special sums and study every term separately. Of decisive importance here are the substitution (1.17) and inequalities (1.18), (1.26).

By virtue of (1.22), the estimates of integrals (2.14) are closely connected with the behavior of the function $\lambda(t,\varepsilon)$. Using (1.19), we isolate temporal intervals on which it has different asymptotics and which are convenient for use when we deal with integrals (2.14). Let $\xi_1^{(t_*)} = \Delta_1 \varepsilon^{1/3}$, $\xi_1^{(t_{**})} = \delta_1 \varepsilon^{1/3} \ln(1/\varepsilon)$, where $\delta_1 > 0$ is small and Δ_1 is the same as in (1.18), (1.26). Formula (1.19) gives

$$0 \leq \lambda(t,\varepsilon) \leq N\varepsilon^{1/3}, \qquad \tau_1 \leq t \leq t_*, \qquad (2.15)$$

$$0 \leq \lambda(t,\varepsilon) \leq \delta_2 \varepsilon^{1/3} \ln(1/\varepsilon), \qquad \tau_1 \leq t \leq t_{**}, \qquad (2.16)$$

where $\delta_2 > 0$ is small.

We divide the first integral in (2.14) in two by the point t_*. Inequality (2.15) makes it possible to estimate the first of them from above by the value $N\varepsilon^{2/3}$, and to obtain

$$\int_{t_*}^{\tau_2} K_0(\tau,\tau_1)\, d\tau = \int_{t_*}^{\tau_2} K_0(\tau,t_*)\, d\tau \cdot K_0(t_*,\tau_1) \leq N \int_{t_*}^{\tau_2} K_0(\tau,t_*)\, d\tau \qquad (2.17)$$

for the second integral.

For the integral on the right-hand side of (2.17), with the use of the substitution (1.17) and inequalities (1.18), (1.26), we find the estimate

$$\int_{t_*}^{\tau_2} K_0(\tau,t_*)\, d\tau \leq N_1 \varepsilon \int_{\xi_1^{(t_*)}}^{\xi_1^{(\tau_2)}} \left(\frac{\xi_1^{(\tau)}}{\xi_1^{(t_*)}}\right)^2 \frac{d\xi_1^{(\tau)}}{(\xi_1^{(\tau)})^2} \leq N_2 \varepsilon^{1/3}. \qquad (2.18)$$

Taking into account (2.5) and (2.18), we infer that the first quantity in (2.14) is bounded.

Dividing the integration interval in the second integral of (2.14) by the point t_{**}, we denote by I_1 and I_2 the terms of the resulting sum. Carrying out the substitution (1.17) in I_2 and taking (1.18), (1.26) into account, we infer that

$$I_2 \leq N_1 \varepsilon \int_{\xi_1^{(t_{**})}}^{\xi_1^{(\tau_2)}} \left(\int_{\xi_1^{(\theta)}}^{\xi_1^{(\tau_2)}} \left(\frac{\xi_1^{(\tau)}}{\xi_1^{(\theta)}}\right)^2 \frac{d\xi_1^{(\tau)}}{(\xi_1^{(\tau)})^2} \right) \frac{d\xi_1^{(\theta)}}{(\xi_1^{(\theta)})^2} \leq N_2 / \ln^3 \frac{1}{\varepsilon}.$$

Again using the point t_{**}, we divide the inner integral in I_1, which makes it possible to write it as $I_{11} + I_{12}$. Using inequality (2.16), we arrive at the estimate

$$I_{11} \leq \frac{1}{\varepsilon} \int_{\tau_1}^{t_{**}} \left(\int_{\theta}^{t_{**}} \exp\{\delta_2(\tau - \theta)\varepsilon^{1/3} \ln \frac{1}{\varepsilon}\} \, d\tau \right) d\theta \leq N\varepsilon^{1/3-\delta_3}/\ln^2 \frac{1}{\varepsilon}, \quad (2.19)$$

where $\delta_3 > 0$ is small. Then, in the representation

$$\varepsilon I_{12} = \int_{\tau_1}^{t_{**}} K_0(t_{**}, \theta) \, d\theta \cdot \int_{t_{**}}^{\tau_2} K_0(\tau, t_{**}) \, d\tau \quad (2.20)$$

we write the first factor as

$$\int_{\tau_1}^{t_*} K_0(t_*, \theta) \, d\theta \cdot K_0(t_{**}, t_*) + \int_{t_*}^{t_{**}} K_0(t_{**}, \theta) \, d\theta. \quad (2.21)$$

Using (2.15) and (1.26), we obtain

$$\int_{\tau_1}^{t_*} K_0(t_*, \theta) \, d\theta \leq N\varepsilon^{2/3}, \qquad K_0(t_{**}, t_*) \leq N \ln^2 \frac{1}{\varepsilon}. \quad (2.22)$$

The estimate of the second term in (2.21) and of the second factor in (2.20) are based on substitution (1.17), which leads to the inequalities

$$\int_{t_*}^{t_{**}} K_0(t_{**}, \theta) \, d\theta \leq N_1 \varepsilon \int_{\xi_1^{(t_*)}}^{\xi_1^{(t_{**})}} \left(\frac{\xi_1^{(t_{**})}}{\xi_1^{(\theta)}} \right)^2 \frac{d\xi_1^{(\theta)}}{(\xi_1^{(\theta)})^2} \leq N_2 \varepsilon^{2/3} \ln^2 \frac{1}{\varepsilon}, \quad (2.23)$$

$$\int_{t_{**}}^{\tau_2} K_0(\tau, t_{**}) \, d\tau \leq N_1 \varepsilon \int_{\xi_1^{(t_{**})}}^{\xi_1^{(\tau_2)}} \left(\frac{\xi_1^{(\tau)}}{\xi_1^{(t_{**})}} \right)^2 \frac{d\xi_1^{(\tau)}}{(\xi_1^{(\tau)})^2} \leq N_2 \varepsilon^{1/3}/\ln^2 \frac{1}{\varepsilon}. \quad (2.24)$$

Now the boundedness of I_1 is the union of inequalities (2.19)–(2.24).

We have thus proved the boundedness of $h_2(t, \varepsilon)$ on the interval $[\tau_1, \tau_2]$, i.e., overcome the main difficulty. What remains is much easier. We set $\tau_2' = \tau_2 + \Delta_3 \varepsilon$, where $\Delta_3 > 0$ is sufficiently large, and consider the interval $[\tau_2, \tau_2']$. After the substitution $t \to \varepsilon t$, the dependence on ε in system (1.10) becomes regular and must be integrated on a finite time interval. Therefore,

$$|h_1(t, e)| \leq N|h_1(\tau_2, \varepsilon)|, \qquad \tau_2 \leq t \leq \tau_2'. \quad (2.25)$$

It turns out that in this case
$$|h_1(\tau_2, \varepsilon)| \leq N/\varepsilon. \tag{2.26}$$

Inequality (2.26) can be derived from the estimate
$$|h_1(\tau_2, \varepsilon)| \leq N_1 + N_2 \frac{1}{\varepsilon} \int_0^{\tau_2} K_0(\tau_2, \tau) \, d\tau,$$
resulting from (2.6) (with due account of Lemma 1.3), as follows: we divide the integral on the right-hand side in two by the point t_*. To estimate the first integral we represent $K_0(\tau_2, \tau) = K_0(\tau_2, t_*) K_0(t_*, \tau)$ and then use inequalities (1.26) and (2.15). In the case of the second integral, we have

$$\frac{1}{\varepsilon} \int_{t_*}^{\tau_2} K_0(\tau_2, \tau) \, d\tau \leq N_1 \int_{\xi_1^{(t_*)}}^{\xi_1^{(\tau_2)}} \left(\frac{\xi_1^{(\tau_2)}}{\xi_1^{(\tau)}} \right)^2 \frac{d\xi_1^{(\tau)}}{(\xi_1^{(\tau)})^2} \leq \frac{N_2}{\varepsilon}.$$

Inequality (2.26) allows us to infer that $h_2(t, \varepsilon)$ is bounded for $\tau_2 \leq t \leq \tau_2'$. It remains to consider the interval $\tau_2' \leq t \leq \tau_3$. By virtue of the choice of τ_2', for these t the points of the trajectory of relaxation system (1.9) lie in a small (independent of ε) neighborhood of the drop point. Therefore, the elements of the matrix $A_{11}(t, \varepsilon)$ are close to constants and its eigenvalues lie on the left of the imaginary axis. It follows that when $\tau_2' \leq \tau \leq t \leq \tau_3$, we have
$$|K(t, \tau)| \leq N_1 \exp \left\{ -\frac{N_2}{\varepsilon}(t - \tau) \right\}, \tag{2.27}$$
where, as we recall, $K(t, \tau)$ is the Cauchy matrix from (1.21). For $t = \tau_2'$, inequality (2.27) and estimate (2.25) ensure the boundedness of $h_2(t, \varepsilon)$ throughout the interval $[\tau_1, \tau_3]$.

At the end of 2.1, we pointed out that to prove the validity of an analog of Theorem 2.1 for suitable $t \geq \tau_3$, it is necessary to find out whether $h_1(\tau_3, \varepsilon)$ is bounded. We can carry out the proof as follows: we rewrite the integral equation (2.6) for the initial moment τ_2'; then, in order to estimate its right-hand side, use the boundedness of $h_2(t, \varepsilon)$ that we have already proved and inequalities (2.25)–(2.27). This chain leads to the estimate

$$|h_1(\tau_3, \varepsilon)| \leq N_1 \frac{1}{\varepsilon} \exp \left\{ -\frac{N_2}{\varepsilon}(\tau_3 - \tau_2') \right\} + N_3 \frac{1}{\varepsilon} \int_{\tau_2'}^{\tau_3} \exp \left\{ -\frac{N_4}{\varepsilon}(\tau_3 - \tau) \right\} d\tau,$$

which completes the proof.

Relation (2.6) implies the boundedness of $h_1(t,\varepsilon)$ outside of arbitrarily small neighborhoods of the points t_p. From this and from the second equation of (1.10) follows the convergence of $h_2(t,\varepsilon)$ to the solutions of (1.4), which is uniform with respect to t outside of intervals of the form $[\tau_1, \tau_3]$, i.e., fast-motion sections of the relaxation system trajectory. Thus, in order to find $h_{20}(t)$ for $t_0 < t < t_1$ (by virtue of Theorem 2.1, $h_{20}(t)$ is known for $0 \le t \le t_0$), we must calculate $h_{20}(t_0 + 0) = \lim_{\varepsilon \to 0} h_2(\tau_3, \varepsilon)$, and so on.

2.3. Calculating the Limit of $h_2(\tau_3, \varepsilon)$ as $\varepsilon \to 0$

As we have pointed out, the problem reduces to the analysis of the first term on the right-hand side of (2.13) which we denote by S. First of all, we isolate from S parts that tend to zero as $\varepsilon \to 0$. We have to perfect somewhat the analytic apparatus we shall use since for negative ξ_1 inequality (1.18) is no longer valid. As a matter of fact, by virtue of (1.19) and Lemma 1.3, only the case of asymptotically small values of ξ_1 is of interest. We shall show later that of main importance is the interval $\tau_1 - \delta(\varepsilon) \le t \le \tau_1$, where $\delta(\varepsilon) = \varepsilon^{2/3} \ln^2(1/\varepsilon)$. Note that, by virtue of relations $\xi_1 = O(\sqrt{-\eta_1})$ for $\eta_1 \le 0$ and $\eta_1(\tau_1 - \delta(\varepsilon), \varepsilon) = -\delta(\varepsilon) + O(\varepsilon^{2/3})$, we have

$$\xi_1^{(\tau_1 - \delta(\varepsilon))} = O(\varepsilon^{1/3} \ln(1/\varepsilon)). \tag{2.28}$$

Next, for $-N_1 \varepsilon^{1/3} \ln(1/\varepsilon) \le \xi_1 \le N_2 \varepsilon^{1/3}$, we have the equality

$$F_1(\xi, \eta) = \xi_1^2 + \varepsilon^{2/3} v_0(\xi_1/\varepsilon^{1/3}) + O(\varepsilon \ln^3(1/\varepsilon)), \tag{2.29}$$

where the function v_0 is defined in [77]. Relation (2.29) makes the substitution (1.17) constructive.

We set $S = S_1 + S_2$, dividing for this purpose the inner integral by the point $\tau_1 - \delta(\varepsilon)$. We infer from (2.13) that

$$|S_1| \le N \frac{1}{\varepsilon} \int_{\tau_1}^{\tau_3} \left(\int_0^{\tau_1 - \delta(\varepsilon)} |K(\tau, \theta)| \, d\theta \right) d\tau. \tag{2.30}$$

Representing $K(\tau, \theta) = K(\tau, \tau_1) K(\tau_1, \tau_1 - \delta(\varepsilon)) K(\tau_1 - \delta(\varepsilon), \theta)$ and using Lemma 1.3, we pass from (2.30) to the estimate

$$|S_1| \le N \frac{1}{\varepsilon} \int_{\tau_1}^{\tau_3} |K(\tau, \tau_1)| \, d\tau \int_0^{\tau_1 - \delta(\varepsilon)} K_0(\tau_1 - \delta(\varepsilon), \theta) \, d\theta \cdot K_0(\tau_1, \tau_1 - \delta(\varepsilon)). \tag{2.31}$$

Theorem on the C^1-Proximity

We represent the first factor on the right-hand side of (2.31) as the sum of integrals along the intervals $[\tau_1, \tau_2]$ and $[\tau_2, \tau_3]$. We use Lemma 1.3 to estimate the first integral and then divide the resulting integral of K_0 in two by the point t_*. In the first integral that results, we use inequality (2.15) and in the second, the substitution (1.17) with due account of inequalities (1.18) and (1.26). As a result we obtain

$$\int_{\tau_1}^{\tau_2} |K(\tau, \tau_1)| \, d\tau \leq N_1 \int_{\tau_1}^{\tau_2} K_0(\tau, \tau_1) \, d\tau \leq N_2 \varepsilon^{1/3}. \tag{2.32}$$

For the second integral we have

$$\int_{\tau_2}^{\tau_3} |K(\tau, \tau_1)| \, d\tau \leq N_1 \int_{\tau_2}^{\tau_3} |K(\tau, \tau_2)| \, d\tau \cdot K_0(\tau_2, \tau_1) \leq$$

$$\leq N_2 \int_{\tau_2}^{\tau_3} |K(\tau, \tau_2)| \, d\tau \cdot K_0(\tau_2, t_*) \leq N_3 \varepsilon^{-2/3} \int_{\tau_2}^{\tau_3} |K(\tau, \tau_2)| \, d\tau. \tag{2.33}$$

The first inequality in (2.33) is a consequence of Lemma 1.3, the second inequality follows from (2.15) and the third from (1.26). The last integral in (2.33) is estimated from above by a quantity of order ε by virtue of (2.27) and by virtue of the estimate $|K(\tau, \tau_2)| \leq N$, $\tau_2 \leq \tau \leq \tau_2'$, since $\tau_2' - \tau_2 = O(\varepsilon)$. From this, with due account of (2.32), we get the inequality

$$\int_{\tau_1}^{\tau_3} |K(\tau, \tau_1)| \, d\tau \leq N \varepsilon^{1/3} \tag{2.34}$$

for the first factor in (2.31).

We use the Stieltjes integral to write the second factor in (2.31) and then employ the asymptotic equality $\lambda(\tau_1 - \delta(\varepsilon), \varepsilon) = O(\varepsilon^{1/3} \ln(1/\varepsilon))$ that follows from (1.19) and (2.28). As a result we make sure that

$$\int_{0}^{\tau_1 - \delta(\varepsilon)} K_0(\tau_1 - \delta(\varepsilon), \theta) \, d\theta \leq$$

$$\leq N_1 \varepsilon^{2/3} \left(\ln \frac{1}{\varepsilon} \right)^{-1} \int_{0}^{\tau_1 - \delta(\varepsilon)} d_\theta K_0(\tau_1 - \delta(\varepsilon), \theta) \leq N_2 \varepsilon^{2/3} / \ln \frac{1}{\varepsilon}. \tag{2.35}$$

It remains to estimate the last factor in (2.31). Substitution (1.17) and relation (2.29) lead to the inequality

$$K_0(\tau_1, \tau_1 - \delta(\varepsilon)) \leq N_1 \exp\left\{\int_{-N_2 \ln(1/\varepsilon)}^{0} \frac{2u}{u^2 + v_0(u)} du\right\} \leq$$
$$\leq N_3 \exp\{-N_4 \ln^3(1/\varepsilon)\}. \qquad (2.36)$$

Combining (2.34)–(2.36), we finally obtain

$$|S_1| \leq N_1 \exp\{-N_2 \ln^3(1/\varepsilon)\}/\ln(1/\varepsilon).$$

We set $S_2 = S_{21} + S_{22}$, dividing the integration interval in the inner integral by the point τ_1. Then we represent S_{22} as the sum $S_{22}^1 + S_{22}^2$ of the integrals along the intervals $[\tau_1, t_{**}]$ and $[t_{**}, \tau_3]$ respectively. Using Lemma 1.3 and estimate (2.16), we infer that

$$|S_{22}^1| \leq N_1 \frac{1}{\varepsilon} \int_{\tau_1}^{t_{**}} \left(\int_{\tau_1}^{\tau} K_0(\tau, \theta) d\theta\right) d\tau \leq N_2 \varepsilon^{1/3 - \delta_4}/\ln^2(1/\varepsilon),$$

where $\delta_4 > 0$ is small. Dividing the inner integral in S_{22}^2 by the point t_{**}, we write it as the sum $S_{22}^{21} + S_{22}^{22}$. Substitution (1.17) and inequalities (1.18), (1.26), (2.27) lead to the conclusion that $S_{22}^{22} \to 0$ as $\varepsilon \to 0$. Indeed, dividing the inner and outer integrals in the inequality

$$|S_{22}^{22}| \leq N \frac{1}{\varepsilon} \int_{t_{**}}^{\tau_3} \left(\int_{t_{**}}^{\tau} |K(\tau, \theta)| d\theta\right) d\tau$$

by the point τ_2, we represent its right-hand side as a sum. In the resulting integrals we make substitution (1.17) and use inequalities (1.18), (1.26) for $t_{**} \leq \tau$, $\theta \leq \tau_2$ and use the estimate $|K(\tau, \theta)| \leq N$ already mentioned and inequality (2.27), respectively, for $\tau_2 \leq \tau$, $\theta \leq \tau_2'$ and $\tau_2' \leq \tau$, $\theta \leq \tau_3$. Then we find that $|S_{22}^{22}| \leq N/\ln^3(1/\varepsilon)$.

Thus, the integrals S_{21} and S_{22}^{21} make an essential contribution to S as $\varepsilon \to 0$, i.e.,

$$S_{21} = \frac{1}{\varepsilon} \int_{\tau_1}^{\tau_3} A_{21}(\tau, \varepsilon) K(\tau, \tau_1) d\tau \int_{\tau_1 - \delta(\varepsilon)}^{\tau_1} K(\tau_1, \theta) A_{12}(\theta, \varepsilon) h_2(\theta, \varepsilon) d\theta, \qquad (2.37)$$

$$S_{22}^{21} = \frac{1}{\varepsilon} \int_{t_{**}}^{\tau_3} A_{21}(\tau, \varepsilon) K(\tau, t_{**}) d\tau \int_{\tau_1}^{t_{**}} K(t_{**}, \theta) A_{12}(\theta, \varepsilon) h_2(\theta, \varepsilon) d\theta. \qquad (2.38)$$

Theorem on the C^1-Proximity

Let us prove the boundedness of expressions (2.37) and (2.38). Relations (2.37) give the estimate

$$|S_{21}| \leq N\frac{1}{\varepsilon}\int_{\tau_1}^{\tau_3}|K(\tau,\tau_1)|\,d\tau \int_{\tau_1-\delta(\varepsilon)}^{\tau_1} K_0(\tau_1,\theta)\,d\theta.$$

We have encountered the first integral on the right-hand side when we considered inequality (2.31), where we showed that it did not exceed $N\varepsilon^{1/3}$. The second integral does not exceed $N\varepsilon^{2/3}$, which fact we can easily verify if we divide it by the point $\tau_1 - \tau_{10}\varepsilon^{2/3}$. Indeed, we can estimate the second of the resulting integrals from above by the quantity $N\varepsilon^{2/3}$, since $K_0 \leq 1$, and get the estimate we need for the first integral when we write it in Stieltjes form and take into account the relation $\lambda(\tau_1 - \tau_{10}\varepsilon^{2/3}, \varepsilon) = O(\varepsilon^{1/3})$ that we encountered above. It follows from (2.38) that

$$|S_{22}^{21}| \leq N\frac{1}{\varepsilon}\int_{\tau_1}^{t_{**}} K_0(t_{**},\theta)\,d\theta \int_{t_{**}}^{\tau_3}|K(\tau,t_{**})|\,d\tau.$$

Comparing it with (2.20), we infer that when we substitute τ_2 for τ_3, we get a product which we have already studied. Therefore, the boundedness of $|S_{22}^{21}|$ is a consequence of the chain of inequalities

$$\int_{\tau_2}^{\tau_3}|K(\tau,t_{**})|\,d\tau \leq N_1 K_0(\tau_2,t_{**})\int_{\tau_2}^{\tau_3}|K(\tau,\tau_2)|\,d\tau \leq$$

$$\leq N_2\varepsilon^{-2/3}(\ln(\frac{1}{\varepsilon}))^{-2}\int_{\tau_2}^{\tau_3}|K(\tau,\tau_2)|\,d\tau.$$

We have made the first transition here with the use of the obvious representation $K(\tau,t_{**}) = K(\tau,\tau_2)K(\tau_2,t_{**})$ and Lemma 1.3 and the second transition with use of inequality (1.26). It remains to note that the last integral appearing in this relation was encountered in inequality (2.33).

It is not so easy to carry out a direct computation of the limits of S_{21} and S_{22}^{21} as $\varepsilon \to 0$. Below we shall show a technique of calculating the limit of the sum $S_{21} + S_{22}^{21}$ which rests only on the fact of its convergence as $\varepsilon \to 0$. For this purpose, in addition to Theorem 2.1 we must also know the limit of $h_2(t,\varepsilon)$ as $\varepsilon \to 0$ on the interval $[\tau_1, t_{**}]$.

THEOREM 2.2. *We have* $h_{22}(t,\varepsilon) \to h_{20}(t_0 - 0)$ *uniformly with respect to* $\tau_1 \leq t \leq t_{**}$.

PROOF. From the integral equation (2.7), rewritten for the initial moment τ_1, we derive the inequality

$$\max_{\tau_1 \leq t \leq t_{**}} |h_2(t,\varepsilon) - h_2(\tau_1,\varepsilon)| \leq N_1 |h_1(\tau_1,\varepsilon)| \int_{\tau_1}^{t_{**}} K_0(\tau,\tau_1)\,d\tau +$$

$$+ N_2 \frac{1}{\varepsilon} \int_{\tau_1}^{t_{**}} \left(\int_{\tau_1}^{\tau} K_0(\tau,\theta)\,d\theta \right) d\tau. \tag{2.39}$$

Let us first estimate the first term on the right-hand side of (2.39). We have

$$\int_{\tau_1}^{t_{**}} K_0(\tau,\tau_1)\,d\tau = \int_{\tau_1}^{t_*} K_0(\tau,\tau_1)\,d\tau + K_0(t_*,\tau_1) \int_{t_*}^{t_{**}} K_0(\tau,t_*)\,d\tau.$$

Here we estimate the first term and the factor before the second term by means of inequality (2.15), and in the second integral we make the substitution (1.17) with due account of inequalities (1.18) and (1.26). As a result we obtain

$$\int_{\tau_1}^{t_{**}} K_0(\tau,\tau_1)\,d\tau \leq N\varepsilon^{2/3} \ln(1/\varepsilon).$$

Recalling inequality (2.5), we make sure that the first term on the right-hand side of (2.39) does not exceed $N\varepsilon^{1/3} \ln(1/\varepsilon)$. Changing the order of integration in the second integral of (2.39), we note that we have already considered it in 2.2, where we have obtained estimate (2.19) for it. Thus, in order to complete the proof of the theorem, we must take into account that, by virtue of Theorem 2.1, $h_2(\tau_1,\varepsilon) \to h_{20}(t_0 - 0)$ as $\varepsilon \to 0$.

Theorems 2.1 and 2.2 allow us to replace $h_2(\theta,\varepsilon)$ by $h_{20}(t_0 - 0)$ in (2.37), (2.38). Therefore, as $\varepsilon \to 0$, relation (2.13) is transformed into the relation

$$h_{20}(t_0 + 0) - h_{20}(t_0 - 0) = H h_{20}(t_0 - 0), \tag{2.40}$$

where H is a matrix whose structure is clear on the intuitive level (see Lemma 1.3 and the technique of the choice of coordinates ξ, η indicated in 1.4). By virtue of (2.37), (2.38), for a strict consideration of the problem, it is necessary to substantiate the "almost decomposability" of $K(t,\tau)$ for $\tau_1 - \delta(\varepsilon) \leq \tau < t \leq t_{**}$.

We write Eq. (1.21) as the system

$$\varepsilon \dot{h}^1 = A_{11}^{11}(t,\varepsilon) h^1 + A_{11}^{12}(t,\varepsilon) h^2,$$
$$\varepsilon \dot{h}^2 = A_{11}^{21}(t,\varepsilon) h^1 + A_{11}^{22}(t,\varepsilon) h^2, \tag{2.41}$$

setting $h = (h^1, h^2)$, where $h^1 \in R$, $h^2 \in R^{k-1}$. We have $A_{11}(t,\varepsilon) = F'_\xi$ on the time interval of interest to us, and therefore the coefficients of system (2.41) are rather specific. First, the function $A_{11}^{11}(t,\varepsilon)$ has the same asymptotics as $\lambda(t,\varepsilon)$. Second, the eigenvalues of the matrix $A_{11}^{22}(t,\varepsilon)$ lie in the left complex half-plane and its elements are close to constants uniformly with respect to ε. Third, it follows from (2.28) and from the choice of t_{**} that

$$|A_{11}^{12}(t,\varepsilon)| \leq N_1 \varepsilon^{1/3} \ln(1/\varepsilon), \qquad |A_{11}^{21}(t,\varepsilon)| \leq N_2 \varepsilon^{2/3} \ln^2(1/\varepsilon). \qquad (2.42)$$

THEOREM 2.3. *Let $K_*(t,\tau) = \text{diag } \{K_1(t,\tau), K_2(t,\tau)\}$, where $K_1(t,\tau) = \exp\{\frac{1}{\varepsilon} \int_\tau^t A_{11}^{11}(s,\varepsilon) ds\}$ and $K_2(t,\tau)$ is the Cauchy matrix for the system $\varepsilon \dot{h}^2 = A_{11}^{22}(t,\varepsilon) h^2$. Then*

$$K(t,\tau) = K_*(t,\tau) + O(\varepsilon^{1/3} \ln^3(1/\varepsilon))$$

uniformly with respect to $\tau_1 - \delta(\varepsilon) \leq \tau < t \leq t_{**}$.

PROOF. We denote by $h^1(t,\tau)$, $h^2(t,\tau)$ the solutions of system (2.41) with the initial conditions $h^1(\tau,\tau) = h_0^1$, $h^2(\tau,\tau) = h_0^2$, independent of ε. We have

$$h^1(t,\tau) = K_1(t,\tau) h_0^1 + \frac{1}{\varepsilon} \int_\tau^t K_1(t,s) A_{11}^{12}(s,\varepsilon) h^2(s,\tau) ds, \qquad (2.43)$$

$$h^2(t,\tau) = K_2(t,\tau) h_0^2 + \frac{1}{\varepsilon} \int_\tau^t K_2(t,s) A_{11}^{21}(s,\varepsilon) h^1(s,\tau) ds. \qquad (2.44)$$

Because of the inequality $|K_2(t,\tau)| \leq N_1 \exp\{\frac{-N_2}{\varepsilon}(t-\tau)\}$ and the second inequality in (2.42) we derive from (2.44)

$$\sup_{t,\tau} |h^2(t,\tau)| \leq N_1 + N_2 \varepsilon^{2/3} \ln^2(1/\varepsilon) \sup_{t,\tau} |h^1(t,\tau)| \qquad (2.45)$$

and get the estimate

$$\sup_{t,\tau} |h^1(t,\tau)| \leq (N_1 + N_2 \ln^3(1/\varepsilon) \sup_{t,\tau} |h^2(t,\tau)|) \sup_{t,\tau} K_1(t,\tau) \qquad (2.46)$$

from the first inequality of (2.42) and relation (2.43). Since K_1 is asymptotically equivalent to K_0, we have $K_1(t,\tau) \leq N_1 K_0(t,\tau) \leq N_2 K_0(t_{**},\tau_1) \leq$ $\leq N_3 \ln^2(1/\varepsilon)$ for t and τ being considered. From (2.45) and (2.46) we get

$$\sup_{t,\tau} |h^1(t,\tau)| \leq N \ln^5(1/\varepsilon). \qquad (2.47)$$

Proceeding from (2.47), we find from (2.44) that

$$h^2(t,\tau) = K_2(t,\tau)h_0^2 + O(\varepsilon^{2/3}\ln^7(1/\varepsilon)). \tag{2.48}$$

Finally, substituting (2.48) into (2.43), we arrive at the relation

$$h^1(t,\tau) = K_1(t,\tau)h_0^1 + O(\varepsilon^{1/3}\ln^3(1/\varepsilon)). \tag{2.49}$$

Combining the asymptotic representations (2.48) and (2.49), we make sure of the validity of the theorem.

We apply Theorem 2.3 to (2.37) and (2.38) with due account of the properties of $A_{12}(\theta,\varepsilon)$ (in the first row, as $\varepsilon \to 0$, the first element tends to 1 and the other elements tend to zero). These properties reflect the singularities of the function $F_1(\xi,\eta)$ (it does not contain terms of the form $c_j\eta_j$, where $j = 2,\ldots,m$). We come to the conclusion that the nonzero elements of the matrix H can only be in the first column. Since in this case relation (2.40) is satisfied by the function $\dot{y}_0(t)$, it follows that the first column of H is the resolution of the vector $\dot{y}_0(t_0+0) - \dot{y}_0(t_0-0)$ into components along the coordinates $(\eta_1,\ldots,\eta_m)$. We have come to condition (1.5) for $p = 0$.

Thus, the function $h_{20}(t)$ is the solution of system (1.4), (1.5) with the initial condition $h_{20}(0) = h_2^0$, and the function $h_{10}(t)$ is defined by the relation $h_{10}(t) = -A^{-1}(t)B(t)h_{20}(t)$. This completes the proof of Theorem 1.1.

In conclusion, we want to make the following remark. The technique used to prove the theorem on the C^1-proximity was elaborated in connection with consideration of the problem concerning the stability of a two-dimensional relaxation cycle in a medium with diffusion [24, 25]. Its detailed exposition can be found in [26] by A. Yu. Kolesov.

3. Principle of Reduction in the Neighborhood of a Junction Point

3.1. The First Procedure of the Reduction Principle

Here we consider a trivial procedure of passing from an arbitrary relaxation system in the neighborhood of a junction point to a system with one fast variable. Recall that, in the special coordinates described in 1.4, a relaxation system in the neighborhood of a junction point has the form

$$\varepsilon\dot{\xi}_j = f_j(\xi_1,\xi_2,\eta_1,\eta_2), \qquad \dot{\eta}_j = g_j(\xi_1,\xi_2,\eta_1,\eta_2), \qquad j = 1,2. \tag{3.1}$$

Here the smooth functions f_j, g_j of the variables $\xi_1, \eta_1 \in R$, $\xi_2 \in R^{k-1}$, $\eta_2 \in R^{m-1}$, $k, m \geq 1$, are defined in a sufficiently small neighborhood of zero, and

$$f_2(0,0,0,0) = 0, \quad g_2(0,0,0,0) = 0, \quad g_1(0,0,0,0) = 1,$$
$$\partial_{\xi_1} f_2(0,0,0,0) = 0, \quad f_1(\xi_1, \xi_2, \eta_1, \eta_2) = \xi_1^2 + \eta_1 + \ldots,$$

where the dots stand for quadratic terms different from ξ_1^2 and for terms of a higher order of smallness. We also suppose that all eigenvalues of the matrix $\partial_{\xi_2} f_2(0,0,0,0)$ lie on the left of the imaginary axis.

LEMMA 3.1. *System (3.1) has an exponentially stable (with the coefficient of the exponent of order ε^{-1}) invariant integral manifold*

$$\xi_2 = V(\xi_1, \eta_1, \eta_2, \varepsilon), \tag{3.2}$$

where the function V smoothly depends on all its arguments.

To prove this lemma, we must note that after the substitution $t \to \varepsilon t$, we arrive at the well-known case of the central integral manifold [71].

On the integral manifold (3.2) we obtain the relaxation system

$$\varepsilon \dot{\xi} = F(\xi, \eta_1, \eta_2, \varepsilon), \quad \dot{\eta}_j = G_j(\xi, \eta_1, \eta_2, \varepsilon), \quad j = 1, 2, \tag{3.3}$$

with one fast variable $\xi = \xi_1$, whose right-hand sides possess the same properties as before for $\varepsilon = 0$.

3.2. The Second Procedure of the Reduction Principle

To describe the main idea of the principle, we first use the equation $F|_{\varepsilon=0} = 0$ to determine the smooth function $\eta_1 = \Phi(\xi, \eta_2)$, $\Phi(0,0) = 0$, in the neighborhood of zero. Then we substitute the equality $\eta_1 = \Phi(\xi, \eta_2)$ into the equation $\partial_\xi F|_{\varepsilon=0} = 0$ and find the smooth function $\xi = \varphi(\eta_2)$, $\varphi(0) = 0$, from this equality. Now we set

$$H_0(\xi, \eta_2) = \begin{cases} \Phi(\xi, \eta_2) & \text{for } -q \leq \xi \leq \varphi(\eta_2), \\ \Phi(\varphi(\eta_2), \eta_2) & \text{for } \varphi(\eta_2) \leq \xi \leq q, \end{cases}$$

where the positive constant q is sufficiently small.

THEOREM 3.1. *System (3.3) has an invariant integral manifold*

$$\eta_1 = H(\xi, \eta_2, \varepsilon); \quad H(\xi, \eta_2, 0) = H_0(\xi, \eta_2). \tag{3.4}$$

Here the function H and its derivatives with respect to ξ, η_2 are continuous in all the variables for $|\xi|, \|\eta_2\| \leq q$, $0 \leq \varepsilon \leq \varepsilon_0$, where ε_0 is sufficiently small. In addition, it is G_1-Lipschitzian. i.e.,

$$|H_{G_1+\Delta G_1} - H_{G_1}| \leq M\|\Delta G_1\|_C, \qquad (3.5)$$

where M is a constant and ΔG_1 is measured in the metric of the functions continuous in ξ, η_1, η_2 in the domain $|\xi|, |\eta_1|, \|\eta_2\| \leq q$. Furthermore, the solutions of system (3.3) exponentially approach the integral manifold (3.4) with the coefficient of the exponent of order $\varepsilon^{\lambda-1}$, $0 < \lambda < 1/3$ as long as $-q \leq \xi \leq -\varepsilon^\lambda$, and remain in its neighborhood for $-\varepsilon^\lambda \leq \xi \leq q$.

For $-q \leq \xi \leq -q_1 < 0$, the statement of the theorem follows from the results of [71]. Then we extend the integral manifold (3.4) along the trajectory of system (3.3) up to the point $\xi = q$ (this is possible since we have $\dot\xi > 0$, $\dot\eta_1 > 0$ on the trajectories of system (3.3)). It follows from this technique of extension that it is sufficient to verify the validity of inequality (3.5) on each trajectory of system (3.3) separately. The last problem reduces to the integration of system (1.10) in which, to the right-hand side of the second equation, we add the term

$$\Delta G_1|_{\xi=\xi(t,\varepsilon),\eta_1=\eta_1(t,\varepsilon),\eta_2=\eta_2(t,\varepsilon)},$$

where $\xi(t,\varepsilon)$, $\eta_1(t,\varepsilon)$, $\eta_2(t,\varepsilon)$ are the components of the trajectories of system (3.3). From this and from the results of Sec. 2, the fact that we need follows trivially.

Thus, the problem of constructing the asymptotics of the solutions of system (3.1) in the neighborhood of the junction point stated in [77] reduces to the integration of the equation

$$\frac{d\eta_2}{d\xi} = \partial_\xi H \cdot G_2(\xi, H, \eta_2, \varepsilon)/(G_1 - \partial_{\eta_2} H \cdot G_2) \qquad (3.6)$$

in R^{m-1}.

LEMMA 3.2. *The right-hand side of (3.6) satisfies the Lipschitz condition with a certain universal constant for $0 \leq \varepsilon \leq \varepsilon_0$ with respect to the variables ξ, η_2.*

From the asymptotic properties of the trajectories of system (3.3) given earlier (see (1.15), (1.16) and (1.25)), we obtain

$$H = \begin{cases} H_0 + O(\varepsilon/\xi) & \text{for } -q \leq \xi \leq -\varepsilon^\lambda, \\ H_0 + O(\varepsilon^{2/3}) & \text{for } -\varepsilon^\lambda \leq \xi \leq q; \end{cases} \qquad (3.7)$$

$$\partial_\xi H = \begin{cases} \partial_\xi H_0 + O(\varepsilon/\xi^2) & \text{for } -q \le \xi \le -\varepsilon^\lambda, \\ \partial_\xi H_0 + O(\varepsilon^{1/3}) & \text{for } -\varepsilon^\lambda \le \xi \le q. \end{cases} \quad (3.8)$$

Then, dividing the second and third equations of system (3.3) by the first equation and investigating the system of variational equations, we make sure that

$$\partial_{\eta_2} H = \begin{cases} \partial_{\eta_2} H_0 + O(\varepsilon/\xi) & \text{for } -q \le \xi \le -\varepsilon^\lambda, \\ \partial_{\eta_2} H_0 + O(\varepsilon^{2/3}) & \text{for } -\varepsilon^\lambda \le \xi \le q. \end{cases} \quad (3.9)$$

Writing now (3.6) in terms of system (3.3) (i.e., representing its right-hand side as $\varepsilon G_2/F$) and differentiating its right-hand side with respect to ξ or η_2, and then taking (3.7)–(3.9) into account, we get the statement of the lemma.

We shall show later that Theorem 3.1 makes it possible to answer some essential questions that arise when we analyze the asymptotic expansions of the function H on different intervals of variation of ξ. For instance, proceeding from the theorem on the C^1-proximity and taking into account that H satisfies the G_1-Lipschitz condition, we shall later solve the problem of sewing together the asymptotic formulas for $\xi = -\varepsilon^\lambda$.

It is easy to see that, in general, the technique of finding H from the equation

$$\partial_\xi H \cdot F + \varepsilon(\partial_{\eta_2} H, G_2) = \varepsilon G_1$$

is similar to the construction of the trajectories of a relaxation system in the plane. It turns out that the asymptotic behavior of the function H is the most symmetric and simple when system (3.3) is reduced to the so-called normal form [91], which we shall describe at the beginning of the next section.

4. Asymptotics of Multidimensional Relaxation Oscillations in the Neighborhood of a Junction Point

4.1. Normal Form

As we stated in 3.1, under standard constraints an arbitrary relaxation system in the neighborhood of a junction point reduces to the following system with one fast variable,

$$\varepsilon \dot\xi = f(\xi, \eta_1, \eta_2), \qquad \dot\eta_j = g_j(\xi, \eta_1, \eta_2), \qquad j = 1, 2. \quad (4.1)$$

Here $0 < \varepsilon \ll 1$ and the functions f, g_j, smooth in the neighborhood of the origin, of the variables $\xi, \eta_1 \in R$, $\eta_2 \in R^{m-1}$, $m \ge 2$, are such that

$$g_2(0,0,0) = 0, \qquad g_1(0,0,0) = 1, \qquad f(\xi, \eta_1, \eta_2) = \xi^2 + \eta_1 + \ldots,$$

where the dots stand for quadratic terms different from ξ^2 and terms of a higher order of smallness.

System (4.1) can be further simplified as follows. We find the function $\xi = \varphi(\eta_1, \eta_2)$ from the equation $f'_\xi = 0$ and the function $\eta_1 = \psi(\eta_2)$ from the equation $f(\varphi, \eta_1, \eta_2) = 0$. Then, in (4.1), we successively carry out the substitution

$$(\xi, \eta_1, \eta_2) \to (\xi - \varphi(\psi(\eta_2), \eta_2), \eta_1 - \psi(\eta_2), \eta_2),$$

which rectifies the $(m-1)$-dimensional manifold of breakoff, and, for a fixed η_2, the substitutions with respect to ξ and η_1, described in [77], which transform the surface $f = 0$ into $\eta_1 = -\xi^2$. As a result we get the system

$$\varepsilon\dot\xi = \xi^2 + \eta_1 + \varepsilon\theta(\xi, \eta_1, \eta_2), \qquad \dot\eta_j = g_j(\xi, \eta_1, \eta_2), \qquad j = 1, 2, \qquad (4.2)$$

where the smooth functions θ, g_j are such that $\theta(0,0,0) = 0$, $g_2(0,0,0) = 0$, $g_1(0,0,0) = 1$. In addition we can suppose that

$$g_2(0, 0, \eta_2) = 0, \qquad g_1(0, 0, \eta_2) = 1. \qquad (4.3)$$

Indeed, the first of relations (4.3) will be satisfied after the substitution $\eta_2 \to \eta_2 + \mathfrak{G}(\eta_2)\eta_1$, where $\mathfrak{G}(\eta_2) = -g_2(0, 0, \eta_2)/g_1(0, 0, \eta_2)$; the second relation will be satisfied after the substitution $(\xi, \eta_1, \eta_2) \to (\gamma^{-1/3}\xi, \gamma^{-2/3}\eta_1, \eta_2)$, where $\gamma = g_1(0, 0, \eta_2)$, and the division of the right-hand sides by $\gamma^{1/3}$. Also make note of the fact that we can remove the regular component of order ε on the right-hand side of the first relation in (4.2) since, as we shall see later, it does not affect the structure of the asymptotics of the solution.

Without loss of generality, we can restrict our consideration to the case $m = 2$ since, although for $m > 2$ in all of the asymptotic expansions that we shall encounter below the scalar coefficients are replaced by vector ones, this is inessential. Because of what we said above, this leads to the standard three-dimensional system

$$\varepsilon\dot\xi = \xi^2 + \eta_1, \qquad \dot\eta_1 = g_1(\xi, \eta_1, \eta_2), \qquad \dot\eta_2 = g_2(\xi, \eta_1, \eta_2), \qquad (4.4)$$

where the smooth functions g_1, g_2 satisfy identities (4.3), which we shall call the normal form of the multidimensional relaxation system in the neighborhood of a junction point.

Recall (see Theorem 3.1) that system (4.4) has an invariant integral manifold

$$\eta_1 = H(\xi, \eta_2, \varepsilon); \qquad H(\xi, \eta_2, 0) = H_0(\xi, \eta_2), \qquad (4.5)$$

where the function H and its derivatives with respect to ξ, η_2 are continuous in all the variables in $|\xi|, |\eta_2| \leq q$, $0 \leq \varepsilon \leq \varepsilon_0$, and for sufficiently small $q, \varepsilon_0 > 0$,

$$H_0(\xi, \eta_2) = \begin{cases} -\xi^2 & \text{for} \quad -q \leq \xi \leq 0, \\ 0 & \text{for} \quad 0 \leq \xi \leq q. \end{cases}$$

In addition, function (4.5) is g_1-Lipschitzian, i.e.,

$$|H_{g_1 + \Delta g_1} - H_{g_1}| \leq M \|\Delta g_1\|_C,$$

where M is a positive constant and Δg_1 is measured in the metric of continuous (in ξ, η_1, η_2) functions in the domain $|\xi|, |\eta_1|, |\eta_2| \leq q$. Finally, the solutions of system (4.4) exponentially approach the integral manifold (4.5) with the coefficient of the exponent of order $\varepsilon^{\lambda-1}$, $0 < \lambda < 1/3$, as long as $-q \leq \xi \leq -\varepsilon^\lambda$, and remain in its neighborhood when $-\varepsilon^\lambda \leq \xi \leq q$.

Thus, two problems naturally arise, namely, the construction of the asymptotics of the integral manifold (4.5) and the asymptotic integration of the equation

$$\frac{d\eta_2}{d\xi} = \frac{\partial H}{\partial \xi} g_2 / \left(g_1 - \frac{\partial H}{\partial \eta_2} g_2\right) \quad (= \varepsilon g_2/(\xi^2 + H)) \tag{4.6}$$

on the integral manifold. As is shown in 3.2, for $0 \leq \varepsilon \leq \varepsilon_0$ the right-hand side of (4.6) satisfies the Lipschitz condition with respect to the variables ξ, η_2 with a certain universal constant, i.e., Eq. (4.6) regularly depends on ε.

4.2. Asymptotics of an Integral Manifold at the Beginning and in the Middle of a Breakoff Section

In order to find H from the equation

$$\frac{\partial H}{\partial \xi}(\xi^2 + H) + \varepsilon \frac{\partial H}{\partial \eta_2} g_2(\xi, H, \eta_2) = \varepsilon g_1(\xi, H, \eta_2) \tag{4.7}$$

we shall consider the following three sections in succession: $-q \leq \xi \leq -\varepsilon^{\lambda_1}$, $-\varepsilon^{\lambda_1} \leq \xi \leq \varepsilon^{\lambda_2}$, $\varepsilon^{\lambda_2} \leq \xi \leq q$, where $0 < \lambda_1, \lambda_2 < 1/3$. In this subsection we shall construct the asymptotics of H on the first two sections.

For $-q \leq \xi \leq -\varepsilon^{\lambda_1}$, we seek the integral manifold of system (4.4) in the form

$$H = -\xi^2 + \sum_{n=1}^{\infty} \varepsilon^n a_n(\xi, \eta_2). \tag{4.8}$$

Substituting expansion (4.8) into (4.7), we find that $a_1 = -g_1(\xi, -\xi^2, \eta_2)/2\xi$, and then the functions a_n can be expressed as $a_1, \ldots, a_{n-1}$. As in [77], using induction, we establish the asymptotic expansions, as $\xi \to -0$,

$$a_n(\xi, \eta_2) = \frac{1}{\xi^{3n-2}} \sum_{k=0}^{\infty} a_{k,n}(\eta_2) \xi^k, \quad n \geq 1, \qquad (4.9)$$

where the smooth functions $a_{k,n}(\eta_2)$ are defined recurrently. For instance, for the leading coefficients $a_{0,n}$ of series (4.9) we have the formulas

$$a_{0,1} = -1/2, \qquad a_{0,n} = -\frac{3n-4}{4} \sum_{k=1}^{n-1} a_{o,k} a_{0,n-k}, \quad n \geq 2,$$

which exactly coincide with those given in [77].

It follows from expansion (4.9) and the choice of λ_1 that

$$\varepsilon^n a_n(-\varepsilon^{\lambda_1}, \eta_2) = O(\varepsilon^{n(1-3\lambda_1)+2\lambda_1}) \qquad (4.10)$$

uniformly with respect to η_2. Relation (4.10) makes it possible to reveal the asymptotic properties of series (4.8) on the interval $-q \leq \xi \leq -\varepsilon^{\lambda_1}$. We use a special technique, which we shall also use in what follows. We denote by $H_N(\xi, \eta_2, \varepsilon)$ a segment of series (4.8) of length N and perturb the function g_1 in Eq. (4.7) and system (4.4) by the addition

$$\Delta g_1 = \varepsilon^{-1} \left[\frac{\partial H_N}{\partial \xi}(\xi^2 + H_N) + \varepsilon \frac{\partial H_N}{\partial \eta_2} g_2(\xi, H_N, \eta_2) - \varepsilon g_1(\xi, H_N, \eta_2) \right]. \qquad (4.11)$$

From the method of definition of H_N and relation (4.10) it follows that, in the metric of the space of continuous functions, Δg_1 becomes arbitrarily small as N increases. It is significant that the perturbed system (4.4) has an integral manifold $\eta_1 = H_N$. This circumstance allows us to use the g_1-Lipschitzian property and obtain the following statement.

THEOREM 4.1. *For* $-q \leq \xi \leq -\varepsilon^{\lambda_1}$ *uniformly with respect to* $-q \leq \eta_2 \leq q$ *the asymptotic formula*

$$H(\xi, \eta_2, \varepsilon) = H_{N-1}(\xi, \eta_2, \varepsilon) + O(\varepsilon^{N(1-3\lambda_1)+2\lambda_1}), \qquad (4.12)$$

where N is an arbitrary natural number and $0 < \lambda_1 < 1/3$, is valid for function (4.5).

Let $-\varepsilon^{\lambda_1} \leq \xi \leq \varepsilon^{\lambda_2}$. Carrying out the normalizations $\eta_1 = \mu^2 v$, $\xi = \mu u$, $\mu^3 = \varepsilon$ in (4.7), we obtain

$$\frac{\partial v}{\partial u}(u^2 + v) + \mu^2 g_2(\mu u, \mu^2 v, \eta_2) \frac{\partial v}{\partial \eta_2} = g_1(\mu u, \mu^2 v, \eta_2). \qquad (4.13)$$

Theorem on the C^1-Proximity

Setting $\mu = 0$ in (4.13) and taking into account property (4.3) of the function g_1, we make sure that the zero approximation of the function $v = v(u, \eta_2, \mu)$ is the solution of the Riccati differential equation

$$\frac{dv}{du} = \frac{1}{u^2 + v}.$$

It was shown in [77] that it has a unique solution $v_0(u)$ that possesses the properties

$$\begin{aligned} v_0(u) &= -u^2 - \tfrac{1}{2u} + O\!\left(\tfrac{1}{u^4}\right), & u &\to -\infty, \\ v_0(u) &= \Omega_0 - \tfrac{1}{u} + O\!\left(\tfrac{1}{u^3}\right), & \Omega_0 > 0, \quad u &\to +\infty. \end{aligned} \qquad (4.14)$$

Relations (4.4) mean that the solution $v_0(u)$ deviates to the least degree from the function

$$H_0(u) = \begin{cases} -u^2 & \text{for } u \leq 0, \\ 0 & \text{for } u \geq 0, \end{cases}$$

which is an approximation of the integral manifold H in the new variables. Therefore, it is expedient to seek the function $v(u, \eta_2, \mu)$ in the form

$$v(u, \eta_2, \mu) = v_0(u) + \sum_{n=1}^{\infty} \mu^n v_n(u, \eta_2). \qquad (4.15)$$

Substituting (4.15) into (4.13) and equating the coefficients in like powers of μ, we have the sequence of differential equations

$$\frac{\partial v_n}{\partial u} + \frac{v_n}{[u^2 + v_0(u)]^2} = \Phi_n(u, \eta_2), \qquad (4.16)$$

where the right-hand sides Φ_n depend only on $v_0, v_1, \ldots, v_{n-1}$, which defines the functions v_n. Regarding (4.16) as an ordinary differential equation, we choose as v_n its solution which is the closest to the bounded solution. It is unique and is written in terms of the Green's function in the form

$$v_n(u, \eta_2) = M(u) \int_{-\infty}^{u} \Phi_n(\theta, \eta_2) \frac{d\theta}{M(\theta)}, \qquad M(u) = \exp \int_{u}^{\infty} \frac{d\theta}{[\theta^2 + v_0(\theta)]^2}.$$

Suppose that v_n^+ and v_n^- are asymptotic expansions of the function v_n as $u \to +\infty$ and $u \to -\infty$ respectively. As in [77], we use induction to derive the equalities

$$v_n^- = u^{n-1} \sum_{k=0}^{\infty} b_{k,n}(\eta_2)/u^{3k}, \qquad n \geq 1, \qquad (4.17)$$

$$v_n^+ = u^{n-1} \sum_{\nu=0}^{n} \sum_{k=3\nu}^{\infty}{}^* \frac{b_{k,\nu,n}(\eta_2)}{u^k} \ln^\nu u +$$
$$+ b_{0,\pi(n),n}(\eta_2) \ln^{\pi(n)} u, \qquad n \geq 0, \qquad (4.18)$$

where all the coefficients smoothly depend on η_2 and the symbol $\sum\limits_{k=m}^{\infty}{}^*$ stands for the sum $\sum\limits_{k=m, k\neq m+1}^{\infty}$ and

$$\pi(n) = \left[\frac{n}{3}\right] + \begin{cases} 0 & \text{if } n \not\equiv 1 \pmod 3, \\ 1 & \text{if } n \equiv 1 \pmod 3. \end{cases}$$

These relations are important for the sequel.

Let us elucidate the asymptotic properties of series (4.15) under the condition that $-\varepsilon^{\lambda_1 - 1/3} \leq u \leq \varepsilon^{\lambda_2 - 1/3}$ and sew together the formulas resulting here with (4.12) for $\xi = -\varepsilon^{\lambda_1}$. This is the central feature in the construction of the asymptotics. It follows from expansions (4.17) and (4.18) that

$$\begin{aligned}\mu^{n+2} v_n(-\varepsilon^{\lambda_1 - 1/3}, \eta_2) &= O(\varepsilon^{1+\lambda_1(n-1)}), \\ \mu^{n+2} v_n(\varepsilon^{\lambda_2 - 1/3}, \eta_2) &= O(\varepsilon^{1+\lambda_2(n-1)}),\end{aligned} \qquad (4.19)$$

uniformly with respect to $-q \leq \eta_2 \leq q$. We denote by $V_N(u, \eta_2, \mu)$ a segment of series (4.15) of length N and perturb the function g_1 in system (4.4) and Eq. (4.13) by adding Δg_1 defined by (4.11), where now $H_N = \mu^2 V_N(\frac{\xi}{\mu}, \eta_2, \mu)$. As before, formulas (4.19) guarantee an arbitrary smallness of Δg_1 for a sufficiently large N uniformly with respect to $-2\varepsilon^{\lambda_1} \leq \xi \leq \varepsilon^{\lambda_2}$, $|\eta_2| \leq q$. In addition, for the ξ being considered, the perturbed system (4.4) has an integral manifold $\eta_1 = H_N$, and, since it is g_1-Lipschitzian, it also has an integral manifold H with asymptotics (4.12) for $-2\varepsilon^{\lambda_1} \leq \xi \leq -\varepsilon^{\lambda_1}$. Let us prove that H coincides with H_N for $\xi = -\varepsilon^{\lambda_1}$ with an accuracy to within ε^α, where $\alpha > 0$ can be arbitrary.

Assuming the contrary, we take, as the initial condition of system (4.4), a point which lies at the intersection of the manifold H_N with the plane $\xi = -\varepsilon^{\lambda_1}$ and does not belong to the manifold H. Let ε^{α_1}, $\alpha_1 > 0$, be the distance from this point to H. Integrating in the direction of decrease of time and using the fact that for $-2\varepsilon^{\lambda_1} \leq \xi \leq -\varepsilon^{\lambda_1}$ the manifold H is stable with the coefficient of the exponent of order $\varepsilon^{\lambda_1 - 1}$ and the time of traverse of the interval $-2\varepsilon^{\lambda_1} \leq \xi \leq -\varepsilon^{\lambda_1}$ has order $\varepsilon^{2\lambda_1}$ (this follows from the equation $\dot\xi = -g_1(\xi, -\xi^2, \eta_2)/2\xi$ for the zero approximation of the trajectory), we infer that the solution being considered a fortiori leaves the

Theorem on the C^1-Proximity

neighborhood of order ε^{α_2} for any $\alpha_2 > 0$ of the manifold H. Indeed, at the point $\xi = -2\varepsilon^{\lambda_1}$ the deviation from manifold (4.5) calculated with the use of the linear approximation is a quantity of order $\varepsilon^{\alpha_1} \exp(M\varepsilon^{3\lambda_1-1})$, $M > 0$. On the other hand, the solution lies on the surface H_N and, hence, is at a distance of order ε^{α_2} from H for a certain $\alpha_2 > 0$ throughout the interval $-2\varepsilon^{\lambda_1} \leq \xi \leq -\varepsilon^{\lambda_1}$. We have obtained a contradiction.

Thus, the integral manifold H_N of the perturbed system (4.4) is a continuation of the manifold H to the interval $-\varepsilon^{\lambda_1} \leq \xi \leq \varepsilon^{\lambda_2}$. Proceeding now from the g_1-Lipschitzian property, we arrive at the following proposition.

THEOREM 4.2. *For $-\varepsilon^{\lambda_1} \leq \xi \leq 0$ uniformly with respect to $|\eta_2| \leq q$ the asymptotic formula*

$$H(\xi, \eta_2, \varepsilon) = \mu^2 V_N\left(\frac{\xi}{\mu}, \eta_2, \mu\right) + O(\varepsilon^{1+\lambda_1 N}) \qquad (4.20)$$

is valid for the integral manifold (4.5) and the formula

$$H(\xi, \eta_2, \varepsilon) = \mu^2 V_N\left(\frac{\xi}{\mu}, \eta_2, \mu\right) + O(\varepsilon^{1+\lambda_2 N}) \qquad (4.21)$$

is valid for $0 \leq \xi \leq \varepsilon^{\lambda_2}$.

4.3. Asymptotics of an Integral Manifold at the End of a Breakoff Section

For $\varepsilon^{\lambda_2} \leq \xi \leq q$, we divide the second and third equations of (4.4) by the first equation. As a result we get the system

$$\frac{d\eta_1}{d\xi} = \frac{\mu^3 g_1(\xi, \eta_1, \eta_2)}{\xi^2 + \eta_1}, \quad \frac{d\eta_2}{d\xi} = \frac{\mu^3 g_2(\xi, \eta_1, \eta_2)}{\xi^2 + \eta_1}, \qquad (4.22)$$

whose right-hand sides are regular for the values of ξ under consideration. Considering the trajectories of system (4.22) that start at the points of the curve which is the intersection of the surface H with the plane $\xi = \varepsilon^{\lambda_2}$, we obtain the integral manifold H for $\xi \geq \varepsilon^{\lambda_2}$. It is clear that the solutions of system (4.22) can be expanded in an asymptotic series whose structure is defined by the initial conditions for $\xi = \varepsilon^{\lambda_2}$. Setting $\xi = \varepsilon^{\lambda_2}$ in (4.21) and taking into account the asymptotic relations (4.18), we infer that for $\xi \geq \varepsilon^{\lambda_2}$ the solution of (4.7) should be sought in the form

$$H = \sum_{n=2}^{\infty} \mu^n h_n(\xi, \eta_2, \mu), \quad h_n(\xi, \eta_2, \mu) = \sum_{\nu=0}^{\pi(n-2)} W_{n,\nu}(\xi, \eta_2) \ln^\nu \frac{1}{\mu}. \qquad (4.23)$$

Substituting (4.23) into (4.7) and equating the coefficients in like powers of μ, we arrive at the equations

$$\frac{\partial h_n}{\partial \xi} = \sum_{\nu=0}^{\pi(n-2)-1} V_{n,\nu}(\xi, \eta_2) \ln^\nu \frac{1}{\mu}, \qquad (4.24)$$

where the functions $V_{n,\nu}$ depend only on $W_{m,k}$ with subscripts $m \leq n-1$. Thus, for instance,

$$\frac{\partial h_2}{\partial \xi} = 0, \qquad \frac{\partial h_3}{\partial \xi} = \frac{g_1(\xi, 0, \eta_2)}{\xi^2}, \qquad \frac{\partial h_4}{\partial \xi} = 0. \qquad (4.25)$$

Equations (4.24), (4.25) give a series of equations with variables separable,

$$\frac{\partial W_{2,0}}{\partial \xi} = 0, \qquad \frac{\partial W_{3,0}}{\partial \xi} = \frac{g_1(\xi, 0, \eta_2)}{\xi^2}, \qquad \frac{\partial W_{3,1}}{\partial \xi} = 0,$$

$$\frac{\partial W_{4,0}}{\partial \xi} = 0, \qquad \frac{\partial W_{n,\nu}}{\partial \xi} = V_{n,\nu}, \qquad \frac{\partial W_{n,\pi(n-2)}}{\partial \xi} = 0,$$

$$\nu = 0, 1, \ldots, \pi(n-2) - 1, \qquad n \geq 5.$$

As in [77], we choose their particular solutions according to the following law. We set

$$W_{2,0} = b_{0,0,0}, \qquad W_{3,1} = b_{0,1,1}, \qquad W_{4,0} = b_{0,0,2},$$

$$W_{3,0} = b_{0,0,1} + (H)\int_0^\xi \frac{g_1(s, 0, \eta_2)}{s^2}\, ds,$$

$$W_{n,\nu} = b_{0,\nu,n-2} + (H)\int_0^\xi V_{n,\nu}(s, \eta_2)\, ds,$$

$$W_{n,\pi(n-2)} = b_{0,\pi(n-2),n-2}, \qquad n \geq 5.$$

Here $b_{0,\nu,m}$ are the same as in expansion (4.18) and $(H)\int$ is the Hadamard regularization. Recall [77] that the Hadamard regularization is the choice of a particular solution of an equation with variables separable whose expansion does not include a constant term as $\xi \to +0$. Note that as $\xi \to +0$, the asymptotics of the functions $W_{n,\nu}$ results when we set $u = \xi/\mu$ in (4.15) and take into account expansions (4.18). In this way we arrive at the relations (cf. [77])

$$W_{n,\nu} = \sum_{\varkappa=0}^{\pi(n-2)-\nu-1} \sum_{k=2-n+\varkappa+\nu}^{\infty} d_{k,\varkappa}^{n,\nu}(\eta_2)\xi^k \ln^\varkappa \xi +$$

$$+ d^{n,\nu}_{0,\pi(n-2)-\nu}(\eta_2) \ln^{\pi(n-2)-\nu} \xi,$$
$$\nu = 0, 1, \ldots, \pi(n-2), \quad n \geq 2, \tag{4.26}$$

where the coefficients are expressed in terms of the coefficients of expansions (4.18) with the use of the formulas

$$d^{n,\nu}_{k,\varkappa} = C^{\nu}_{\nu+\varkappa} b_{n+k-2, -k, \nu+\varkappa}, n = 3, 5, 6, \ldots,$$
$$\nu = 0, 1, \ldots, \pi(n-2) - 1,$$
$$\varkappa = 0, 1, \ldots, \pi(n-2) - 1 - \nu, \quad k \geq 2 - n + \nu + \varkappa;$$
$$d^{n,\nu}_{0,\pi(n-2)-\nu} = C^{\nu}_{\pi(n-2)} b_{n-2, 0, \pi(n-2)},$$
$$n \geq 2, \quad \nu = 0, 1, \ldots, \pi(n-2).$$

It follows from the regularity of system (4.22) that the partial sums of series (4.23) are asymptotic approximations of the integral manifold H, i.e., we have the following statement.

THEOREM 4.3. *For $\varepsilon^{\lambda_2} \leq \xi \leq q$ uniformly with respect to $|\eta_2| \leq q$ the asymptotic representation*

$$H(\xi, \eta_2, \varepsilon) = \sum_{n=2}^{N} \mu^n h_n(\xi, \eta_2, \mu) + O(\varepsilon^{\frac{N+1}{3} - \lambda_2(N-1)}), \tag{4.27}$$

where N is an arbitrary natural number, is valid for the function H.

4.4. Asymptotics of Trajectories at the Beginning of a Breakoff Section

Thus, the construction of the asymptotics of the trajectories of system (4.4) have reduced to the asymptotic integration of Eq. (4.6). This is connected with the multidimensionality of the system, since in the case of a plane this equation does not have an analog. It is sufficient to construct the asymptotics of the Cauchy problem for (4.6) with the initial condition

$$\eta_2(-q, \varepsilon) = \eta_2^0, \quad |\eta_2^0| \leq q, \tag{4.28}$$

independent of ε. It follows from the regularity properties of (4.6) indicated in 4.1 that on each of the three introduced intervals of variation of ξ the asymptotics of the solution is dictated by the asymptotics of the right-hand side and of the initial condition, the problem of the substantiation of formulas with a remainder being trivial.

For $-q \leq \xi \leq -\varepsilon^{\lambda_1}$, we substitute the asymptotic relation (4.8) into (4.6). Then it is natural to seek the solution of problem (4.6), (4.28) in the form

$$\eta_2(\xi, \varepsilon) = \eta_{2,0}(\xi) + \sum_{n=1}^{\infty} \varepsilon^n \eta_{2,n}(\xi). \tag{4.29}$$

Substituting (4.29) into (4.26) we get the Cauchy problem

$$\frac{d\eta_{2,0}}{d\xi} = -\frac{2\xi g_2(\xi, -\xi^2, \eta_{2,0})}{g_1(\xi, -\xi^2, \eta_{2,0})}, \qquad \eta_{2,0}(-q) = \eta_2^0 \tag{4.30}$$

for determining $\eta_{2,0}(\xi)$. After elementary but rather cumbersome computations we find that

$$\frac{d\eta_{2,n}}{d\xi} = M_n, \qquad \eta_{2,n}(-q) = 0, \qquad n \geq 1, \tag{4.31}$$

where

$$M_n = \frac{\Lambda_n}{\gamma_1} + \sum_{l=1}^{n} \Lambda_{n-l} \sum_{\nu=1}^{l} \frac{(-1)^\nu}{\gamma_1^{\nu+1}} \sum_{\substack{i_1+\ldots+i_\nu=l+\nu \\ i_j \geq 2}} \gamma_{i_1} \cdots \gamma_{i_\nu}, \tag{4.32}$$

$$\gamma_n = a_n(\xi, \eta_{2,0}) + \sum_{m=1}^{n-1} \sum_{k=1}^{n-m} \frac{a_{m\eta_2}^{(k)} a_{m\eta_2}^{(k)}(\xi, \eta_{2,0})}{k!} \times$$

$$\times \sum_{\substack{i_1+\ldots+i_k=n-m \\ i_j \geq 1}} \eta_{2,i_1} \cdots \eta_{2,i_k}, \qquad n \geq 1, \tag{4.33}$$

$$\Lambda_n = \sum_{0 < k+m \leq n} \frac{g_{2\eta_1\eta_2}^{(k)(m)}(\xi, -\xi^2, \eta_{2,0})}{k! \, m!} A_{k,m}^n, \qquad n \geq 1, \tag{4.34}$$

$$\Lambda_0 = g_2(\xi, -\xi^2, \eta_{2,0}), \qquad A_{k,m}^n = \sum_{s=0}^{n-k-m} v_s^k W_{n-k-m-s}^m, \tag{4.35}$$

$$v_\nu^k = \sum_{\substack{i_1+\ldots+i_k=k+\nu \\ i_j \geq 1}} \gamma_{i_1} \cdots \gamma_{i_k}, \qquad W_\nu^m = \sum_{\substack{i_1+\ldots+i_m=m+\nu \\ i_j \geq 1}} \eta_{2,i_1} \cdots \eta_{2,i_m},$$

$$v_0^0 = W_0^0 = 1, \qquad v_\nu^0 = W_\nu^0 = 0, \qquad \nu = 1, 2, \ldots \tag{4.36}$$

Here and in what follows the symbol $F_{\eta_1\eta_2\ldots}^{(m)(k)\ldots}$ is used to denote the derivative of the function F of the form $\partial^{m+k+\ldots}/\partial \eta_1^m \partial \eta_2^k \ldots$. It follows from the structure of formulas (4.32)–(4.36) that Eq. (4.31) can be written as

$$\frac{d\eta_{2,n}}{d\xi} = \varkappa(\xi)\eta_{2,n} + \varphi_n(\xi), \tag{4.37}$$

where $\varkappa(\xi)$ is the derivative of the right-hand side of (4.30) with respect to $\eta_{2,0}$ for $\eta_{2,0} = \eta_{2,0}(\xi)$ and the functions $\varphi_n(\xi)$ depend only on $\eta_{2,s}$ with $s \leq n-1$. Thus the functions $\eta_{2,n}$ are defined recurrently.

Proceeding from (4.32)–(4.36), we shall show that the asymptotic expansions

$$\eta_{2,0}(\xi) = \eta_{2,0}(0) + \sum_{n=3}^{\infty} \varkappa_n \xi^n, \qquad (4.38)$$

$$\eta_{2,n}(\xi) = \frac{1}{\xi^{3n-3}} \sum_{k=0}^{n} \sum_{\nu=3k}^{\infty} c_{k,\nu,n} \xi^\nu \ln^k |\xi| + c_n \ln^n |\xi|, \qquad n \geq 1, \qquad (4.39)$$

hold true as $\xi \to -0$. It is clear that only the last of them needs to be proved, since (4.38) is a consequence of the regularity of problem (4.30).

The substantiation of (4.39) rests on certain algebraic properties of its right-hand side. We denote by P_n an arbitrary polynomial of order n of the variable $\xi^3 \ln |\xi|$ with coefficients which can be expanded in series by the integral powers of ξ. Let $\tilde{P}_n = \frac{1}{\xi^{3n-3}} P_n + c_n \ln^n |\xi|$, where c_n is a constant. In the new notations (4.39) assumes the form

$$\eta_{2,n} = \tilde{P}_n. \qquad (4.40)$$

It is easy to verify that

$$\tilde{P}_n \tilde{P}_m = \tilde{P}_{n+m}, \qquad \tilde{P}_n P_m = \frac{P_{n+m}}{\xi^{3n}}, \qquad \int \frac{P_n}{\xi^{3n-2}} d\xi = \tilde{P}_n, \qquad (4.41)$$

where the same letters P_n and $\tilde{P}_n$ denote different representatives of the classes of functions we have introduced.

For $\eta_{2,1}$, (4.40) follows from the equation

$$\frac{d\eta_{2,1}}{d\xi} = \varkappa(\xi)\eta_{2,1} + g'_{2\eta_1}(\xi, -\xi^2, \eta_{2,0}) - \frac{g_2(\xi, -\xi^2, \eta_{2,0})}{[a_1(\xi, \eta_{2,0})]^2} a_2(\xi, \eta_{2,0}),$$

where use is made of properties (4.9) of the functions a_1, a_2. Let us suppose that (4.40) is valid for all $\eta_{2,m}$ with subscripts $m \leq n-1$. Then, with the aid of (4.41) and properties (4.9) of the functions a_n, we successively obtain

$$\gamma_n = \frac{P_{n-1}}{\xi^{3n-2}}, \qquad v_\nu^k = \frac{P_\nu}{\xi^{3\nu+k}}, \qquad W_\nu^m = \tilde{P}_{m+\nu},$$

$$A_{k,m}^n = \frac{P_{n-k}}{\xi^{3n-2k}}, \qquad \Lambda_n = \frac{P_{n-1}}{\xi^{3n-2}}, \qquad \varphi_n = \frac{P_n}{\xi^{3n-2}},$$

where, as before, the same notations are used for the functions P_m and $\tilde{P}_n$ with the same structure. From these relations and from (4.37), by virtue of the last property of (4.41), it follows that

$$\eta_{2,n} = \int_{-q}^{\xi} \exp\left\{\int_{s}^{\xi} \varkappa(\sigma)\,d\sigma\right\} \varphi_n(s)\,ds = \int \frac{P_n}{s^{3n-2}}\,ds = \tilde{P}_n.$$

We have completely proved (4.40) (and, hence, (4.39)), and now we can infer that the following statement is valid.

THEOREM 4.4. *For* $-q \leq \xi \leq -\varepsilon^{\lambda_1}$ *the asymptotic relation*

$$\eta_2(\xi,\varepsilon) = \eta_{2,0}(\xi) + \sum_{n=1}^{N} \varepsilon^n \eta_{2,n}(\xi) + O(\varepsilon^{(1-3\lambda_1)N+1}) \qquad (4.42)$$

holds true for the solution of the Cauchy problem (4.6), (4.26).

Taking this relation into account in representation (4.12), we get the asymptotics of $\eta_1(\xi,\varepsilon)$.

PROPOSITION 4.1. *For* $-q \leq \xi \leq -\varepsilon^{\lambda_1}$, *for the coordinate* η_1 *of system* (4.4) *we have the asymptotic formula*

$$\eta_1 = -\xi^2 + \sum_{n=1}^{N} \varepsilon^n \gamma_n(\xi) + O(\varepsilon^{N(1-3\lambda_1)+1-\lambda_1}), \qquad (4.43)$$

where N is an arbitrary natural number, and for the functions γ_n defined by (4.33), *as $\xi \to -0$, the asymptotic expansions*

$$\gamma_n(\xi) = \frac{1}{\xi^{3n-2}} \sum_{k=0}^{n-1} \sum_{\nu=3k}^{\infty} c_{k,\nu}^n \xi^\nu \ln^k |\xi|, \qquad n \geq 1, \qquad (4.44)$$

hold true.

4.5. Asymptotics of Trajectories in the Middle of a Breakoff Section

We carry out the substitutions $\xi = \mu u$, $\eta_1 = \mu^2 v$, $\eta_2 = \eta_{2,0}(0) + \mu^3 \omega$, $\varepsilon = \mu^3$ in (4.6) for $-\varepsilon^{\lambda_1} \leq \xi \leq \varepsilon^{\lambda_2}$. Taking (4.15) into account, we arrive at the equation

$$\frac{d\omega}{du} = \frac{1}{\mu}\frac{g_2(\mu u, \mu^2 v, \eta_{2,0}(0) + \mu^3 \omega)}{u^2 + v}, \qquad (4.45)$$

whose right-hand side can be expanded in an asymptotic series in integer powers of μ. We are interested in the solution of (4.45) with an initial condition dependent on ε whose asymptotic behavior is derived from (4.38), (4.39), and (4.42), setting $\xi = -\varepsilon^{\lambda_1}$ in them. Analyzing the resulting asymptotic expansions, we make sure that the solution we need should be sought in the form

$$\omega(u,\mu) = \sum_{n=0}^{\infty} \mu^n \omega_n(u,\mu), \qquad \omega_n(u,\mu) = \sum_{k=0}^{[\frac{n}{3}]+1} \omega_{n,k}(u) \ln^k \mu. \qquad (4.46)$$

After substituting (4.46) into (4.45) and equating the coefficients in like powers of μ, we arrive at the sequence of differential equations

$$\frac{d\omega_n}{du} = M_{n+1}, \qquad n \geq 0, \qquad (4.47)$$

where

$$M_n = \frac{\Lambda_n}{u^2 + v_0(u)} + \sum_{l=1}^{n} \Lambda_{n-l} \sum_{\nu=1}^{l} \frac{(-1)^\nu}{[u^2 + v_0(u)]^{\nu+1}} \times$$

$$\times \sum_{\substack{i_1+\ldots+i_\nu=l \\ i_j \geq 1}} \gamma_{i_1} \ldots \gamma_{i_\nu}, \qquad n \geq 1, \qquad (4.48)$$

$$\Lambda_n = \sum_{\substack{0 < m+2k+3\nu \leq n \\ m+k \neq 0}} \frac{g_{2\xi\eta_1\eta_2}^{(m)(k)(\nu)}(0,0,\eta_{2,0}(0))}{m!\, k!\, \nu!} A_{m,k,\nu}^n, \qquad \Lambda_0 = 0, \qquad (4.49)$$

$$\gamma_n = v_n(u,\eta_{2,0}(0)) + \sum_{m=1}^{[\frac{n}{3}]} \sum_{k=1}^{m} \frac{v_{n-3m\,\eta_2}^{(k)}(u,\eta_{2,0}(0))}{k!} \times$$

$$\times \sum_{\substack{i_1+\ldots+i_k=3m-3k \\ i_j \geq 0}} \omega_{i_1} \ldots \omega_{i_k}, \qquad n \geq 0, \qquad (4.50)$$

$$A_{m,k,\nu}^n = u^m \sum_{s=0}^{n-m-2k-3\nu} z_s^k \lambda_{n-m-2k-3\nu-s}^\nu, \qquad (4.51)$$

$$z_r^k = \sum_{\substack{i_1+\ldots+i_k=r \\ i_j \geq 0}} \gamma_{i_1} \ldots \gamma_{i_k}, \qquad \lambda_r^\nu = \sum_{\substack{i_1+\ldots+i_\nu=r \\ i_j \geq 0}} \omega_{i_1} \ldots \omega_{i_\nu},$$

$$z_0^0 = \lambda_0^0 = 1, \qquad z_r^0 = \lambda_r^0 = 0, \qquad r = 1, 2, \ldots \qquad (4.52)$$

Using the relation

$$\max_{\substack{i_1+\ldots+i_\nu=r \\ i_j \geq 0}} \left\{ \left[\frac{i_1}{3}\right] + \ldots + \left[\frac{i_\nu}{3}\right] \right\} = \left[\frac{r}{3}\right]$$

to estimate the maximum power in which $\ln\mu$ enters into expressions (4.48)–(4.52), we infer that

$$M_{n+1} = \sum_{k=0}^{[\frac{n}{3}]} M_{n,k}(u)\ln^k\mu, \qquad (4.53)$$

where $M_{n,k}$ depends only on ω_m with subscripts $m \leq n - 3$. Therefore, the sequence of differential equations is recurrent and disintegrates into the following series of equations with variables separable:

$$\frac{d\omega_{n,k}}{du} = M_{n,k}(u), \qquad k = 0, 1, \ldots, \left[\frac{n}{3}\right],$$

$$\frac{d\omega_{n,[\frac{n}{3}]+1}}{du} = 0, \qquad n \geq 0.$$

We choose solutions of these equations such that for $\xi = -\varepsilon^{\lambda_1}$ all members of expansions (4.42) and (4.46) independent of λ_1 coincide. Hence

$$\omega_{n,k} = \int_0^u M_{n,k}(s)\,ds + (H)\int_{-\infty}^0 M_{n,k}(s)\,ds + \beta(n)c_{k,n,[\frac{n}{3}]+1},$$

$$k = 0, 1, \ldots, \left[\frac{n}{3}\right], \qquad (4.54)$$

$$\omega_{n,[\frac{n}{3}]+1} = \beta(n)c_{[\frac{n}{3}]+1}, \qquad \beta(n) = \begin{cases} 1 & \text{if } n \equiv 0\pmod{3}, \\ 0 & \text{if } n \not\equiv 0\pmod{3}, \end{cases} \qquad (4.55)$$

where $c_{k,n,[\frac{n}{3}]+1}$, $c_{[\frac{n}{3}]+1}$ are the same as in (4.39). Using, say, (4.54), (4.55), we obtain

$$\omega_0(u) = g'_{2\xi}(0, 0, \eta_{2,0}(0))\left[\int_0^u \frac{s\,ds}{s^2 + v_0(s)} + \right.$$

$$\left. + (H)\int_{-\infty}^0 \frac{s\,ds}{s^2 + v_0(s)}\right] + c_{0,0,1} + c_1\ln\mu. \qquad (4.56)$$

Note that in [77], where the asymptotics of the trajectories is constructed with an accuracy to within quantities of order ε, the second and third terms in a formula similar to (4.56) are not taken into account. As we said above, M_{n+1} depends only on ω_m with subscripts $m \leq n - 3$. Therefore $M_{n,k} = 0$ for $n = 0, 1, \ldots, 3k - 1$. Hence, by virtue of (4.54), (4.55), we have

$$\omega_{n,k} = 0, \qquad n = 0, 1, \ldots, 3k - 4,$$

$$\omega_{3k-3,k} = c_k, \qquad \omega_{3k-2,k} = \omega_{3k-1,k} = 0, \qquad k = 1, 2, \ldots$$

Theorem on the C^1-Proximity

It is easy to obtain the asymptotic expansions of $\omega_{n,k}$ as $u \to -\infty$: we have to set $\xi = \mu u$, $\eta_2 = \eta_{2,0}(0) + \mu^3 \omega$ in (4.29) and regroup the members of the asymptotic series. In this way we get

$$\omega_{n,0}^- = \varkappa_{n+3} u^{n+3} + u^n \sum_{m=0}^{[\frac{n}{3}]} \sum_{\nu=m}^{\infty} \frac{c_{m,n,\nu}}{u^{3\nu-3}} \ln^m |u| +$$
$$+ \beta(n) c_{[\frac{n}{3}]+1} \ln^{[\frac{n}{3}]+1} |u|, \qquad (4.57)$$

$$\omega_{n,k}^- = u^n \sum_{m=k}^{[\frac{n}{3}]} \sum_{\nu=m}^{\infty} C_m^k \frac{c_{m,n,\nu}}{u^{3\nu-3}} \ln^{m-k} |u| +$$
$$+ \beta(n) c_{[\frac{n}{3}]+1} C_{[\frac{n}{3}]+1}^k \ln^{[\frac{n}{3}]+1-k} |u|, \qquad n \geq 3k, \qquad (4.58)$$

where $\varkappa_{n+3}, c_{m,n,\nu}$ and $c_{[\frac{n}{3}]+1}$ are the same as in (4.38), (4.39). Analyzing now (4.48)–(4.52) and using induction with respect to n, we make sure that

$$\omega_{n,k}^+ = u^{n-3k} \sum_{\nu=0}^{n-3k} \sum_{l=3\nu}^{\infty} {}^* g_{\nu,l}^{n,k} \frac{\ln^\nu u}{u^l} +$$
$$+ g_{\pi(n+1)-k,0}^{n,k} \ln^{\pi(n+1)-k} u, \qquad n \geq 3k. \qquad (4.59)$$

By way of example, we give the proof for $k = 0$. As in the proof of (4.39), we use the same symbol P_n for different functions from the class of polynomials of degree n of the variable $\frac{\ln u}{u^3}$ with coefficients which can be expanded in series in the integer powers of $1/u$ and which do not contain members of the first power. We also introduce a class of functions $\tilde{P}_n = u^n P_n + g \ln^{\pi(n+1)} u$, where g is an arbitrary number. Then, for $k = 0$, (4.59) assumes the form

$$\omega_{n,0}^+ = \tilde{P}_n. \qquad (4.60)$$

For $\omega_{0,0}^+$ relation (4.60) immediately follows from (4.56) and from the asymptotic properties of the function $v_0(u)$ (see (4.18) for $n = 0$). Let us suppose that it holds true for all subscripts not exceeding $n - 1$. Then, taking into account the equalities

$$P_n \cdot P_m = P_{n+m}, \qquad \tilde{P}_{i_1} \cdot \ldots \cdot \tilde{P}_{i_k} = u^{i_1+\ldots+i_k+3k} P_{i_1+\ldots+i_k},$$

which can be easily verified, we infer from (4.48)–(4.52) that $M_{n,0} = u^{n-1} P_n$, whence we find that

$$\omega_{n,0}^+ = \int u^{n-1} P_n \, du = u^n P_n + g \ln^{\pi(n+1)} u = \tilde{P}_n.$$

We have proved (4.59) for $k = 0$. The proof of (4.59) for $k \geq 1$ is similar but somewhat more cumbersome.

The asymptotic representations (4.58) and (4.59) allows us to infer that the following statement is valid.

THEOREM 4.5. *For solving the Cauchy problem* (4.6), (4.28), *we have the asymptotic relation*

$$\eta_2(\xi,\varepsilon) = \eta_{2,0}(0) + \mu^3 \sum_{n=0}^{N} \mu^n \omega_n\left(\frac{\xi}{\mu},\mu\right) + O(\varepsilon^{\lambda_1(N+4)}) \qquad (4.61)$$

for $-\varepsilon^{\lambda_1} \leq \xi \leq 0$ *and the relation*

$$\eta_2(\xi,\varepsilon) = \eta_{2,0}(0) + \mu^3 \sum_{n=0}^{N} \mu^n \omega_n\left(\frac{\xi}{\mu},\mu\right) + O(\varepsilon^{\lambda_2(N+1)+1}) \qquad (4.62)$$

for $0 \leq \xi \leq \varepsilon^{\lambda_2}$.

We substitute formulas (4.61), (4.62) into (4.20), (4.21), which define the asymptotics of the integral manifold H for the corresponding values of ξ. As a result we get the asymptotic representation

$$\eta_1 = \mu^2 \sum_{n=0}^{\infty} \mu^n \gamma_n \qquad (4.63)$$

of the coordinate η_1 of system (4.4). Here the functions $\gamma_n = \gamma_n(u,\mu)$ defined by (4.50) have the structure

$$\gamma_n = \sum_{k=0}^{[\frac{n}{3}]} \gamma_{n,k}(u) \ln^k \mu. \qquad (4.64)$$

Note that $\gamma_{3k,k} = 0$ for $k \geq 1$, since the function $v_0(u)$ does not depend on η_2 and the corresponding derivatives entering into (4.50) are zero. As $u \to -\infty$, we have the asymptotic representations

$$\gamma_{0,0}^- = v_0(u)^- = -u^2 + \sum_{k=0}^{\infty} \frac{\alpha_k}{u^{3k+1}}, \qquad (4.65)$$

$$\gamma_{n,k}^- = u^{n-1} \sum_{m=k}^{[\frac{n}{3}]} \sum_{\nu=m}^{\infty} \frac{\alpha_{m,\nu}^{n,k}}{u^{3\nu}} \ln^{m-k} |u|, \qquad n \geq 1, \qquad (4.66)$$

for the functions $\gamma_{n,k}$. These representations result from (4.43), (4.44) after the substitutions $\xi = \mu u$, $\eta_1 = \mu^2 v$ and the re-expansion in the power of μ.

As $u \to +\infty$, we infer from (4.50) and expansions (4.59) that

$$\gamma_{n,k}^+ = u^{n-1-3k} \sum_{\nu=0}^{n-3k} \sum_{l=3\nu}^{\infty} {}^*\frac{\beta_{\nu,l}^{n,k}}{u^l} \ln^\nu u +$$
$$+ \beta_{\pi(n)-k,0}^{n,k} \ln^{\pi(n)-k} u, \qquad n \geq 0. \tag{4.67}$$

Relations (4.63)–(4.67) lead to the following statement.

PROPOSITION 4.2. *For* $-\varepsilon^{\lambda_1} \leq \xi \leq 0$ *we have*

$$\eta_1 = \mu^2 \sum_{n=0}^{N} \mu^n \gamma_n\left(\frac{\xi}{\mu}, \mu\right) + O(\varepsilon^{\lambda_1 N+1}) \tag{4.68}$$

and for $0 \leq \xi \leq \varepsilon^{\lambda_2}$ *the relation*

$$\eta_1 = \mu^2 \sum_{n=0}^{N} \mu^n \gamma_n\left(\frac{\xi}{\mu}, \mu\right) + O(\varepsilon^{\lambda_2 N+1}) \tag{4.69}$$

holds true.

4.6. Asymptotics of Trajectories at the End of a Breakoff Section

For $\varepsilon^{\lambda_2} \leq \xi \leq q$, we consider the equation

$$\frac{d\eta_2}{d\xi} = \frac{\mu^3 g_2(\xi, \eta_1, \eta_2)}{\xi^2 + \eta_1}, \tag{4.70}$$

which results if we set $\varepsilon = \mu^3$ in (4.6) and substitute the asymptotic series (4.23) for $\eta_1 = H$. We are interested in its solution with the initial condition (4.69) for $\xi = \varepsilon^{\lambda_2}$. We seek this solution in the form

$$\eta_2(\xi, \mu) = \eta_{2,0}(0) + \sum_{n=3}^{\infty} \mu^n \eta_{2,n}(\xi, \mu), \qquad \eta_{2,n}(\xi, \mu) = \sum_{k=0}^{\pi(n-2)} \eta_{2,n,k}(\xi) \ln^k \mu. \tag{4.71}$$

Substituting (4.71) into (4.70) and equating the coefficients in like powers of μ, we obtain

$$\frac{d\eta_{2,n}}{d\xi} = B_{n-3}, \qquad n \geq 3, \tag{4.72}$$

where

$$B_n = \sum_{\varkappa=0}^{n} \Lambda_{n-\varkappa} \sum_{k=0}^{[\frac{\varkappa}{2}]} \frac{(-1)^k}{\xi^{2k+2}} \lambda_{\varkappa-2k}^k, \qquad n \geq 0, \tag{4.73}$$

$$\Lambda_n = \sum_{0<2k+3m\leq n} \frac{g_{2\eta_1\eta_2}^{(k)(m)}(\xi,0,\eta_{2,0}(0))}{k!\,m!} A^n_{k,m}, \qquad n \geq 1,$$

$$\Lambda_0 = g_2(\xi,0,\eta_{2,0}(0)), \qquad (4.74)$$

$$\gamma_n = h_n(\xi,\eta_{2,0}(0)) + \sum_{m=1}^{[\frac{n}{3}]}\sum_{k=1}^{m} \frac{h^{(k)}_{n-3m\,\eta_2}(\xi,\eta_{2,0}(0))}{k!} \times$$

$$\times \sum_{\substack{i_1+\ldots+i_k=3m \\ i_j\geq 3}} \eta_{2,i_1}\ldots\eta_{2,i_k}, \qquad n \geq 2, \qquad (4.75)$$

$$A^n_{k,m} = \sum_{r=0}^{n-2k-3m} \lambda_r^k \delta^m_{n-2k-3m-r}, \qquad (4.76)$$

$$\lambda^k_\nu = \sum_{i_1+\ldots+i_k=2k+\nu} \gamma_{i_1}\ldots\gamma_{i_k}, \qquad \delta^m_\nu = \sum_{\substack{i_1+\ldots+i_m=3m+\nu \\ i_j\geq 3}} \eta_{2,i_1}\ldots\eta_{2,i_m},$$

$$\lambda^0_0 = \delta^0_0 = 1, \qquad \delta^0_\nu = \lambda^0_\nu = 0, \qquad \nu \geq 1. \qquad (4.77)$$

Analyzing (4.72)–(4.77) and taking into account the relations

$$\max_{\substack{i_1+\ldots+i_k=l \\ i_j\geq 2}} \{\pi(i_1-2)+\ldots+\pi(i_k-2)\} =$$

$$= \max_{\substack{i_1+\ldots+i_k=l \\ i_j\geq 3}} \{\pi(i_1-2)+\ldots+\pi(i_k-2)\} = \pi(l-2),$$

we infer that

$$B_{n-3} = \sum_{\nu=0}^{\pi(n-2)-1} B_{n-3,\nu}\ln^\nu\mu,$$

where $B_{n-3,\nu}$ depends only on $\eta_{2,m}$ with subscripts $m \leq n-3$. From this and from (4.71) we get the following series of differential equations, dependent on n, with variables separable:

$$\frac{d\eta_{2,n,\nu}}{d\xi} = B_{n-3,\nu}, \qquad \nu = 0,1,\ldots,\pi(n-2)-1,$$

$$\frac{d\eta_{2,n,\pi(n-2)}}{d\xi} = 0. \qquad (4.78)$$

As in 4.5, we choose the solutions of (4.78) such that for $\xi = \varepsilon^{\lambda_2}$ all members of expansions (4.71) and (4.62), independent of λ_2, coincide. In this way we

Theorem on the C^1-Proximity

make sure that

$$\eta_{2,n,\nu} = (H)\int_0^\xi B_{n-3,\nu}(s)\,ds + \sum_{\{k:2k+\nu\leq n-3, k\leq \nu\}} (-1)^{\nu-k} g_{\nu-k,n-3k-3}^{n-3,k},$$

$$\nu = 0, 1, \ldots, \pi(n-2) - 1,$$

$$\eta_{2,n,\pi(n-2)} = \beta(n)c_{[\frac{n}{3}]} + \sum_{k=0}^{[\frac{n}{3}]-1} g_{\pi(n-2)-k,0}^{n-3,k}(-1)^{\pi(n-2)-k}, \quad (4.79)$$

where all the coefficients are taken from expansions (4.59) of the functions $\omega_{n,k}$ for $u \to +\infty$.

As $\xi \to +0$, the asymptotic expansions of functions (4.79) result from (4.46), (4.59), where we must set $u = \xi/\mu$ and carry out a re-expansion with respect to μ. Having performed these operations, we infer that

$$\eta_{2,n,s} = \frac{1}{\xi^{n-3}} \sum_{\varkappa=0}^{\pi(n-2)-s-1} \sum_{p=\varkappa+s}^{\infty} z_{p,\varkappa}^{n,s} \xi^p \ln^\varkappa \xi +$$

$$+ z_{0,\pi(n-2)-s}^{n,s} \ln^{\pi(n-2)-s} \xi, \quad (4.80)$$

where

$$z_{p,\varkappa}^{n,s} = \sum_{\{k:2k+\varkappa+s\leq p, k\leq s\}} (-1)^{s-k} C_{\varkappa+s-k}^{s-k} g_{\varkappa+s-k,n-3k-3}^{p,k},$$

$$n = 3, 5, 6, \ldots, \quad 0 \leq s \leq \pi(n-2) - 1,$$
$$0 \leq \varkappa \leq \pi(n-2) - s - 1, \quad p \geq \varkappa + s;$$

$$z_{0,\pi(n-2)-s}^{n,s} = \sum_{k=0}^{s} (-1)^{s-k} C_{\pi(n-2)-k}^{s-k} g_{\pi(n-2)-k,0}^{n-3,k},$$

$$s = 0, 1, \ldots, \pi(n-2) - 1;$$

$$z_{0,0}^{n,\pi(n-2)} = \beta(n)c_{[\frac{n}{3}]} + \sum_{k=0}^{[\frac{n}{3}]-1} (-1)^{\pi(n-2)-k} g_{\pi(n-2)-k,0}^{n-3,k}.$$

Relations (4.80) allow us to formulate the following statement.

THEOREM 4.6. *For $\varepsilon^{\lambda_2} \leq \xi \leq q$ we have*

$$\eta_2(\xi,\mu) = \eta_{2,0}(0) + \sum_{n=3}^{N} \mu^n \eta_{2,n}(\xi,\mu) + O(\varepsilon^{\frac{N+1}{3}-\lambda_2(N-2)}). \quad (4.81)$$

By virtue of (4.27) and (4.81), we obtain the asymptotic series

$$\eta_1 = \sum_{n=2}^{\infty} \mu^n \gamma_n(\xi,\mu), \quad \gamma_n(\xi,\mu) = \sum_{s=0}^{\pi(n-2)} \gamma_{n,s}(\xi) \ln^s \mu \quad (4.82)$$

for the coordinate η_1 of system (4.4). Here the functions γ_n are defined by (4.75) and, as $\xi \to 0$, $\gamma_{n,s}$ can be expanded in the asymptotic series

$$\gamma_{n,s} = \frac{1}{\xi^{n-2}} \sum_{\varkappa=0}^{\pi(n-2)-s-1} \sum_{p=\varkappa+s}^{\infty} \varphi_{p,\varkappa}^{n,s} \xi^p \ln^{\varkappa} \xi +$$

$$+ \varphi_{0,\pi(n-2)-s}^{n,s} \ln^{\pi(n-2)-s} \xi, \qquad n \geq 3, \qquad \gamma_2 = \Omega_0, \qquad (4.83)$$

whose coefficients are expressed in terms of the coefficients of series (4.67):

$$\varphi_{p,\varkappa}^{n,s} = \sum_{\{k:2k+\varkappa+s \leq p, k \leq s\}} (-1)^{s-k} C_{\varkappa+s-k}^{s-k} \beta_{\varkappa+s-k, n-3k-3}^{p,k},$$

$$n = 3, 5, 6, \ldots, \qquad s = 0, 1, \ldots, \pi(n-2) - 1,$$

$$\varkappa = 0, 1, \ldots, \pi(n-2) - s - 1, \qquad p \geq \varkappa + s;$$

$$\varphi_{0,\pi(n-2)-1}^{n,s} = \sum_{k=0}^{s} (-1)^{s-k} C_{\pi(n-2)-k}^{s-k} \beta_{\pi(n-2)-k,0}^{n-2,k},$$

$$s = 0, 1, \ldots, \pi(n-2) - 1;$$

$$\varphi_{0,0}^{n,\pi(n-2)} = \sum_{k=0}^{[\frac{n}{3}]-1} (-1)^{\pi(n-2)-k} \beta_{\pi(n-2)-k,0}^{n-2,k}.$$

It is interesting that series (4.82) and (4.83) are similar in their structure to the corresponding series in the case of a relaxation system in the plane [77].

We sum up what we said above as the following statement.

PROPOSITION 4.3. *For* $\varepsilon^{\lambda_2} \leq \xi \leq q$ *we have*

$$\eta_1 = \sum_{n=2}^{N} \mu^n \gamma_n(\xi, \mu) + O(\varepsilon^{\frac{N+1}{3} - \lambda_2(N-1)}). \qquad (4.84)$$

4.7. Conclusion

The statements we have formulated make it possible to trivialize the results of [76, 77], which substantiate the asymptotic formulas of the relaxation system trajectories with an accuracy up to ε. Indeed, let us divide the second and third equations of (4.4) by the first equation and linearize the resulting systems on the formal trajectory. The results we established in Secs. 1 and 2 allow us to state that the Cauchy matrix of the system obtained is bounded uniformly with respect to ε. The subsequent reasoning is standard, namely, taking into account the residual and the nonlinear terms, we use the Lagrange formula to pass to an integral equation and apply the contraction mapping principle in the space of functions continuous in $\xi \in [-q, q]$.

Thus, although we cited the results of [77] when proving Theorem 1.2, in actual fact our presentation is sufficiently self-contained, i.e., the technique that we have developed makes it possible to obtain all the necessary facts independently. In particular, the scheme we have suggested will be later used in 10.3 when we consider relaxation parabolic systems.

5. Uniqueness, Asymptotics, and Stability of a Relaxation Cycle

5.1. Asymptotics of a Relaxation Cycle

Suppose that the relay system (1.1) has a dichotomous periodic solution $y_0(t)$ with period T_0 and let, for definiteness, $(\Phi_1(y_0(0)), y_0(0)) \in \Gamma_{1-}$. As in 1.2, we denote by Z_0 the cycle of the hybrid system (1.7) obtained by means of a "rise" of a cycle of the relay system from R^m to R^n. Then, when Conditions 1.1–1.3 from Sec. 1 are satisfied, the relaxation system (1.9) has, by virtue of Theorem 1.1, a unique cycle $Z(\varepsilon)$ with period $T(\varepsilon)$ for all sufficiently small values of ε in a certain neighborhood independent of ε, with $Z(\varepsilon) \to Z_0$, $T(\varepsilon) \to T_0$ as $\varepsilon \to 0$. We denote by $(x(\varepsilon), y(\varepsilon))$, where $x(0) = \Phi_1(y_0(0))$, $y(0) = y_0(0)$, its initial condition defined for $t = 0$.

THEOREM 5.1. *We have the asymptotic representations*

$$T(\varepsilon) = T_0 + \sum_{n=2}^{\infty} \varepsilon^{n/3} \sum_{\nu=0}^{\pi(n-2)} T_{n,\nu} \ln^\nu \frac{1}{\varepsilon}, \qquad (5.1)$$

$$x(\varepsilon) = \Phi_1(y(\varepsilon)) + \sum_{n=1}^{\infty} \varepsilon^n F_n(y(\varepsilon)), \qquad (5.2)$$

$$y(\varepsilon) = y_0(0) + \sum_{n=2}^{\infty} \varepsilon^{n/3} \sum_{\nu=0}^{\pi(n-2)} y_{n,\nu} \ln^\nu \frac{1}{\varepsilon}, \qquad (5.3)$$

where (5.2) is the result of the substitution of (5.3) into the series for the integral manifold of system (1.9) in the neighborhood of $(\Phi_1(y_0(0)), y_0(0))$.

5.2. Proof of Theorem 5.1

Let us generically draw an intersecting hyperplane L through the point $(\Phi_1(y_0(0)), y_0(0))$ belonging to the domain Γ_{1-} and consider the Poincaré operator $\Pi(\varepsilon)$ in its neighborhood.

It is convenient to represent it as a superposition of mappings, namely, $\Pi_1(\varepsilon)$ acts from L into the hyperplane $\xi = -q$, $\Pi_2(\varepsilon)$ acts from $\xi = -q$

into $\xi = q$, $\Pi_3(\varepsilon)$ acts from $\xi = q$ into the hyperplane $\xi = -q$ on the second domain Γ_{2-}, and so on. By virtue of the results from [7], some of these operators can be expanded in asymptotic series in integer powers of ε with smooth coefficients. The asymptotics of the operator $\Pi_2(\varepsilon)$ (and other, similar, operators) was found in Sec. 4. Therefore, for the components x and y of the operator $\Pi(\varepsilon)$ we have asymptotic series with smooth coefficients similar to (5.2) and (5.3).

It follows that the coordinates of its fixed point can be expanded in formal series (5.2) and (5.3), the character of whose asymptotic properties can be revealed thanks to the following two facts. First, in accordance with Theorem 1.1, $\Pi'(\varepsilon)|_{x=x(\varepsilon),y=y(\varepsilon)} \to B$ as $\varepsilon \to 0$, where the matrix B does not have eigenvalues on the unit circle. Second, if the finite intervals of series (5.2), (5.3) are substituted into the equation $\Pi(\varepsilon)(x(\varepsilon),y(\varepsilon)) = (x(\varepsilon),y(\varepsilon))$, the residue is of a higher order of smallness. It remains to use the contraction mapping principle.

Relation (5.1) is a simple consequence of the structure of the operator $\Pi(\varepsilon)$.

5.3. Proof of the Pontryagin–Mishchenko Theorem

We shall consider here the proof of this theorem, which we have repeatedly mentioned, in order to make more clear the reasons why it is valid.

For $\varepsilon = 0$, the operator $\Pi(\varepsilon)$ introduced above turns into the Poincaré operator Π_0 along the trajectory of the hybrid system (1.7). We change the latter in a definite way, namely, we suppose that when moving along the trajectory of the hybrid system, we get to the hyperplane $\xi = -q$. We have introduced the operator $\Pi_2(\varepsilon)$ acting from $\xi = -q$ into $\xi = q$. In the asymptotic series in which the operator $\Pi_2(\varepsilon)$ is expanded, we preserve the zero-order terms and the terms of orders $\varepsilon^{2/3}$ and $\varepsilon \ln \frac{1}{\varepsilon}$ and neglect the other orders of smallness. We assume that for $-q \leq \xi \leq q$ the hybrid system trajectory has been changed by means of the operator $\Pi_2(\varepsilon)$ shortened by the technique described above. We act in a similar way when the hybrid system trajectory approaches the other breakoff manifold. We denote the operator of the hybrid system changed in this way by $\Pi_0(\varepsilon)$.

Recall that it is supposed in the Pontryagin–Mishchenko theorem that all eigenvalues of the matrix Π'_0, calculated at a fixed point, are smaller than unity in absolute value. Therefore, the operator $\Pi_0(\varepsilon)$ has a unique exponentially stable equilibrium position $M_0(\varepsilon)$, which passes, for $\varepsilon = 0$, into a similar equilibrium position of the operator $\Pi_0(0)$.

Theorem on the C^1-Proximity

We set $M = M_0(\varepsilon) + h$ and introduce into consideration a ball $\|h\| \leq c_0\varepsilon$. We have

$$\|\Pi(\varepsilon)M - M_0(\varepsilon)\| \leq \|\Pi_0(\varepsilon)M - M_0(\varepsilon)\| + \|\Pi(\varepsilon)M - \Pi_0(\varepsilon)M\|.$$

By virtue of the asymptotic estimates from [75, 76, 78, 79], the second term in this inequality does not exceed the quantity $K\varepsilon$, where the constant K does not depend on c_0. Also make note of the fact that

$$\|\Pi_0(\varepsilon)[M_0(\varepsilon) + h] - M_0(\varepsilon)\| \leq qc_0\varepsilon,$$

where $q < 1$. Therefore,

$$\|\Pi(\varepsilon)M - M_0(\varepsilon)\| \leq c_0\varepsilon$$

if $c_0 = K/(1-q)$. Thus, the operator $\Pi(\varepsilon)$ maps into itself a ball whose radius is of order ε, and this makes it possible to construct the cycles of the relaxation system with an accuracy to within ε.

5.4. Example

Since relaxation systems are now widely used for the simulation of various physical and biophysical processes, it is not necessary to make a special analysis of the corresponding examples. We refer those who want to study them in greater detail to [2] (see also Sec. 12). By the way, the collection of mathematical facts that we have established gives the necessary strictness to the analysis of different mathematical models of electronic devices carried out in [2].

Here we shall only mention Example I.2 (see [77]):

$$\varepsilon\dot{x}_1 = -\alpha(y_1 - y_2) + \varphi(x_1) - x_2,$$
$$\varepsilon\dot{x}_2 = \alpha(y_1 - y_2) + \varphi(x_2) - x_1,$$
$$\dot{y}_1 = x_1, \qquad \dot{y}_2 = x_2,$$

where $\alpha > 0$ is a constant and the smooth monotonically decreasing function $\varphi(u)$ is defined for $-1 < u < 1$, and

$$\lim_{u \to -1+0} \varphi(u) = \infty, \qquad \lim_{u \to 1-0} \varphi(u) = -\infty.$$

In the domain $-1 < x_1, x_2 < 1$, the system being considered, which appears in the description of a number of electronic devices, has a unique orbitally exponentially stable cycle for $0 < \varepsilon \ll 1$ (see III.2).

6. Stochastic Character of the Behavior of Trajectories in Three-Dimensional Relaxation Systems

6.1. Base Theorem

Let us apply the relaxation oscillation theory presented in Secs. 1–4 to the problem of the existence of a strange attractor in the system

$$\varepsilon \dot{x} = f(x,y), \qquad \dot{y} = g(x,y), \qquad 0 < \varepsilon \ll 1, \qquad (6.1)$$

where $x \in R$, $y \in R^2$, and f and g are sufficiently smooth functions. As usual, we suppose that the equation $f = 0$ defines in R^3 a smooth surface Γ divided into the parts Γ_{1-}, Γ_+, Γ_{2-} by the curves Γ_{10}, Γ_{20}. Recall that $(x,y) \in \Gamma_{1-} + \Gamma_{2-}$ if $f'_x < 0$; $(x,y) \in \Gamma_+$ if $f'_x > 0$; $(x,y) \in \Gamma_{10} + \Gamma_{20}$ if $f'_x = 0$. We suppose, in addition, that the domains Γ_{1-} and Γ_{2-} are defined by the equations $x = \Phi_1(y)$ and $x = \Phi_2(y)$, respectively, where Φ_1 and Φ_2 are smooth functions. Finally, we suppose that

$$f''_{xx} \neq 0, \qquad (\mathrm{grad}_y\, f, g) \neq 0 \qquad (6.2)$$

for $(x,y) \in \Gamma_{10} + \Gamma_{20}$.

As was shown in Secs. 1, 2, under the indicated conditions the relaxation system (6.1) is a special small perturbation of the relay system

$$\dot{y} = g(x,y), \qquad x = \begin{cases} \Phi_1(y), \\ \Phi_2(y) \end{cases} \qquad (6.3)$$

about which we shall assume the following.

CONDITION 6.1. The global secant l with parametrization $0 \leq s \leq 1$ is defined in the domain Γ_{1-} and the trajectories of the relay system (6.3) with the initial conditions on l generically cut it after a complete traversal.

Obviously, under Condition 6.1 the Poincaré continuously differentiable function is defined,

$$s \longrightarrow F_0(s). \qquad (6.4)$$

As is known (see [71]), outside of a certain neighborhood of the curve Γ_{10} the relaxation system (6.1) has a smooth integral manifold $x = H(y,\varepsilon)$, which is at a distance not exceeding ε from Γ_{1-}. We take a smooth surface Γ_* which has a generic intersection with Γ_{1-} along l. It cuts the integer surface $x = H(y,\varepsilon)$ along a certain smooth curve $l(\varepsilon)$, the parametrization on which we denote by $0 \leq s \leq 1$ as before. We use σ to denote the coordinate along a certain smooth curve, belonging to Γ_*, which has a generic intersection with

$l(\varepsilon)$. It is clear that for $0 \le s \le 1$, $-\sigma_1 \le \sigma \le \sigma_2$, where the constants σ_1 and σ_2 are sufficiently small, the Poincaré operator Π along the trajectories of the relaxation system (6.1) is defined.

THEOREM 6.1. *Under Condition 6.1 we have*

$$\Pi : \begin{array}{l} s \longrightarrow F_1(s,\sigma,\varepsilon), \\ \sigma \longrightarrow \Delta(s,\sigma,\varepsilon)\exp(-\alpha/\varepsilon), \end{array} \qquad (6.5)$$

where the constant $\alpha > 0$, the functions F_1, Δ are continuously differentiable with respect to s, σ and their derivatives are uniformly bounded with respect to ε, in the C^1-metric $F_1(s,0,\varepsilon)$ differs from $F_0(s)$ from (6.4) by a quantity of order $\varepsilon^{2/3}$.

The proof follows immediately from the results given in Secs. 1–4.

6.2. Complicated Dynamics of System (6.1)

Under additional restrictions, the base Theorem 6.1 makes it possible to formulate interesting propositions concerning the dynamics of system (6.1). Suppose that mapping (6.4) has an exponentially unstable cycle

$$s_1, \ldots, s_k \qquad (6.6)$$

with $F_0'(s_i) \ne 0$, $i = 1, \ldots, k$. We say that it has a structurally stable homoclinic trajectory

$$\{p_n, -\infty < n < \infty\} \qquad (6.7)$$

if the numbers p_n coincide with one of the numbers (6.6) for $n \ge n_0$, the sequence (p_n) disintegrates into subsequences as $n \to -\infty$, and each of the subsequences converges to one of the numbers (6.6), and, finally, $F_0'(p_n) \ne 0$ for all integers n.

CONDITION 6.2. Mapping (6.4) has an exponentially unstable cycle (6.6) with a structurally stable homoclinic trajectory (6.7).

THEOREM 6.2. *Under Conditions 6.1, 6.2, system (6.1) has a countable number of dichotomous cycles.*

PROOF. Passing to the necessary iteration, we can suppose that mapping (6.4) has an equilibrium position s_0 with a structurally stable homoclinic trajectory. Then, by virtue of (6.5), the operator Π is in a saddle equilibrium position $s_0(\varepsilon) \to s_0$, $\sigma_0(\varepsilon) \to 0$ as $\varepsilon \to 0$. We denote its stable and unstable states as $\gamma_-(\varepsilon)$ and $\gamma_+(\varepsilon)$ respectively. On $\gamma_+(\varepsilon)$ we take an open set $\omega_+(\varepsilon)$ containing $(s_0(\varepsilon), \sigma_0(\varepsilon))$ and consider the sets $\omega_m(\varepsilon) = \Pi^m(\omega_+(\varepsilon))$, $m = 1, 2, \ldots$. It follows from Condition 6.2 that $\omega_{m_0}(\varepsilon)$ contains the points

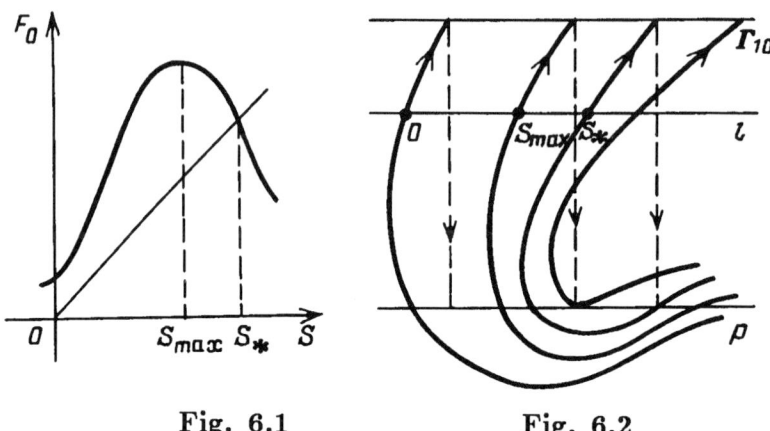

Fig. 6.1　　　　　Fig. 6.2

$\gamma_{-}(\varepsilon)$ for a certain m_0 since, if we assume the contrary, it would follow that the equilibrium position s_0 of mapping (6.4) can only have a structurally stable homoclinic trajectory. It remains to use the theorem from [89].

It is clear that the nonmonotonicity of mapping (6.4) is a necessary prerequisite for the validity of Condition 6.2, and it is valid, indeed (see Fig. 6.1), if, for instance, on Γ_{1-} the trajectory has the form shown in Fig. 6.2, where the lower straight line P conventionally shows the drop curve and the dashed line shows the character of motion on Γ_{2-}.

Naturally, it does not yet follow from Theorem 6.2 that system (6.1) has a strange attractor. However, if mapping (6.4) has Sharkovsky's mixing attractor, then, according to Theorems 6.1, 6.2, and the results of [88], the behavior of the solutions of system (6.1) is chaotic.

7. Pontryagin's Problem on the Tori of Singularly Perturbed Systems

7.1. Statement of the Problem

Here we shall consider the system

$$\varepsilon \dot{x} = f(x,y), \qquad \dot{y} = g(x,y), \qquad 0 < \varepsilon \ll 1, \qquad (7.1)$$

supposing that $x \in R^k$, $y \in R^m$, $k, m \geq 2$, and the sufficiently smooth right-hand sides of f and g are defined throughout the space R^n, $n = k + m$.

Theorem on the C^1-Proximity

We are interested in the auto-oscillations of system (7.1) under the following two constraints.

CONDITION 7.1. For every fixed y, the fast-motion system

$$\dot{x} = f(x,y) \tag{7.2}$$

has an exponentially orbitally stable cycle $x_0(\tau, y)$ with period $T(y)$. Here $x_0(\tau+1, y) \equiv x_0(\tau, y)$, $\tau = t/T(y)$.

CONDITION 7.2. The averaged slow-motion system

$$\dot{y} = \mathfrak{G}_0(y), \qquad \mathfrak{G}_0(y) = T(y) \int_0^1 g(x_0(\tau, y), y)\, d\tau \tag{7.3}$$

has an exponentially orbitally stable cycle $y_0(t)$ with period T_0.

The multidimensional singularly perturbed systems under Condition 7.1 were considered for the first time in [80, 81], where it was shown that an exponentially stable equilibrium position of system (7.3) was associated with a periodic solution of system (7.1) of the same stability. This statement constitutes the content of the well-known Pontryagin–Rodygin theorem. The review [79] points out the possibility of generalizing this theorem to the case of the cycle in system (7.3) and gives the formulation of the corresponding hypothesis, which is substantiated below with the use of the apparatus of integer manifolds [71].

7.2. Auxiliary Lemma

Carrying out the substitution $t \to \varepsilon t$ in (7.1), we arrive at the system

$$\dot{x} = f(x,y), \qquad \dot{y} = \varepsilon g(x,y), \tag{7.4}$$

which, for $\varepsilon = 0$, has an exponentially stable invariant integer manifold $x = x_0(\tau, y)$ with the equations

$$\dot{y} = 0, \qquad \dot{\tau} = 1/T(y) \tag{7.5}$$

on it. According to [71], this manifold is preserved for $\varepsilon > 0$, but now it is defined by the relation

$$x = x_0(\tau, y) + X(\tau, y, \varepsilon), \qquad X(\tau, y, 0) \equiv 0, \tag{7.6}$$

and instead of system (7.5) we have

$$\dot{y} = \varepsilon g(x_0 + X, y), \qquad \dot{\tau} = [1 + \Phi(\tau, y, \varepsilon)]/T(y). \tag{7.7}$$

Here X, Φ are some functions smooth with respect to all the variables and periodic with respect to τ with period 1, and $\Phi(\tau, y, 0) \equiv 0$.

Here we shall describe the algorithm of the asymptotic computation of the right-hand sides of (7.6), (7.7). We set

$$X = \sum_{k=0}^{\infty} \varepsilon^k x_k(\tau, y), \qquad \Phi = \sum_{k=1}^{\infty} \varepsilon^k \alpha_k(y)$$

and determine the unknown coefficients of the expansion of these series from the identity

$$\left(\frac{\partial x_0}{\partial \tau} + \frac{\partial X}{\partial \tau}\right)(1 + \Phi)/T(y) + \varepsilon\left(\frac{\partial x_0}{\partial y} + \frac{\partial X}{\partial y}\right)g(x_0 + X, y) =$$
$$= f(x_0 + X, y). \tag{7.8}$$

Equating in (7.8) the coefficients in like powers of ε, we arrive at the recurrent sequence of differential equations

$$\frac{\partial x_k}{\partial \tau} + \alpha_k(y)\frac{\partial x_0}{\partial \tau} = A_0(\tau, y)x_k + F_k(\tau, y), \tag{7.9}$$

where

$$A_0(\tau, y) = T(y)\frac{\partial f}{\partial x}\bigg|_{x=x_0(\tau, y)},$$

and the functions F_k depend only on x_r, α_r with subscripts $r \leq k - 1$. From the condition of solvability of Eq. (7.9) in the class of functions periodic with respect to τ, we find that

$$\alpha_k(y) = \int_0^1 (F_k(\tau, y), g_0(\tau, y))\, d\tau,$$

where g_0 is a periodic solution of the equation

$$\frac{dg}{d\tau} = -A_0^*(\tau, y)g, \qquad \left(\frac{\partial x_0}{\partial \tau}, g_0\right) = 1.$$

Now we can uniquely determine from (7.9) the function $x_k(\tau, y)$ which is periodic with respect to τ and which satisfies the condition

$$\int_0^1 (x_k(\tau, y), g_0(\tau, y))\, d\tau \equiv 0.$$

Thus, the results given in [71], in conjunction with the computations presented, allow us to formulate the following statement.

Theorem on the C^1-Proximity

LEMMA 7.1 *Under Condition 7.1, system (7.4) has an exponentially orbitally stable invariant integer manifold (7.6) with the system of equations (7.7) on it. In addition, for any natural N we have the asymptotic representations*

$$X = \sum_{k=1}^{N} \varepsilon^k x_k + \varepsilon^{N+1}\Delta_1(\tau, y, \varepsilon), \qquad (7.10)$$

$$\Phi = \sum_{k=1}^{N} \varepsilon^k \alpha_k + \varepsilon^{N+1}\Delta_2(\tau, y, \varepsilon), \qquad (7.11)$$

where Δ_1, Δ_2 are functions smooth with respect to all the variables and periodic with respect to τ.

7.3. The Main Result

Let us consider system (7.7), which is the result of the reduction of system (7.4) to the integer manifold (7.6). Dividing the first equation of (7.7) by the second, we obtain the following nonautonomous system equivalent to it and written in standard form [71]:

$$\frac{dy}{d\tau} = \varepsilon \mathfrak{G}(\tau, y, \varepsilon). \qquad (7.12)$$

By virtue of Condition 7.2, the averaged equation of the first approximation corresponding to (7.12) has an exponentially stable cycle. The results of [71] allow us to infer that in this case system (7.12) has an exponentially stable toroidal integral manifold

$$y = y_0(T_0 \sigma) + H(\tau, \sigma, \varepsilon), \qquad H(\tau, \sigma, 0) \equiv 0, \qquad (7.13)$$

with the equation

$$\frac{d\sigma}{d\tau} = \frac{\varepsilon}{T_0}[1 + \Psi(\tau, \sigma, \varepsilon)], \qquad \Psi(\tau, \sigma, 0) \equiv 0, \qquad (7.14)$$

on it, where the smooth functions H, Ψ periodically depend on τ, σ with period 1.

To find the asymptotics of H, Ψ, we fix an arbitrary natural N and make the substitution

$$y = \xi + \sum_{k=1}^{N} \varepsilon^k Y_k(\tau, \xi) \qquad (7.15)$$

in system (7.12), which reduces it to the form

$$\frac{d\xi}{d\tau} = \varepsilon\mathfrak{G}_0(\xi) + \sum_{k=1}^{N-1} \varepsilon^{k+1}\mathfrak{G}_k(\xi) + \varepsilon^{N+1}\Delta_3(\tau,\xi,\varepsilon), \quad (7.16)$$

where all of the functions are smooth and Y_k and Δ_3 are periodic with respect to τ. The algorithm of the construction of substitution (7.15) and the right-hand side of (7.16) can be found in [6], for instance. Next we consider the abridged system

$$\frac{d\xi}{d\tau} = \varepsilon\mathfrak{G}_0(\xi) + \sum_{k=1}^{N-1} \varepsilon^{k+1}\mathfrak{G}_k(\xi). \quad (7.17)$$

On the basis of the condition of structural stability of the periodic solution $y_0(t)$ of (7.3), we infer that system (7.17) has a periodic solution $\xi_0(\sigma,\varepsilon)$ with period $T_0(\varepsilon)/\varepsilon$, where $\xi_0(\sigma+1,\varepsilon) \equiv \xi_0(\sigma,\varepsilon)$, $\sigma = \varepsilon\tau/T_0(\varepsilon)$, and that the asymptotic expansions

$$\xi_0(\sigma,\varepsilon) = y_0(T_0\sigma) + \sum_{k=1}^{\infty} \varepsilon^k \xi_k(\sigma), \quad T_0(\varepsilon) = T_0 + \sum_{k=1}^{\infty} \varepsilon^k T_k \quad (7.18)$$

are valid. Since the difference between the right-hand sides of (7.16), (7.17) is of the order ε^{N+1}, it follows that because of the smooth dependence of the integer manifold (7.13) on the right-hand side of $\mathfrak{G}$, the corresponding manifolds of these equations also differ only by the order ε^{N+1}. It follows from the latter fact and from (7.18) that as a result of substitution (7.15) relation (7.13) reduces to the form

$$\xi = y_0(T_0\sigma) + \sum_{k=1}^{N} \varepsilon^k \xi_k(\sigma) + \varepsilon^{N+1}\Delta_4(\tau,\sigma,\varepsilon) \quad (7.19)$$

and Eq. (7.14) assumes the form

$$\frac{d\sigma}{d\tau} = \varepsilon\left[1\bigg/\left(T_0 + \sum_{k=1}^{N} \varepsilon^k T_k\right) + \varepsilon^{N+1}\Delta_5(\tau,\sigma,\varepsilon)\right]. \quad (7.20)$$

LEMMA 7.2. *Under Condition 7.2, system* (7.12) *has an exponentially stable invariant integer manifold* (7.13) *with Eq.* (7.14) *on it, and the asymptotics of the right-hand sides of* (7.13), (7.14) *is defined by* (7.15), (7.19), (7.20).

Returning to system (7.4), we derive the following statement from Lemmas 7.1 and 7.2.

THEOREM 7.1. *When Conditions 7.1 and 7.2 are satisfied, system (7.4) has an exponentially orbitally stable two-dimensional invariant torus (7.6), (7.13) the equations on which are the second equation of (7.7) and equation (7.14).*

8. Stable Relaxation Tori in Three-Dimensional Systems

8.1. Statement of the Problem and Description of the Result

Here we formulate conditions under which the singularly perturbed system

$$\varepsilon \dot{x} = f(x, y), \qquad \dot{y} = g(x, y), \qquad x \in R^2, \qquad y \in R, \qquad (8.1)$$

where f and g are sufficiently smooth functions, has a relaxation torus, i.e., a torus on which the trajectory successively passes with time from one stable integer manifold of form (7.6) to another. It should be emphasized that here the connection with the results of the preceding section is the same as the connection between the results given in [3], concerning the cycle of the slow-motion system, and the results given in [77] concerning a relaxation cycle. It should also be noted that our restrictions, which exclude the possibility of applying the Pontryagin–Rodygin theorem [81], constitute a natural generalization of the conditions given in Sec. 1, under which it is customary to study relaxation oscillations.

CONDITION 8.1. The dynamic properties of the fast-motion equation

$$\dot{x} = f(x, y), \quad x \in R^2, \qquad (8.2)$$

depend on the parameter y as follows: for all $y < y_1$ Eq. (8.2) has a unique global exponentially stable cycle, and when y passes through y_1 in the domain bounded by this cycle, the contraction of the trajectories generates another two cycles, a stable cycle and an unstable one; for $y_1 < y < y_2$ Eq. (8.2) has exactly three structurally stable cycles and for $y = y_2$ there occurs a bifurcation of the merging of the stable and the unstable cycle; for $y > y_2$ the situation is the same as for $y < y_1$.

We denote by $x_1(\tau, y)$, $x_2(\tau, y)$ the stable cycles of Eq. (8.2) with periods $T_1(y)$ and $T_2(y)$ respectively. Here

$$x_1(\tau + 1, y) \equiv x_1(\tau, y), \qquad \tau = t/T_1(y) \qquad \text{if } y \leq y_2;$$

$$x_2(\tau + 1, y) = x_2(\tau, y), \qquad \tau = t/T_2(y) \qquad \text{if } y \geq y_1.$$

CONDITION 8.2. We suppose that

$$\mathfrak{G}_1(y) = T_1(y) \int_0^1 g(x_1(\tau, y), y) \, d\tau > 0 \qquad \text{for } y \leq y_2,$$

$$\mathfrak{G}_2(y) = T_2(y) \int_0^1 g(x_2(\tau, y), y) \, d\tau < 0 \qquad \text{for } y \geq y_1.$$

For $y = y_1$, the cycle $x_2(t) = x_2(t/T_2(y_1), y_1)$ of Eq. (8.2) becomes multiple. Therefore, the equation in variations $\dot{h} = A_0(t)h$ corresponding to it has linearly independent periodic solutions $h_1(t) = dx_2/dt$, $h_2(t)$ such that $(h_j(t), g_k(t)) \equiv \delta_{jk}$, $j, k = 1, 2$, where g_1, g_2 are linearly independent solutions of the conjugate equation. According to the results of [35], by means of the substitution

$$x = x_2(s) + \xi h_2(s) + \xi^2 h_3(s),$$

where $h_3(s)$ is a certain $T_2(y_1)$-periodic function, we transform Eq. (8.2), in the neighborhood of the cycle x_2, with an accuracy to within the terms of order ξ^3, into the system

$$\dot{\xi} = \alpha \xi^2, \qquad \dot{s} = 1 + \beta \xi^2.$$

CONDITION 8.3. The conditions of nondegeneracy

$$\int_0^{T_2(y_1)} (a(t), g_2(t)) \, dt \neq 0, \qquad \alpha \neq 0, \tag{8.3}$$

where

$$a(t) = \left. \frac{\partial f}{\partial y} \right|_{x=x_0(t), y=y_1}$$

are valid. We suppose that inequalities similar to (8.3) are also satisfied for $y = y_2$.

Condition 8.3 characterizes a kind of generic position and Conditions 8.1, 8.2 define the global behavior of the solutions. When they are satisfied, the trajectories of (8.1) behave as follows.

Suppose, for instance, that the initial conditions $x(0, \varepsilon) = x_0$, $y(0, \varepsilon) = y_0$ are such that $y_0 < y_1$. Then the component $x(t, \varepsilon)$ quickly (in the time of order ε) gets into the neighborhood of the cycle $x_1(\tau, y_0)$ and the component $y(t, \varepsilon)$ becomes close to $y_0(t)$, i.e., to the solution of the equation $\dot{y} = \mathfrak{G}_1(y)$,

Theorem on the C^1-Proximity

with the initial condition $y(0) = y_0$, and with a further increase in t, the component $x(t,\varepsilon)$ performs rapid oscillations along the cycles $x_1(\tau,y)$ approximately according to the law $x_1(t/\varepsilon T_1(y_0(t)), y_0(t))$.

It is clear that at a certain moment of time $y_0(t)$ falls in a sufficiently small neighborhood of the value y_2. Then there occurs a breakoff to the cycle $x_2(\tau, y_2)$ and this leads to a motion similar to the already described motion along the cycles $x_2(\tau, y)$, and $y(t,\varepsilon)$ decreases now. In the neighborhood of $y = y_1$ there occurs a breakoff to $x_1(\tau, y_1)$, and so on. Therefore, with time, any trajectory of system (8.1) becomes close to a two-frequency trajectory with frequencies of order unity and order ε^{-1}.

THEOREM 8.1. *Under Conditions 8.1–8.3, system (8.1) has an exponentially orbitally stable two-dimensional invariant torus, whose zero approximation is a closed surface Γ_0 consisting of two lateral surfaces*

$$x = x_1(\tau, y), \qquad x = x_2(\tau, y), \qquad 0 \leq \tau \leq 1, \qquad y_1 \leq y \leq y_2$$

and two bases $y = y_1$ and $y = y_2$, where x varies in annular domains bounded by the cycles $x_1(\tau, y_1)$, $x_2(\tau, y_1)$ and $x_1(\tau, y_2)$, $x_2(\tau, y_2)$ respectively.

To prove Theorem 8.1, we first use the technique developed in Secs. 1, 2 in order to investigate the system in variations on the arbitrary trajectory of system (8.1) and then use the method of point-to-point mapping.

The statement of Theorem 8.1 admits a generalization to the multidimensional case. However, in this case not only are the notations of Conditions 8.1–8.3 more cumbersome but also the relaxation invariant torus can be dichotomous. Naturally, in a multidimensional case, Eq. (8.2) can have tori of different dimensions instead of cycles.

8.2. Example

We shall describe a class of systems (8.1) for which Conditions 8.1–8.3 are satisfied. We define Eq. (8.2) in polar coordinates:

$$\frac{dr}{dt} = r[f(r^2) - y], \qquad \frac{d\varphi}{dt} = 1,$$

where $f(z)$, $z > 0$, is a smooth function whose graph is shown in Fig. 8.1. We suppose that the equation $f'(z) = 0$ has only two roots $z_1, z_2 > 0$ with

$$f'(z) < 0 \quad \text{for} \quad 0 < z < z_1 \quad \text{and} \quad z > z_2,$$

$$f'(z) > 0 \quad \text{for} \quad z_1 < z < z_2,$$

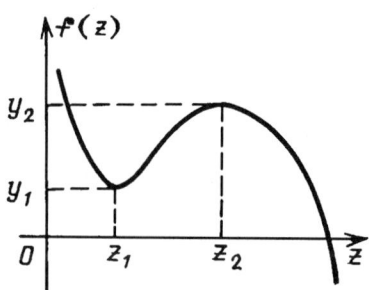

Fig. 8.1

$$f''(z_1)f''(z_2) \neq 0, \qquad f(z_1) > 0, \ f(z_2) > 0,$$
$$\lim_{z \to 0} f(z) = +\infty, \qquad \lim_{z \to +\infty} f(z) = -\infty.$$

It is easy to see that for such a choice of the right-hand side $f(x,y)$ of system (8.1), Conditions 8.1 and 8.3 are satisfied. If we now choose a function $g(x,y)$ satisfying Condition 8.2, we get a three-dimensional relaxation system for which Theorem 8.1 holds true.

Chapter 2

Relaxation Oscillations in a Medium with Diffusion

It is well known that relaxation boundary-value problems of the reaction–diffusion type serve as base models for a large number of biophysical problems [1, 17]. Therefore, the description of their properties is timely. In what follows, the results of the Chapter 1 are extended to this class of differential equations. In particular, a limit parabolic relay system is introduced and a theorem on the C^1-proximity, similar to Theorem 1.1, is formulated.

Recall that systems of the reaction–diffusion type are usually used for simulating the process of origination of spatially nonhomogeneous structures in homogeneous media, and this is important for many problems of biophysics, ecology, and chemistry. Most often this phenomenon is described with the use of the Turing–Prigogine theorem, which is based on the phenomenon of the diffuse instability of a spatially homogeneous equilibrium position discovered by Turing [95] (for a certain variation of the diffusion coefficient the homogeneous equilibrium position loses stability, and spatially nonhomogeneous equilibrium positions, the so-called dissipative structures, appear in its neighborhood). However, a majority of mathematical models of biophysical and ecological processes have a spatially homogeneous (independent of the spatial variables) cycle. Therefore, the extension of the Turing–Prigogine theorem to the case of oscillations is of a certain interest.

In this chapter we consider the problem of stability of a homogeneous relaxation cycle. The results obtained are used to analyze a number of mathematical models of biophysics.

The presentation is based primarily on [24, 25, 28, 39].

9. On a certain Class of Parabolic Relay Systems

9.1. The General Statement of the Problem

Let us consider a parabolic boundary-value problem

$$\varepsilon \frac{\partial x}{\partial t} = D_1 \Delta x + f(x,y), \qquad \left.\frac{\partial x}{\partial \nu}\right|_{\partial \Omega} = 0, \qquad (9.1)$$

$$\frac{\partial y}{\partial t} = D_2 \Delta y + g(x,y), \qquad \left.\frac{\partial y}{\partial \nu}\right|_{\partial \Omega} = 0, \qquad (9.2)$$

where $0 < \varepsilon \ll 1$, $x \in R$, $y \in R^m$, $D_1 > 0$, the eigenvalues of the matrix D_2 lie on the right of the imaginary axis, Δ is the Laplace operator, ν is the direction of the outer normal to a sufficiently smooth boundary $\partial \Omega$ of the bounded domain $\Omega \subset R^N$. We suppose that the functions f and g, smooth with respect to the set of variables, are defined for all $(x,y) \in R^{m+1}$, with $f > 0$ for $x \leq -M$ and $f < 0$ for $x \geq M$, where $M = M(y) > 0$. We also assume that the function f is analytic with respect to the variable x.

As a phase space (the space of initial conditions $(x(s), y(s))$) of the boundary-value problem (9.1), (9.2), we take the space $E = C^\alpha \times C^\alpha$, where C^α consists of continuous functions satisfying in $\bar{\Omega}$ Hölder's condition with the exponent $0 < \alpha < 1$. We consider the norm in E to be introduced in an ordinary way.

We pose the following question: what boundary-value problems of type (9.1), (9.2) should be classed with relaxation problems? In order to answer this question, it is convenient to proceed from the approach, developed in Ch. 1, which relates the properties of the original system and the properties of the limit relay system.

9.2. Properties of Parabolic Relay Systems

In our case the analog of the relay system (1.1), introduced in Sec. 1, is the boundary-value problem

$$\frac{\partial y}{\partial t} = D_2 \Delta y + g(\Phi(y), y), \qquad \left.\frac{\partial y}{\partial \nu}\right|_{\partial \Omega} = 0, \qquad (9.3)$$

where the relay operator $\Phi(y)$ will be defined by construction.

We arbitrarily fix a function $y_0(s) \in C^\alpha$ and consider the boundary-value problem

$$\frac{\partial x}{\partial t} = D_1 \Delta x + f(x,y), \qquad \left.\frac{\partial x}{\partial \nu}\right|_{\partial \Omega} = 0 \qquad (9.4)$$

for $y = y_0(s)$. It follows from the constraints imposed on f and the results given in [44, 59] that the boundary-value problem (9.4) has at least one asymptotically stable equilibrium position $x_0(s)$. In the general case it is exponentially stable, and we assume the truth of this fact. It follows that for all $y(s)$ close to $y_0(s)$ in the norm C^α, the manifold $\Gamma_1 = \{(\Phi_1(y), y)\}$ is defined in the space E, where the Fréchet smooth operator $\Phi_1(y)$, $\Phi_1(y_0(s)) = x_0(s)$, acting from C^α into $C^{2+\alpha}$, puts the function $y(s)$ into correspondence with the exponentially stable equilibrium position $x(s)$ of the boundary-value problem (9.4). Let us extend this manifold, including into it the points $(x(s), y(s))$, where $x(s)$ is the equilibrium position of the boundary-value problem (9.4) for $y = y(s)$, whose second coordinate $y(s)$ can be connected with $y_0(s)$ by a curve continuous in C^α so that the original equilibrium position $x_0(s)$ is continuously extended along this curve up to $x(s)$, the structural stability being preserved. As a result, we get a locally smooth manifold with the boundary $\Gamma_{10} = \{(\Phi_1(y), y), y \in \Lambda_{10}\}$. Here the set Λ_{10} is defined by the relation $\lambda_0 = 0$, where $\lambda_0 = \lambda_0(y(s))$ is the largest eigenvalue of the operator

$$Lh = D_1 \Delta h + f'_x(\Phi_1(y), y)h, \quad \left.\frac{\partial h}{\partial \nu}\right|_{\partial \Omega} = 0. \quad (9.5)$$

The constructions that we have carried out make it possible first to define the solution of the relay system (9.3) as the solution of the mixed problem

$$\frac{\partial y}{\partial t} = D_2 \Delta y + g(\Phi_1(y), y), \quad \left.\frac{\partial y}{\partial \nu}\right|_{\partial \Omega} = 0, \quad y|_{t=0} = y_0(s). \quad (9.6)$$

Let us suppose that at some first moment $t_0 > 0$ the point $(\Phi_1(y(t, s)), y(t, s))$ falls in Γ_{10}, the nondegeneracy condition

$$\int_\Omega f''_x(\Phi_1(y(t_0, s)), y(t_0, s))e_0^3(s)ds \neq 0 \quad (9.7)$$

and the normal switching condition

$$l_0\left(\left.\frac{\partial y}{\partial t}\right|_{t=t_0-0}\right) \equiv \int_\Omega f'_y(\Phi_1(y(t_0, s)), y(t_0, s))\left.\frac{\partial y}{\partial t}\right|_{t=t_0-0} \cdot e_0(s)ds \neq 0 \quad (9.8)$$

being satisfied. Here $e_0(s)$ is an eigenvalue of the operator (9.5) for $y = y(t_0, s)$, which corresponds to the zero eigenvalue.

Let us consider the boundary-value problem (9.4) for $y = y(t_0, s)$. In the neighborhood of the multiple equilibrium position of $\Phi_1(y(t_0, s))$ it has an exponentially stable one-dimensional invariant manifold and the equation on it begins with a quadratic term by virtue of inequality (9.7). Let us take the initial condition of the solution of Eq. (9.4) lying on the one-dimensional "whiskers" emanating from $\Phi_1(y(t_0, s))$. According to the results of [44, 45], as $t \to \infty$ the solution with this initial condition tends to the equilibrium position which, in the general case, is exponentially stable. Therefore, in the same way as we did above, we begin by introducing a manifold $\Gamma_2 = \{(\Phi_2(y), y)\}$ with the boundary Γ_{20} and then, for $t \geq t_0$, define the solution of the relay system (9.3) as the solution of the mixed problem

$$\frac{\partial y}{\partial t} = D_2 \Delta y + g(\Phi_2(y), y), \quad \frac{\partial y}{\partial \nu}\bigg|_{\partial \Omega} = 0, \quad y|_{t=t_0} = y(t_0, s). \qquad (9.9)$$

We suppose that among the moments t at some first time moment $t_1 > t_0$ a new switching occurs and conditions similar to (9.7), (9.8) are satisfied. In this case we act as we did above.

Without making stipulations, we assume everywhere in what follows that conditions similar to (9.7), (9.8) are satisfied for $t = t_p$, $p = 0, 1, \ldots$. Then, for $t \neq t_p$, the solutions of the relay system (9.3) smoothly depend on the initial conditions, and for the first variation $h(t, s)$ of the fixed solution $y(t, s)$ we have the boundary-value problem

$$\frac{\partial h}{\partial t} = D_2 \Delta h + A(t)h, \quad \frac{\partial h}{\partial \nu}\bigg|_{\partial \Omega} = 0, \qquad (9.10)$$

$$h(t_p + 0, s) = h(t_p - 0, s) +$$

$$+ \left(\frac{\partial y}{\partial t}\bigg|_{t=t_p+0} - \frac{\partial y}{\partial t}\bigg|_{t=t_p-0}\right) l_p(h(t_p - 0, s)) / l_p\left(\frac{\partial y}{\partial t}\bigg|_{t=t_p-0}\right). \qquad (9.11)$$

Here the bounded operator $A(t)$ is the result of the ordinary linearization on the trajectory $y(t, s)$ of the nonlinear term on the right-hand side of system (9.3) between the neighboring switching moments, and the linear functionals l_p are defined for $p > 0$ in the same way as l_0 in (9.8).

We have the following analog of the Andronov–Witt theorem.

LEMMA 9.1. *Suppose that $y(t, s)$ is a periodic solution of the relay system (9.3) with period T and all multipliers (except for a simple unit multiplier) of the T-periodic linear boundary-value problem (9.10), (9.11) lie in the interior of a unit circle. Then $y(t, s)$ is exponentially orbitally stable.*

The proof follows immediately from the results of [46] (see also 1.1).

10. Theorem on the C^1-Proximity for Relaxation Parabolic Systems

10.1. Description of the Result

Suppose that $S(r_0)$ is a ball in E of a sufficiently small radius r_0 with center at the point $(x_0(s), y_0(s)) \in \Gamma_1$. We denote by $(x(t,s,\varepsilon), y(t,s,\varepsilon))$ the solution of the parabolic problem (9.1), (9.2) with the initial condition $(x(s), y(s)) \in S(r_0)$ defined for $t = 0$ and by $h_1(t,s,\varepsilon)$, $h_2(t,s,\varepsilon)$ its first variations with the initial conditions $h_1(0,s,\varepsilon) = h_1^0(s)$, $h_2(0,s,\varepsilon) = h_2^0(s)$. In the statement being formulated, $y(t,s)$ is the solution of the relay system (9.3) with the initial condition $y(0,s) = y(s)$ and the switching moments $t_0, t_1, \ldots$.

THEOREM 10.1. *For the values of t which do not belong to the small fixed neighborhoods of the points $0, t_0, t_1, \ldots$, the limiting relations*

$$\lim_{\varepsilon \to 0}(x(t,s,\varepsilon), y(t,s,\varepsilon)) = (\Phi(y(t,s)), y(t,s)), \qquad (10.1)$$

$$\lim_{\varepsilon \to 0}(h_1(t,s,\varepsilon), h_2(t,s,\varepsilon)) = (h_{10}(t,s), h_{20}(t,s)) \qquad (10.2)$$

are uniform with respect to $(x(s), y(s)) \in S(r_0)$ and $(h_1^0(s), h_2^0(s))$ from a unit ball with center at zero in the metric of E. Here $h_{20}(t,s)$ is the solution of the mixed problem for (9.10), (9.11) with the initial condition $h_{20}(0,s) = h_2^0(s)$, and $h_{10}(t,s) = \Phi'(y(t,s))h_{20}(t,s)$. The convergence in (10.1), (10.2) is uniform with respect to the indicated values of t from a bounded interval.

Theorem 10.1 generalizes the statement obtained in Secs. 1, 2 on the C^1-proximity of the solutions of the limiting and original systems to the case being considered and allows us to judge the stationary conditions of the boundary-value problem (9.1), (9.2) from the structurally stable stationary behavior of system (9.3). Since the proof of Theorem 10.1 almost verbatim (with some natural changes connected with parabolicity) repeats the proof of Theorem 1.1, we shall discuss only its singularities.

We say that the boundary-value problem (9.1), (9.2) is a *relaxation problem* if the relay system (9.3) has solutions with an infinite number of switchings.

Suppose that for $y \in R$ the functions f and g are subjected to constraints such that the boundary-value problem (9.1), (9.2) has a spatially homogeneous cycle $(x(t,\varepsilon), y(t,\varepsilon))$. In this case the boundary-value problem (9.1), (9.2) is obviously relaxational, and Lemma 9.1 and Theorem 10.1 make it possible to describe the structure of the neighborhood of the cycle

$(x(t,\varepsilon), y(t,\varepsilon))$, independent of ε, when the cycle loses stability. However, we may have a situation where the system of ordinary differential equations obtained from (9.1), (9.2) for $D_1 = 0$, $D_2 = 0$ is not relaxational, but for $D_1 \neq 0$, $D_2 \neq 0$ the parabolic system is relaxational.

The representations that we have developed can obviously be extended to more general classes of singularly perturbed parabolic systems with classical boundary conditions.

10.2. Analysis of Problem (9.1), (9.2) Outside of the Neighborhoods of Switching Points

Let us first consider an interval $\delta \leq t \leq t_0 - \delta$, where $\delta > 0$ is sufficiently small. For these t we shall seek a formal integer manifold of system (9.1), (9.2) as a series

$$x = \Phi_1(y) + \sum_{k=1}^{\infty} \varepsilon^k F_k(y). \tag{10.3}$$

Substituting (10.3) into (9.1), (9.2) and equating the coefficients in like powers of ε, we recurrently obtain the sequence of equations for determining the operators F_k. For instance, for determining F_1 we have

$$LF_1 = \Phi_1'(y)(D_2 \Delta y + g(\Phi_1(y), y)), \tag{10.4}$$

where L is operator (9.5). From (10.4) we find F_1 as a Fréchet smooth operator acting from $C^{2+\alpha}$ into $C^{4+\alpha}$. In the same way we successively determine the Fréchet smooth operators $F_k(y)$ acting from $C^{2k+\alpha}$ into $C^{2k+2+\alpha}$.

We can assign a strict meaning to the formal integer manifold (10.3) as follows. We fix the initial condition $(x(s), y(s)) \in S(r_0)$ of system (9.1), (9.2) and denote by $y_0(t, s)$ the solution of (9.6) with the initial condition $y(s)$ defined for $t = 0$. Then we substitute series (10.3) into (9.2) and seek the solution of the resulting equation as a series

$$y(t, s, \varepsilon) = y_0(t, s) + \varepsilon y_1(t, s) + \varepsilon^2 y_2(t, s) + \ldots . \tag{10.5}$$

In this way we obtain a recurrent sequence of linear nonhomogeneous boundary-value problems of the form

$$\frac{\partial y_k}{\partial t} = D_2 \Delta y_k + A(t) y_k + f_k(t, s), \quad \left.\frac{\partial y_k}{\partial \nu}\right|_{\partial \Omega} = 0 \tag{10.6}$$

with the initial condition $y_k|_{t=0} = 0$ for determining the unknown coefficients $y_k(t, s)$ of series (10.5). Here $A(t)$ is a Fréchet derivative of the expression

$g(\Phi_1(y), y)$ for $y = y_0(t, s)$ and the function f_k depends only on y_m with subscripts $m \leq k - 1$ and is smooth with respect to s up to $t = 0$ if the initial condition $y(s) \in C^{2k+\alpha}$. According to the results of [18, 69], it follows that the boundary-value problem (10.6) is solvable in C^α.

LEMMA 10.1. *Suppose that the initial condition* $(x(s), y(s)) \in S(r_0)$ *of the solution of problem* (9.1), (9.2) *is such that* $y(s) \in C^{2k+\alpha}$. *Then, for* $\delta \leq t \leq t_0 - \delta$, *in the metric* C^α *we have*

$$x(t, s, \varepsilon) = \Phi_1(y_0(t, s)) + \sum_{m=1}^{k-1} \varepsilon^m x_m(t, s) + O(\varepsilon^k), \qquad (10.7)$$

$$y(t, s, \varepsilon) = y_0(t, s) + \sum_{m=1}^{k-1} \varepsilon^m y_m(t, s) + O(\varepsilon^k), \qquad (10.8)$$

where (10.8) *is an interval of series* (10.5) *and* (10.7) *is the result of the substitution of* (10.8) *into* (10.3) *and of the re-expansion with respect to* ε. *Asymptotic relations similar to* (10.7), (10.8) *are also valid for the Fréchet derivatives of the functions* $x(t, s, \varepsilon)$, $y(t, s, \varepsilon)$ *with respect to the initial conditions.*

The proof of the lemma is standard and reduces to the analysis of a system of variational equations for (9.1), (9.2) on finite intervals of series (10.7), (10.8). Without giving the corresponding calculations, we only want to point out that the main difficulty lies in obtaining the estimate

$$\|U(t, \tau, \varepsilon)\|_{C^\alpha \to C^\alpha} \leq M \exp\left\{-\frac{\gamma}{\varepsilon}(t - \tau)\right\}, \qquad M, \gamma > 0, \qquad (10.9)$$

of the resolvent operator of the boundary-value problem

$$\varepsilon \frac{\partial h}{\partial t} = D_1 \Delta h + f'_y(\Phi_1(y_0(t, s)), y_0(t, s))h, \qquad \left.\frac{\partial h}{\partial \nu}\right|_{\partial \Omega} = 0.$$

The proof of this estimate completely coincides with the given, say, in [86] substantiation of a similar inequality for the case of ordinary differential equations.

Thus, Lemma 10.1 allows us to find the asymptotic properties of the parabolic system (9.1), (9.2) on the interval $\delta \leq t \leq t_0 - \delta$. In this case the requirement of the smoothness of the component $y(s)$ of the initial condition is not a constraint since, say, for $t = \delta$, the function $x(t, s, \varepsilon)$, $y(t, s, \varepsilon)$ has a sufficiently large (limited by the smoothness of f and g) number of derivatives with respect to s bounded uniformly, with respect to ε (this can be easily found out from the analysis of the system of variational equations corresponding to the trajectory $(x(t, s, \varepsilon), y(t, s, \varepsilon))$ and the estimate (10.9)). Everything that was said is valid for $t_0 + \delta \leq t \leq t_1 - \delta$, and so on.

10.3. Problem (9.1), (9.2) in the Neighborhood of the Switching Point

Just as in Secs. 1 and 2, by virtue of the results of 10.2, the substantiation of Theorem 10.1 reduces to the analysis of the neighborhood of the switching points. Therefore, it is sufficient to consider the neighborhood of the point $(\Phi_1(y(t_0, s)), y(t_0, s))$.

In system (9.1), (9.2) we introduce new coordinates which are completely similar to the coordinates introduced in 1.4. With this aim in view, we shift the zero of the space E to the point $(\Phi_1(y(t_0, s)), y(t_0, s))$ and set $x = (\xi_1, \xi_2)$, $y = (\eta_1, \eta_2)$. Here $\xi_1 \in R$ "measures" distances along the one-dimensional integer manifold of Eq. (9.4) for $y = y(t_0, s)$ and the vector $\xi_2 \in C^\alpha$ is orthogonal to the eigenfunction $e_0(s)$ of operator (9.5) in the $L_2(\Omega)$ sense; $\eta_1 \in R$ corresponds to the direction $\frac{\partial y}{\partial t}(t_0 - 0, s)$ and the vector $\eta_2 \in C^\alpha$ satisfies the relation $l_0(\eta_2) = 0$. After some similarity transformations and time normalization, in the new coordinates the parabolic problem (9.1), (9.2) assumes the form

$$\varepsilon \dot{\xi}_1 = F_1(\xi_1, \xi_2, \eta_1, \eta_2), \tag{10.10}$$

$$\varepsilon \frac{\partial \xi_2}{\partial t} = L\xi_2 + F_2(\xi_1, \xi_2, \eta_1, \eta_2), \quad \left.\frac{\partial \xi_2}{\partial \nu}\right|_{\partial \Omega} = 0, \tag{10.11}$$

$$\dot{\eta}_1 = G_1(\xi_1, \xi_2, \eta_1, \eta_2), \tag{10.12}$$

$$\frac{\partial \eta_2}{\partial t} = D_2 \Delta \eta_2 + G_2(\xi_1, \xi_2, \eta_1, \eta_2), \quad \left.\frac{\partial \eta_2}{\partial \nu}\right|_{\partial \Omega} = 0. \tag{10.13}$$

Here L is operator (9.5) for $y = y(t_0, s)$, the functionals F_1, G_1 and the operators F_2, G_2 are Fréchet smooth in the neighborhood of the origin, and

$$F_1 = \xi_1^2 + \eta_1 + \ldots, \quad F_2(\xi_1, 0, 0, 0) \equiv 0, \quad \partial_{\xi_2} F_2(0, 0, 0, 0) = 0, \tag{10.14}$$

$$G_1(0, 0, 0, 0) = 1, \quad G_2(0, 0, 0, 0) = 0, \tag{10.15}$$

where the dots denote the quadratic terms different from ξ_1^2 and terms of a higher order of smallness.

Thus, in the neighborhood of the switching point we get problem (10.10)–(10.13), which is similar in its properties to system (1.14). If we now construct a formal asymptotics of its trajectory, we can easily verify that we get formulas similar to (1.15), (1.16).

The consequent actions are as follow: we linearize system (10.10)–(10.13) on the formally constructed trajectory, study the properties of the resulting linear system (here all the calculations coincide almost completely with those carried out in Secs. 1, 2), and then, using the scheme described in 4.7, prove the validity of the asymptotic formulas being used. Thus, we simultaneously prove both limiting relations of Theorem 10.1.

11. Criterion of Stability of a Homogeneous Relaxation Cycle

11.1. General Theorem

Suppose that $x \in R^k$ in the boundary-value problem (9.1), (9.2) and the corresponding point pattern

$$\varepsilon \dot{x} = f(x,y), \qquad \dot{y} = g(x,y) \tag{11.1}$$

is such that we can apply Lemma 1.2. Then, according to Theorem 1.2, the boundary-value problem (9.1), (9.2) has a spatially homogeneous cycle $(x(t,\varepsilon), y(t,\varepsilon))$ which is exponentially stable in the framework of system (11.1). Let us establish its stability in the framework of the distributed model (9.1), (9.2).

A theorem analogous to the Andronov–Witt theorem is valid for system (9.1), (9.2) in the phase space E. This theorem reduces the problem of the stability of the homogeneous cycle of this system to the analysis of the system of variational equations corresponding to this cycle. Representing the solutions of the latter as Fourier series in terms of the eigenfunctions of the Laplace operator, we arrive at the relations

$$\varepsilon \dot{h}_1 = [A_{11}(t,\varepsilon) - zD_1]h_1 + A_{12}(t,\varepsilon)h_2, \tag{11.2}$$

$$\dot{h}_2 = A_{21}(t,\varepsilon)h_1 + [A_{22}(t,\varepsilon) - zD_2]h_2, \tag{11.3}$$

where $z = \lambda_k$, $k = 1, 2, \ldots$, λ_k, are eigenvalues of the operator $-\Delta$ under the Neumann boundary condition and the matrices A_{11}, A_{12}, A_{21}, A_{22} are the same as in (1.10). Since we can arbitrarily change the domain Ω and the matrix coefficients of diffusion in (9.1), (9.2), it is convenient to consider the parameter $z \geq 0$ in the system (11.2), (11.3) to be varying continuously. Thus, the linear system (11.2), (11.3) with $T(\varepsilon)$-periodic coefficients and continuous parameter $z \geq 0$ is the object of our analysis.

We shall study the arrangement of the multipliers of system (11.2), (11.3) under the following addition constraint.

LOCALIZATION CONDITION. We suppose that for all $z > 0$ the eigenvalues of the matrix $A(x,y) - zD_1$, where $A(x,y) = f'_x$, lie on the left of the imaginary axis for $(x,y) \in \Gamma_{1-} + \Gamma_{10} + \Gamma_{2-} + \Gamma_{20}$ and the eigenvalue close to zero has the order of smallness z for $(x,y) \in \Gamma_{10} + \Gamma_{20}$ and small values of z.

Suppose that the localization conditions are violated, namely, the matrix $A(x_0, y_0) - z_0 D_1$ has an eigenvalue with a positive real part for a certain $z_0 > 0$ and $(x_0, y_0) \in Z_0$, where Z_0 is a cycle of the hybrid system (1.7). Then system (11.2), (11.3) immediately acquires a multiplier whose absolute value is of the order $\exp(c/\varepsilon)$, $c > 0$.

We set

$$\varphi(\sigma) = \int_{-\infty}^{\infty} \exp\left\{-\int_s^{\infty} \frac{du}{(u^2 + v_0(u))^2} - \sigma(\Omega_0 - v_0(s))\right\} \frac{ds}{(s^2 + v_0(s))^2},$$

$$\sigma \geq 0, \tag{11.4}$$

where the function v_0 and the constant $\Omega_0 > 0$ are the same as in (4.14). Let, furthermore,

$$r = (D_1 \alpha, \beta)(|q_1 q_2|)^{-1/3},$$

where

$$q_1 = (\dot{y}_0(t_p - 0), B^*\beta), \qquad q_2 = \frac{1}{2} \sum_{j_1, j_2, j_3 = 1}^k \beta_{j_1} \frac{\partial^2 f_{j_1}}{\partial x_{j_2} \partial x_{j_3}} \alpha_{j_2} \alpha_{j_3},$$

$$\alpha = (a_1, \ldots, \alpha_k), \qquad \beta = (\beta_1, \ldots, \beta_k), \qquad f = (f_1, \ldots, f_k),$$

the derivatives being calculated on the cycle Z_0 at the switching moments $t_p - 0$ (for notations, see 1.1).

THEOREM 11.1. *Suppose that the Localization Condition is satisfied in addition to the hypothesis of Lemma 1.2. Then k multipliers of system* (11.2), (11.3) *tend to zero uniformly with respect to $z \geq 0$ as $\varepsilon \to 0$ and the other m multipliers are determined, with an accuracy to within $\varepsilon^{1/3} \ln \frac{1}{\varepsilon}$, from the T_0-periodic problem*

$$\dot{h} = [D(t) - zD_2 - C(t)(A(t) - zD_1)^{-1}B(t)]h, \tag{11.5}$$

$$h(t_p + 0) = h(t_p - 0) + \varphi(rz/\varepsilon^{1/3})a(b, h(t_p - 0)). \tag{11.6}$$

It follows from (11.6) that the most interesting case is when the quantity $(D_1 \alpha, \beta)$ is small, to be more precise, is of the order $\varepsilon^{1/3}$. Under this condition, the multipliers of system (11.5), (11.6) vary in a complicated way upon a change in the diffusion coefficients, passing through zero, in particular.

In many respects system (11.2), (11.3) is similar to system (1.10), and it can be analyzed with the use of practically the same scheme. However, if the parameter z is small (of the order $\varepsilon^{1/3}$), a boundary layer appears and the principal parts of the multipliers of system (11.2), (11.3) can be found from (11.5), (11.6) with an accuracy to within $\varepsilon^{1/3} \ln \frac{1}{\varepsilon}$ (for finite positive z the principal parts of the multipliers of system (11.2), (11.3) can be found from (11.5), (11.6) with an accuracy to within $\varepsilon^{2/3}$).

11.2. A Two-Dimensional Case

For $x, y \in R$, i.e., when we consider a relaxation cycle in the plane, we can get an explicit formal answer to the problem on its diffusion instability (this is due to the fact that Eq. (11.6) is scalar). And since it is precisely this case that is encountered in the majority of biophysical problems [1, 17], it is worthwhile to consider it separately.

Suppose that for $x, y \in R$ the smooth functions f and g in (11.1) are such that system (11.1) has an exponentially stable relaxation cycle $(x(t,\varepsilon), y(t,\varepsilon))$. For the sake of simplicity, we shall only consider the case in which it has only two slow-motion sections. Instead of system (11.2), (11.3) with a single parameter z, it is convenient here to consider the two-parameter system

$$\begin{aligned} \varepsilon \dot{h}_1 &= -z_1 h_1 + a(t,\varepsilon)h_1 + b(t,\varepsilon)h_2, \\ \dot{h}_2 &= -z_2 h_2 + c(t,\varepsilon)h_1 + d(t,\varepsilon)h_2, \end{aligned} \tag{11.7}$$

where

$$z_1 = \lambda_k D_1, \qquad z_2 = \lambda_k D_2, \qquad k = 1, 2, \ldots,$$

$$a = \partial f/\partial x, \qquad b = \partial f/\partial y, \qquad c = \partial g/\partial x, \qquad d = \partial g/\partial y,$$

and the derivatives are taken for $x = x(t,\varepsilon)$, $y = y(t,\varepsilon)$. System (11.7), in which the parameters z_1, z_2 are assumed to vary continuously, is the object of our further investigation. The analysis of its stability properties for all $z_1, z_2 \geq 0$ is equivalent to answering the question concerning the stability of a homogeneous cycle of the boundary-value problem (9.1), (9.2) for any diffusion coefficients D_1, D_2 in any domain Ω.

In order to formulate the main result, we set

$$\begin{aligned} \Phi_0(\sigma) = \tfrac{1}{T_0}\Big\{ &\ln\Big|\Big(1 - \tfrac{g(B)}{g(C)}\Big)\varphi(r_1\sigma) + \tfrac{g(B)}{g(C)}\Big| + \\ + \ln&\Big|\Big(1 - \tfrac{g(D)}{g(A)}\Big)\varphi(r_2\sigma) + \tfrac{g(D)}{g(A)}\Big|\Big\}, \qquad \sigma \geq 0, \\ r_1 &= \left(\frac{2}{|f''_{xx}(B) f'_y(B) g(B)|}\right)^{1/3}, \\ r_2 &= \left(\frac{2}{|f''_{xx}(D) f'_y(D) g(D)|}\right)^{1/3}, \end{aligned} \tag{11.8}$$

where T_0 is the zero approximation of the period of the coefficients of system (11.7), the function φ is defined by (11.4), and A, C and B, D are the drop and junction points of the cycle Z_0 (Fig. 11.1). In the general case, function (11.8) has two vertical asymptotes $\sigma = \sigma_1$ and $\sigma = \sigma_2$ and is monotonic outside of the interval $[\sigma_1, \sigma_2]$.

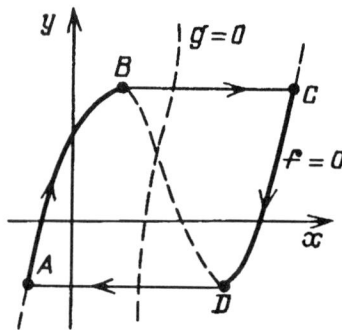

Fig. 11.1

Suppose that μ_0 is a multiplier of system (11.7), the largest in absolute value (the absolute value of the other does not exceed $N_1 \exp(-N_2/\varepsilon)$, where $N_1, N_2 > 0$ and are independent of z_1, z_2, ε). Theorem 11.1 yields the following statement.

THEOREM 11.2. *For any $\delta > 0$ there exist sufficiently small positive numbers $\varepsilon_0 = \varepsilon_0(\delta)$, $\delta_0 = \delta_0(\delta)$, such that inside the "horns"*

$$0 < \varepsilon \leq \varepsilon_0, \qquad \varepsilon^{1/3}(\sigma_j - \delta_0) \leq z_1 \leq \varepsilon^{1/3}(\sigma_j + \delta_0), \qquad j = 1, 2,$$

the inequality $|\mu_0| \leq \delta$ is satisfied and outside of these "horns" μ_0 has order unity as $\varepsilon \to 0$. Higher and lower than the "horns" $\mu_0 > 0$, and between them $\mu_0 < 0$. In addition, there exists a function $F(\varepsilon, z_1)$, $F(\varepsilon, 0) \equiv 0$, such that this multiplier is smaller (larger) than unity in absolute value for $z_2 - F(\varepsilon, z_1) > 0 \; (< 0)$. The representation (asymptotically unimprovable with respect to ε)

$$F(\varepsilon, z_1) = F_0(z_1) + \Phi_0(z_1/\varepsilon^{1/3}) - F_0(+0) + \varepsilon^{1/3} \ln \frac{1}{\varepsilon} \Phi(\varepsilon, z_1), \qquad (11.9)$$

where

$$F_0(z_1) = \frac{1}{T_0} \int_0^{T_0} (d_0(t) - b_0(t)c_0(t)[a_0(t) - z_1]^{-1}) dt \qquad (11.10)$$

is valid for this function (here a_0, b_0, c_0, d_0 are zero approximations of a, b, c, d). The function Φ is bounded uniformly with respect to the set of variables.

We shall describe some properties of function (11.10) that are necessary for the sequel. It is obvious, first of all, that

$$\lim_{z_1 \to \infty} F_0(z_1) = \frac{1}{T_0} \int_0^{T_0} \frac{\partial g}{\partial y} dt. \qquad (11.11)$$

Second, after some transformations we make sure that

$$\lim_{z_1 \to +0} F_0(z_1) = \frac{1}{T_0} \ln \frac{g(B)g(D)}{g(A)g(C)}, \qquad (11.12)$$

where the points A, B, C, D are shown in Fig. 11.1. Note that, in the general case, the integral

$$F_0'(z_1) = -\frac{1}{T_0} \int_0^{T_0} \frac{\partial f}{\partial y} \frac{\partial g}{\partial x} \left(\frac{\partial f}{\partial x} - z_1 \right)^{-2} dt \qquad (11.13)$$

diverges as $z_1 \to +0$.

Theorems 10.1 and 11.2 imply the following statement.

THEOREM 11.3. *Suppose that $\lambda_k D_2 > F_0(\lambda_k D_1)$ for $k = 1, 2, \ldots$ (here λ_k are positive eigenvalues of the operator Δ arranged in the order of increase under the boundary Neumann condition). Then there exists $\varepsilon_0 > 0$ such that for $0 < \varepsilon \leq \varepsilon_0$ the homogeneous relaxation cycle of the boundary-value problem (9.1), (9.2) is orbitally exponentially stable and all solutions of problem (9.1), (9.2) with initial conditions from its certain (independent of ε) neighborhood in the phase space E tend to it as $t \to \infty$. Now if $\lambda_{k_0} D_2 < F_0(\lambda_{k_0} D_1)$ for some k_0, then this cycle is unstable.*

We shall separately consider the case, important for applications [1], in which $D_2 = 0$ and the boundary condition for y is no longer necessary.

THEOREM 11.4. *Suppose that $D_2 = 0$ and the right-hand side of (11.12) is positive. Then there exist a positive number ε_0 and a natural number k_0, independent of ε_0, such that for every $0 < \varepsilon \leq \varepsilon_0$ and every $k \geq k_0$ system (11.7) has an exponentially growing solution.*

This statement also follows from Theorem 11.2. It should be emphasized that under the conditions of Theorem 11.4, the characteristic exponents of the high modes $e_k(s)$ ($e_k(s)$ is an eigenfunction of the operator $-\Delta$ corresponding to the eigenvalue λ_k) have positive real parts of order unity, and this circumstance allows us to speak of an explosive diffusion instability of the homogeneous cycle. It is interesting that a similar situation can occur in the case of harmonic auto-oscillations [22, 50].

It should be pointed out in conclusion that under the conditions of Theorem 11.4, for $D_2 = 0$, in the neighborhood of the homogeneous cycle, the parabolic system (9.1), (9.2) has an exponentially stable infinite-dimensional invariant manifold corresponding to the characteristic indices of the system of variational equations with nonnegative real parts. The proof is based on the knowledge of the spectral properties of the monodromy operator generated by the system of variational equations and on the standard technique of proving the existence of integral manifolds [87] extended to parabolic boundary-value problems [67].

11.3. Asymptotics of the Function $F(\varepsilon, z_1)$ for $z_1 \geq z_{10} > 0$

Here we use the results of Sec. 5 and apply a special technique to reveal all the properties of function (11.9) for $z_1 \geq z_{10}$, where $z_{10} > 0$ is an arbitrarily small fixed number.

Let us consider an auxiliary four-dimensional system

$$\begin{aligned}
\varepsilon \dot{x}_1 &= \tfrac{z_1}{2}(x_2 - x_1) + f(x_1, y_1), \\
\varepsilon \dot{x}_2 &= \tfrac{z_1}{2}(x_1 - x_2) + f(x_2, y_2), \\
\dot{y}_1 &= \tfrac{z_2}{2}(y_2 - y_1) + g(x_1, y_1), \\
\dot{y}_2 &= \tfrac{z_2}{2}(y_1 - y_2) + g(x_2, y_2),
\end{aligned} \qquad (11.14)$$

which we shall call a bilocal model of the boundary-value problem (9.1), (9.2). Obviously, system (11.14) has a relaxation cycle with components $x_1 \equiv x_2 = x(t, \varepsilon)$, $y_1 \equiv y_2 = y(t, \varepsilon)$. After simple verification, we make sure that all the conditions indicated in Sec. 1 are satisfied for $z_1 \geq z_{10} > 0$. It is significant that when we linearize system (11.14) on the indicated cycle and project the difference Laplace operator onto the eigenvectors, we get the linear system (11.7). Therefore, the calculation of the function $F(\varepsilon, z_1)$ reduces to the analysis of the Poincaré operator $\Pi(\varepsilon, z_1, z_2)$ introduced in Sec. 5.

As was pointed out in Sec. 5, formulas similar to the asymptotic formulas (5.2), (5.3) are valid for the components $x = (x_1, x_2)$ and $y = (y_1, y_2)$ of the operator $\Pi(\varepsilon, z_1, z_2)$. Therefore, the function $F(\varepsilon, z_1)$, which can be found from the equation

$$\det(\Pi'|_{x=x(\varepsilon), y=y(\varepsilon)} - I) = 0,$$

where $x(\varepsilon)$, $y(\varepsilon)$ are the initial conditions of the relaxation cycle and I is an identity matrix, admits of an asymptotic expansion

$$F(\varepsilon, z_1) = F_0(z_1) + \sum_{n=2}^{\infty} \varepsilon^{n/3} \sum_{\nu=0}^{\pi(n-2)} F_{n,\nu}(z_1) \ln^{\nu} \frac{1}{\varepsilon}, \qquad (11.15)$$

which is valid uniformly with respect to $z_1 \geq z_{10}$. Moreover, dividing the first equation of (11.7) by z_1 and making z_1 tend to ∞, we infer that

$$\lim_{z_1 \to \infty} F(\varepsilon, z_1) = \frac{1}{T(\varepsilon)} \int_0^{T(\varepsilon)} \frac{\partial g}{\partial y}(x(t, \varepsilon), y(t, \varepsilon)) dt. \tag{11.16}$$

Comparing now relations (11.15) and (11.16), we make sure that as $z_1 \to \infty$, all of the coefficients $F_{n,\nu}(z_1)$ from (11.15) have finite limits equal to the corresponding terms of the expansion of the right-hand side of (11.16).

THEOREM 11.5. *The asymptotic expansion* (11.5) *uniform with respect to* $z_1 \geq z_{10}$, *where* $z_{10} > 0$ *is any fixed number, holds true for the function* $F(\varepsilon, z_1)$.

Note that Theorem 11.2 cannot be obtained by means of the constructions described above, since for $z_1 = 0$ the matrix $A(x, y) = f'_x$ of system (11.14) has a twofold zero eigenvalue at the junction point of the relaxation cycle, i.e., one of the conditions of Sec. 1 is violated.

12. Analysis of Some Mathematical Models of Biophysics

12.1. Stability of a Homogeneous Cycle of a Number of Mathematical Models

The results presented in Sec. 11 allow us to analyze a number of mathematical models of biophysics [1].

Let us suppose that

$$f = y + x - \frac{1}{3}x^3, \qquad g = -x - by + a, \qquad a, b > 0, \tag{12.1}$$

in (9.1), (9.2), i.e., let us consider the Bonhoeffer–van der Pol model. We can easily verify that in the case of (12.1) the right-hand sides in (11.11) and (11.12) are negative and the right-hand side in (11.13) is positive. Therefore, $F_0(z_1)$ increases monotonically and tends to a negative limit as $z_1 \to \infty$. Therefore, $F_0(z_1) < 0$ for all $z_1 > 0$ and, by virtue of Theorem 11.3, the relaxation cycle is stable for any diffusion coefficients.

A similar fact is valid for boundary-value problems of the type (9.1), (9.2) where

$$f = x - x^3 - y, \qquad g = x - B; \tag{12.2}$$

$$f = x - x^3 - y, \qquad g = B - y + \gamma x; \tag{12.3}$$

$$f = A - x(B+1) + x^2(y-x), \qquad g = A - x. \tag{12.4}$$

All parameters are positive here. Note that in the case of (12.2) the corresponding boundary-value problem is called a FitzHugh–Nagumo model, in the case of (12.3) it is called a Nagumo model, and in the case of (12.4) it is a variant of the so-called brusselator.

From the other boundary-value problems (9.1), (9.2) encountered in the articles of collection [1], we choose for our consideration the boundary-value problem suggested by Yakhno:

$$f = x - x^3 - y,$$

$$g = x + 1.5y - 2.7y^2 \quad \text{or} \quad g = x + 3y - 2.7y^2 \quad \text{for} \quad D_2 = D.$$

By virtue of Theorem 11.4, for these f and g we observe an explosive diffusion instability of the homogeneous relaxation cycle. This fact allowed Yakhno to carry out numerical constructions for complicated spatially nonhomogeneous steady states.

Also note that in the case of models (12.1)–(12.4) the inference concerning the stability of the homogeneous cycle is substantiated by Theorem 11.2, but here the analysis is very complicated and can be carried out only with the use of numerical methods.

12.2. Some Inferences

In [52, 57, 58] the parabolic systems with auto-oscillational "ordinary part" are divided, by the nature of their dynamics, into simple physical, complicated physical, and biological systems.

In particular, simple systems include parabolic systems whose homogeneous cycle remains stable upon any change of the diffusion coefficients. This class of parabolic systems cannot have complicated spatially nonhomogeneous steady states since the appearance of the latter is connected with a desynchronization of the auto-oscillations. But this does not prevent them from having well-ordered conditions of the type of "running waves" on the circumference of a circle or "reverberators" in a circle and in domains close to a circle in shape.

In complicated physical and biological parabolic systems, spatially nonhomogeneous steady states are qualitatively more complicated in their structure, and, as is shown in [64], their appearance is connected not only with the loss of stability of the homogeneous cycle upon a decrease in the diffusion coefficients but also with the phenomenon of self-organization when the most complicated and ordered conditions are born "from the air."

As was mentioned in [1, 17], boundary-value problems of the type of (9.1), (9.2) with nonlinearities (12.1)–(12.4) are considered to be base problems in biophysics. However, a contradiction arises with what was presented above. Mainly it is due to the fact that the authors of [1, 17] took modifications of the known models of self-excited oscillators of relaxation oscillations [2] for simulating complicated phenomena, i.e., proceeded from simple physical objects.

At the same time, it is easy to circumvent the use of trivial models: by virtue of Theorem 11.3, it is sufficient for the function $F_0(z_1)$ to be positive for relatively small z_1 and negative for all sufficiently large z_1. Only for $F_0(z_1)$ constructed in this way is the homogeneous cycle stable for all sufficiently large values of D_1, and with a decrease in D_1 it loses stability. This is the way the "normal" mathematical models of biophysics must be constructed.

Chapter 3

Structure of the Neighborhood of a Relaxation Cycle

Recall that the problem concerning the structure of the neighborhood of a cycle for smooth systems of ordinary differential equations and parabolic boundary-value problems can be solved, for instance, by means of the method of constructing a normal form in the neighborhood of a cycle, developed in [29, 35], which is a generalization of the algorithm suggested in [49] on normalization of distributed systems in the neighborhood of an equilibrium position. Its direct extension to relaxation systems causes some difficulties. Below we show that in the relaxation case it is convenient to normalize the Poincaré operator along the trajectories of the system in the neighborhood of the cycle since, as $\varepsilon \to 0$, for this operator we have not only C^1-convergence to the corresponding operator of a relay system but also C^k-convergence for any $k \geq 1$.

In what follows we present the method of extension of the normalization algorithm given in [49, 29, 35] to mappings. Then we indicate applications to relay and relaxation systems. In particular, at the end of the chapter we give a statement which can be regarded as an extension of the Turing–Prigogine bifurcation theorem to the case in which a diffusion instability is characteristic of a spatially homogeneous relaxation cycle.

13. Normal Form of Mapping

13.1. Statement of the Problem

Let us consider the mapping

$$u \mapsto \Pi(u, \varepsilon), \qquad 0 < \varepsilon \ll 1, \tag{13.1}$$

where Π, $\Pi(0, \varepsilon) \equiv 0$, is an operator, smooth with respect to the set of variables, in the real Banach space E. We introduce into consideration a

Structure of the Neighborhood of a Relaxation Cycle 109

linear operator $V_0 = D_u \Pi|_{u=0,\varepsilon=0}$ (D is a Fréchet derivative) and assume that it has $2m$ eigenvalues equal to $\exp(\pm i\omega_k)$, $0 < \omega_k < \pi$, $k = 1, \ldots, m$, and that the other part of its spectrum lies in a circle whose radius is smaller than unity. The indicated eigenvalues are associated with exactly $2m$ linearly independent eigenvectors $a_k = a_k^1 + i a_k^2$, $\bar{a}_k = a_k^1 - i a_k^2$, and

$$l_k(a_n) = \delta_{kn}, \qquad l_k(\bar{a}_n) = 0, \qquad k, n = 1, \ldots, m,$$

where $V_0^* l_k = \exp(-i\omega_k) l_k$, $k = 1, \ldots, m$.

On these assumptions, in the neighborhood of the zero fixed point, mapping (13.1) has a $2m$-dimensional exponentially stable local invariant manifold [87, 67], and this allows us to reduce the problem to a finite-dimensional one. Recall that in [49, 29, 35] the well-known Bogolyubov–Mitropol'skii method is applied to the construction of an integer manifold in the neighborhood of the equilibrium position and in that of a cycle, respectively, and the system of ordinary differential equations is constructed on it directly in the normal form of Poincaré–Dulac. Later on this technique is extended to the case where we are interested in the structure of the neighborhood of the fixed point of the mapping.

We shall seek the equation of the $2m$-dimensional invariant manifold of mapping (13.1) in the form

$$u = \sum_{j=1}^{m} \xi_j [a_j \exp(i\omega_j \tau_j) + \bar{a}_j \exp(-i\omega_j \tau_j) + \varepsilon W_j] +$$

$$+ \sum_{k,j=1}^{m} W_{kj} \xi_k \xi_j + \sum_{k,j,p=1}^{m} W_{kjp} \xi_k \xi_j \xi_p + U(\omega \tau, \xi, \varepsilon) \qquad (13.2)$$

and shall try to define the mapping on it directly in the normal form

$$\xi_k \mapsto \xi_k + \varphi_k, \qquad \tau_k \mapsto \tau_k + 1 + \psi_k, \qquad k = 1, \ldots, m, \qquad (13.3)$$

where

$$\varphi_k = \varepsilon \varphi_k^1 \xi_k + \varepsilon \sum_{\omega_j \in \Omega_k^1} a_{kj} \cos(\omega_j \tau_j - \omega_k \tau_k + \gamma_{kj}) \xi_j +$$

$$+ \sum_{\omega_j, \omega_l \in \Omega_k^2} a_{kjl} \cos(k_j \omega_j \tau_j + k_l \omega_l \tau_l - \omega_k \tau_k + \gamma_{kjl}) \xi_j^{|k_j|} \xi_l^{|k_l|} +$$

$$+ \sum_{\omega_j, \omega_l, \omega_p \in \Omega_k^3} a_{kjlp} \cos(k_j \omega_j \tau_j + k_l \omega_l \tau_l + k_p \omega_p \tau_p - \omega_k \tau_k + \gamma_{kjlp}) \times$$

$$\times \xi_j^{|k_j|} \xi_l^{|k_l|} \xi_p^{|k_p|} + \xi_k \sum_{j=1}^{m} \varphi_{kj}^3 \xi_j^2 + \Phi_k(\omega\tau, \xi, \varepsilon);$$

$$\psi_k = \varepsilon\psi_k^1 + \varepsilon \sum_{\omega_j \in \Omega_k^1} \frac{a_{kj}}{\omega_k} \sin(\omega_j\tau_j - \omega_k\tau_k + \gamma_{kj}) \frac{\xi_j}{\xi_k} +$$

$$+ \sum_{\omega_j,\omega_l \in \Omega_k^2} \frac{a_{kjl}}{\omega_k} \sin(k_j\omega_j\tau_j + k_l\omega_l\tau_l - \omega_k\tau_k + \gamma_{kjl}) \frac{\xi_j^{|k_j|} \xi_l^{|k_l|}}{\xi_k} +$$

$$+ \sum_{\omega_j,\omega_l,\omega_p \in \Omega_k^3} \frac{a_{kjlp}}{\omega_k} \sin(k_j\omega_j\tau_j + k_l\omega_l\tau_l + k_p\omega_p\tau_p - \omega_k\tau_k + \gamma_{kjlp}) \times$$

$$\times \frac{\xi_j^{|k_j|} \xi_l^{|k_l|} \xi_p^{|k_p|}}{\xi_k} + \sum_{j=1}^{m} \psi_{kj}^2 \xi_j^2 + \frac{1}{\xi_k} \Psi_k(\omega\tau, \xi, \varepsilon).$$

Here $\Omega_k^1 = \{\omega_j : \omega_k = \omega_j, j \neq k\}$,

$$\Omega_k^2 = \{\omega_j, \omega_l : \omega_k = k_j\omega_j + k_l\omega_l (\text{mod } 2\pi), |k_j| + |k_l| = 2\},$$

the set Ω_k^3 is defined by analogy (of course, the identical resonances $\omega_k = \omega_k + \omega_j - \omega_j$ are removed); the smooth functions Φ_k, Ψ_k and the vector-function U (with values in E) of the vector argument $\omega\tau = (\omega_1\tau_1, \ldots, \omega_m\tau_m)$ are periodic with period 2π and satisfy the inequalities

$$\|U\|_E, |\Phi_k|, |\Psi_k| \leq M(\varepsilon^2\|\xi\| + \varepsilon\|\xi\|^2 + \|\xi\|^4), \tag{13.4}$$

where M is a constant; $\|*\|_E$, $\|*\|$ is a norm in E and a Euclidean norm; $\xi = (\xi_1, \ldots, \xi_m)$.

13.2. Description of the Algorithm and the Main Result

The principal terms of (13.2), (13.3) described above are sought in parallel when the following algorithm is realized. Let us compose an expression

$$\tilde{u} = F(u, \varepsilon), \tag{13.5}$$

where u is the right-hand side of (13.2) and $\tilde{u}$ is the right-hand side of (13.2) in which ξ_k, τ_k are replaced by $\xi_k + \varphi_k$, $\tau_k + 1 + \psi_k$. Successively equating the coefficients in the power of $\varepsilon\xi_1, \ldots, \varepsilon\xi_m, \xi_1^2, \ldots, \xi_m^3$ in (13.5), we arrive at equations of the form

$$v(\omega(\tau+1)) = V_0 v(\omega\tau) + f(\omega\tau), \tag{13.6}$$

where $\omega(\tau+1) = (\omega_1(\tau_1+1), \ldots, \omega_m(\tau_m+1))$ and f is a trigonometric polynomial with respect to $\omega\tau$ with coefficients from E. We seek the solution of (13.6) with the same dependence on $\omega\tau$ as f. As a result, the problem reduces to the solution of several equations of the form

$$\exp(i\varkappa)z = V_0 z + g, \qquad (13.7)$$

where $\varkappa$ is a linear combination of the numbers ω_k with integer coefficients. From the solvability conditions of (13.7) we determine constants entering into the right-hand sides of (13.3) on which the vector g linearly depends at every step.

The following statement is the main result.

THEOREM 13.1. *Formula* (13.2) *defines a 2m-dimensional exponentially stable local invariant manifold of mapping* (13.1) *in the neighborhood of a zero fixed point,* (13.3) *being the mapping on this manifold.*

The proof of the theorem is based on combining the Bogolyubov–Mitropol'skii reduction principle [71] with the technique of normalizing substitutions. Here are its principal stages.

In (13.1) we set $u = \eta + v$, where $\eta = P(\varepsilon)u$, $v = u - P(\varepsilon)u$, and $P(\varepsilon)$ is the root projector of the operator $V(\varepsilon) = D_u \Pi(0,\varepsilon)$ corresponding to its eigenvalues which turn into $\exp(\pm i\omega_k)$, $k=1,\ldots,m$ for $\varepsilon = 0$. Next, we identify the finite-dimensional space $P(\varepsilon)E$ with R^{2m} and assume that $\eta \in R^{2m}$. In the new coordinates, mapping (13.1) reduces to the form

$$\eta \mapsto B(\varepsilon)\eta + G_1(\eta, v, \varepsilon), \qquad (13.8)$$

$$v \mapsto L(\varepsilon)v + G_2(\eta, v, \varepsilon), \qquad (13.9)$$

where $L(\varepsilon) = (I - P(\varepsilon))V(\varepsilon)(I - P(\varepsilon))$; the matrix $B(0)$ has eigenvalues $\exp(\pm i\omega_k)$, $k=1,\ldots,m$, with a complete set of eigenvectors e_k and $\bar{e}_k$ respectively; the mappings

$$G_1(*,*,\varepsilon): R^{2m} \oplus (I - P(\varepsilon))E \to R^{2m},$$

$$G_2(*,*,\varepsilon): R^{2m} \oplus (I - P(\varepsilon))E \to (I - P(\varepsilon))E,$$

which are smooth with respect to the collection of variables, are defined in a sufficiently small neighborhood of zero of the space $R^{2m} \oplus (I - P(\varepsilon))E$ and vanish together with the first Fréchet derivative with respect to η and v for $\eta = 0$, $v = 0$.

From the results of [87, 67] it follows that the $2m$-dimensional exponentially stable invariant manifold of mapping (13.1) is defined by the relation

$$v = H(\eta, \varepsilon). \qquad (13.10)$$

Here the vector-function H with its values in $(I - P(\varepsilon))E$, which is smooth with respect to the set of variables, is such that

$$H(0,\varepsilon) \equiv 0, \qquad D_\eta H(0,\varepsilon) \equiv 0.$$

Substituting (13.10) into (13.8), we obtain a finite-dimensional mapping on the integer manifold of mapping (13.1)

$$\eta \mapsto B(\varepsilon)\eta + G(\eta,\varepsilon), \qquad (13.11)$$

where the vector-function G has at zero an order of smallness higher than the first.

To complete the proof, we must make a change of variables in (13.11)

$$\eta = \sum_{j=1}^{m} \xi_j [e_j \exp(i\omega_j \tau_j) + \bar{e}_j \exp(-i\omega_j \tau_j) + \varepsilon \eta_j] +$$

$$+ \sum_{k,j=1}^{m} \eta_{kj} \xi_k \xi_j + \sum_{k,j,p=1}^{m} \eta_{kjp} \xi_k \xi_j \xi_p,$$

which can be constructed by means of the algorithm described at the beginning of this subsection and which reduces (13.11) to form (13.3). Returning to mapping (13.1), we obtain the statement of the theorem.

In conclusion, it should be pointed out that Theorem 13.1 can be easily generalized to the case where the operator V_0 has eigenvalues equal to -1 and where there are Jordan blocks. In addition, the parameter ε in (13.1) can be a vector parameter.

13.3. On the Invariants of a Normal Form

Note that the terms written out in (13.3) are invariant with an accuracy up to similitude transformations. We shall consider in greater detail the problem of the number of invariants of the normal form of the mapping in the neighborhood of a zero fixed point which loses stability when passing through its eigenvalue -1 or the pair $\exp(\pm i\omega)$, where $2\pi/\omega$ is irrational.

Suppose that the operator V_0 has a simple eigenvalue equal to -1 which is associated with the eigenvector h_1 and the other part of its spectrum lies in a circle whose radius is smaller than unity. Then, in accordance with Theorem 13.1, for $\varepsilon = 0$, the "abridged" normal form of mapping (13.1) has the form

$$\xi \mapsto -\xi + d_0 \xi^3.$$

The main constraint imposed on mapping (13.1) and characterizing a certain generality of the position is that $d_0 \neq 0$. In this case, on the formally analytic integer manifold

$$u = \xi h_1 + \xi^2 h_2 + \xi^3 h_3 + \ldots \tag{13.12}$$

mapping (13.1), for $\varepsilon = 0$, can be reduced to the form

$$\xi \mapsto -\xi + d_0 \xi^3 + d_1 \xi^5, \tag{13.13}$$

which contains only invariants.

Here is the technique of constructing the indicated manifold. The terms of series (13.12) are determined in succession from the series of linear nonhomogeneous equations

$$(-1)^k h_k = V_0 h_k + f_k, \quad k = 2, 3, \ldots, \tag{13.14}$$

which result when we equate the coefficients in like powers of ξ in an expression similar to (13.5). Note that when k is even, Eq. (13.14) is always solvable and when it is odd, we need a solvability condition. Therefore, we shall choose solutions h_{2m+1}, $m \geq 1$, which include terms $c_m h_1$, where c_m are arbitrary constants. We can find these constants from the conditions of solvability of Eq. (13.14).

Note that at the fifth step, the solvability condition cannot be satisfied at the expense of the choice of the constant c_1, and, therefore, a fifth-power term appears in (13.13). But for $k > 5$ such a situation does not occur and the solvability of Eq. (13.14) is provided by the constants c_2, c_3, and so on.

These facts remain valid when

$$V_0 a = \exp(i\omega) a, \quad V_0 \bar{a} = \exp(-i\omega) \bar{a}, \quad 0 < \omega < \pi;$$

the ratio $2\pi/\omega$ is irrational, and $d_0 \neq 0$. In this case, on the formally analytic invariant manifold (which can be constructed in the same way as (13.12))

$$u = \xi[a \exp(i\omega\tau) + \bar{a} \exp(-i\omega\tau)] + \xi^2 h_2(\omega\tau) + \ldots$$

mapping (13.1), for $\varepsilon = 0$, reduces to the form

$$\xi \mapsto \xi + d_0 \xi^3 + d_1 \xi^5, \quad \tau \mapsto \tau + 1 + c_0 \xi^2,$$

which is a generalization of the well-known result concerning the number of invariants of the normal form in the neighborhood of a weak focus.

14. Intrinsic Resonance upon the Loss of Stability by a Homogeneous Cycle of a Relay System

14.1. The Case of a Closed Interval

Here the results of the preceding section are applied to the relay system (9.3) whose diffusion matrix smoothly depends on the small positive parameter μ and which is considered on the interval $[0, 1]$.

Let us suppose that the relay system

$$\frac{\partial y}{\partial t} = D(\mu)\frac{\partial^2 y}{\partial s^2} + g(\Phi(y), y), \tag{14.1}$$

$$\left.\frac{\partial y}{\partial s}\right|_{s=0} = \left.\frac{\partial y}{\partial s}\right|_{s=1} = 0 \tag{14.2}$$

has a simple homogeneous cycle $y_0(t)$, $(\Phi_1(y_0(0)), y_0(0)) \in \Gamma_1$ (see Sec. 9). Suppose, furthermore, that for $\mu = 0$ all the multipliers of the periodic boundary-value problem (9.10), (9.11), except for the identity multiplier and the pair $\exp(\pm i\omega)$, $0 < \omega < \pi$, which is associated with the Lyapunov–Floquet solutions

$$a(t)\exp(i\omega t)\cos \pi s, \quad \tilde{a}(t)\exp(-i\omega t)\cos \pi s,$$

lie inside a circle of radius 1. Under these constraints, in the phase space $E = C^\alpha$ of the boundary-value problem (14.1), (14.2) we generally draw an intersecting hyperplane to the cycle $y_0(t)$ through the point $y_0(0)$, then shift $y_0(0)$ to the zero of the space E and consider the Poincaré operator Π along the trajectories of system (14.1), (14.2).

By virtue of the constraints imposed on the boundary-value problem (14.1), (14.2), the hypothesis of Theorem 13.1 is satisfied for the operator Π. It follows from this theorem that when there are no strong resonances, the normal form of the mapping of Π in the neighborhood of the zero fixed point has the form

$$\xi \mapsto \xi + \mu \tau_0' \xi + d_0 \xi^2 + \Phi(\omega\tau, \xi, \mu), \tag{14.3}$$

$$\tau \mapsto \tau + 1 + \mu \omega_0' + c_0 \xi^2 + \frac{1}{\xi}\Psi(\omega\tau, \xi, \mu), \tag{14.4}$$

where the functions Φ, Ψ possess the properties described in Sec. 13.

THEOREM 14.1. *Let $\tau_0' > 0$, $d_0 < 0$ in (14.3). Then there exists $\mu_0 > 0$ such that for $0 < \mu \leq \mu_0$ mapping (14.3), (14.4) has an exponentially stable invariant manifold*

$$\xi = \sqrt{-\mu\tau_0'/d_0}\,[1 + \sqrt{\mu}\varphi(\omega\tau, \mu)], \tag{14.5}$$

Structure of the Neighborhood of a Relaxation Cycle

where the function φ, which is smooth with respect to the set of variables and periodic with respect to $\omega\tau$ with period 2π, is uniformly bounded; the mapping on it results from (14.4) when (14.5) is taken into account.

The proof of the theorem immediately follows from the results of [68].

Returning to (14.1), (14.2), we infer from Theorem 14.1 that in the neighborhood of a homogeneous cycle it has an orbitally exponentially stable two-dimensional invariant torus. In the case of a mild loss of stability by a homogeneous cycle, the same is true for a resonance of order 3, i.e., when $\omega = 2\pi/3$. This is due to the fact that here the normal form remains the same, namely, the resonance terms in it disappear by virtue of the relation

$$\int_0^1 \cos^3 \pi s \, ds = 0.$$

For $\omega = \pi/2$ the character of the steady-state behavior of the boundary-value problem (14.1), (14.2), lying at a distance of order $\sqrt{\mu}$ from the cycle $y_0(t)$, can be determined from the analysis of the abridged mapping

$$\xi \mapsto \xi + \mu d_1 \xi + d_2 \xi^3 + d_3 \xi^3 \cos 2\pi(\tau + \gamma),$$

$$\tau \mapsto \tau + 1 + \mu\alpha_0 + \alpha_1 \xi^2 - \frac{2}{\pi} d_3 \xi^2 \sin 2\pi(\tau + \gamma),$$

where all the written-out constants are nonzero in the general case. The results of its investigation are given in [4].

The facts presented above rest on the algorithm of constructing a normal form described in Sec. 13 which makes it possible to take into account the peculiarity of the problem. For instance, at the resonance of order 3, in the case of an arbitrary mapping (13.1), Theorem 13.1 leads to the normal form

$$\begin{aligned}\xi &\mapsto \xi + \mu\alpha_1 \xi + \alpha_2 \xi^2 \cos 2\pi(\tau + \gamma) + \alpha_3 \xi^3 + \Phi(\tau, \xi, \mu),\\ \tau &\mapsto \tau + 1 + \mu\beta_1 - \tfrac{3}{2\pi}\alpha_2 \xi \sin 2\pi(\tau + \gamma) + \beta_2 \xi^2 + \tfrac{1}{\xi}\Psi(\tau, \xi, \mu),\end{aligned} \quad (14.6)$$

whose dynamics is very complicated [4, 9], and in the boundary-value problem (14.1), (14.2) this resonance only affects the behavior of the solutions on the invariant torus.

In conclusion, it should be pointed out that all we have said about the resonance of order 3 remains true for a multidimensional domain, when the least positive eigenvalue λ_1 of the operator $-\Delta$ is simple, the eigenfunction $e_1(s)$ corresponding to it satisfies the relation

$$\int_\Omega e_1^3(s) \, ds = 0, \quad (14.7)$$

and, for $\mu = 0$, the multipliers $\exp(\pm i\omega)$ of the boundary-value problem (9.10), (9.11) are associated with the Lyapunov–Floquet solutions

$$a(t)\exp(i\omega t)e_1(s), \qquad \bar{a}(t)\exp(-i\omega t)e_1(s).$$

Relation (14.7) characterizes a certain symmetry of the domain Ω. When this relation is not valid, the normal form for $\omega = 2\pi/3$ assumes the general form (14.6).

14.2. The Case of a Circle

The results of 14.1 refer to the principal case when the eigenvalue λ_1 of the operator $-\Delta$ is simple. The violation of this requirement entails new dynamic effects. All their peculiarities can be seen from the example of the boundary-value problem

$$\frac{\partial y}{\partial t} = D(\mu)\left[\frac{1}{r}\frac{\partial}{\partial r}\left(r\frac{\partial y}{\partial r}\right) + \frac{1}{r^2}\frac{\partial^2 y}{\partial \varphi^2}\right] + g(\Phi(y), y), \qquad (14.8)$$

$$\left.\frac{\partial y}{\partial r}\right|_{r=1} = 0, \qquad (14.9)$$

where r, φ are polar coordinates, $\Omega = \{(r, \varphi) : r \le 1, 0 \le \varphi \le 2\pi\}$. We assume, as before, that, for $\mu = 0$, the boundary-value problem (9.10), (9.11) has, on the circle of radius 1, a unit multiplier and a pair $\exp(\pm i\omega)$, which is associated with the linearly independent Lyapunov–Floquet solutions

$$a(t)\exp(i\omega t)J_1(x_{11}r)\cos\varphi, \qquad a(t)\exp(i\omega t)J_1(x_{11}r)\sin\varphi,$$
$$\bar{a}(t)\exp(-i\omega t)J_1(x_{11}r)\cos\varphi, \qquad \bar{a}(t)\exp(-i\omega t)J_1(x_{11}r)\sin\varphi,$$

and all its other multipliers are inside the unit circle. Here J_1 is the first-order Bessel function of the first kind and x_{11} is the first positive zero of J_1'.

When there are no strong resonances in this situation, the abridged mapping assumes the form (we do not write out the equations for the phase variables)

$$\begin{aligned}\xi_1 &\mapsto \xi_1 + \varepsilon\alpha_1\xi_1 + \alpha_2(\xi_1^2 + \xi_2^2)\xi_1, \\ \xi_2 &\mapsto \xi_2 + \varepsilon\alpha_1\xi_2 + \alpha_2(\xi_1^2 + \xi_2^2)\xi_2.\end{aligned} \qquad (14.10)$$

Thus, for $\alpha_1 > 0$, $\alpha_2 < 0$, the complete mapping has an exponentially stable two-dimensional torus defined, in the first approximation, by the relation $\xi_1^2 + \xi_2^2 = -\mu\alpha_1/\alpha_2$. We shall use the following technique to elucidate the behavior of the iterations of the mapping on it.

We consider (14.8) in a semicircle assuming that

$$\left.\frac{\partial y}{\partial r}\right|_{r=1} = 0, \quad \left.\frac{\partial y}{\partial \varphi}\right|_{\varphi=0} = \left.\frac{\partial y}{\partial \varphi}\right|_{\varphi=\pi} = 0. \tag{14.11}$$

In order to obtain the abridged normal form for the amplitude variable for the boundary conditions (14.11) from (14.10), we have only to set $\xi_2 = 0$. We evenly continue the solutions that lie on the two-dimensional exponentially stable torus of the boundary-value problem (14.8), (14.11) (which exists by virtue of the inequalities $\alpha_1 > 0$, $\alpha_2 < 0$) to the interval $[-\pi, 0]$ along φ and then to the entire axis by periodicity. As a result, they will also satisfy the boundary-value problem (14.8), (14.9). We infer that the orbitally exponentially stable three-dimensional torus of the boundary-value problem (14.8), (14.9), whose existence we have established, is entirely filled up with two-dimensional tori which are transformed into one another upon the shifts along the variable φ.

It should be emphasized that the properties of a three-dimensional torus that we have described are closely connected with a circular symmetry, i.e., are typical of parabolic boundary-value problems. Somewhat earlier similar effects in different situations were pointed out in [22, 29].

15. Bifurcation of a Relaxation Cycle

15.1. Duplication Bifurcation of a Relaxation Cycle

Let us suppose that system (1.9) has a structurally stable relaxation cycle $x(t, \varepsilon)$, $y(t, \varepsilon)$ and the conditions of Sec. 1 are fulfilled; $\Pi(\varepsilon)$ is the Poincaré operator along the trajectories of system (1.9) introduced in the proof of Theorem 5.1 and Π_0 is the corresponding Poincaré operator along the trajectories of the hybrid system (1.7).

LEMMA 15.1. *Suppose that the right-hand sides f and g of system (1.9) are infinitely differentiable. Then we have the limiting relation*

$$\lim_{\varepsilon \to 0} \Pi(\varepsilon) = \Pi_0 \tag{15.1}$$

in the C^k-metric for any fixed $k \geq 1$.

We represent the operator $\Pi(\varepsilon)$ as the superposition of the maps as we did when proving Theorem 5.1. As a result, the problem reduces to the verification of a relation, similar to (15.1), for the Poincaré operator of mapping from the plane $\xi = -q$ into the plane $\xi = q$ in the neighborhood of the junction point (for other mappings (15.1) follows from the results of [7]).

The relation similar to (15.1) is valid because all asymptotic formulas for the trajectory of the relaxation system in the neighborhood of the junction point from Sec. 4 can be differentiated any number of times with respect to the initial conditions. Indeed, differentiating the trajectory of system (4.22) with respect to the initial conditions, we arrive at linear nonhomogeneous systems which can be integrated, just as system (4.22), by the method presented in Sec. 4.

Lemma 15.1 shows that the Poincaré operator acting along the trajectories of the singularly perturbed system (1.9) regularly depends on ε. Thus, Theorem 13.1 is applicable to it.

Let us suppose, in addition, that the right-hand sides f and g of (1.9) smoothly depend on the parameter $0 < \mu \ll 1$, i.e., we have

$$\varepsilon \dot{x} = f(x, y, \mu), \qquad \dot{y} = g(x, y, \mu). \tag{15.2}$$

Moreover, we suppose that for $\mu = 0$ the periodic problem (1.4), (1.5) has a simple multiplier -1 and its other multipliers (except for the simple unit one) are smaller than unity in absolute value. Then it follows from the results of Sec. 4 and Lemma 15.1 that the linear system (1.10) has a multiplier

$$\lambda(\mu, \varepsilon) = \lambda_0(\mu) + \sum_{n=2}^{\infty} \varepsilon^{n/3} \sum_{k=0}^{\pi(n-2)} \lambda_{n,k}(\mu) \ln^k \frac{1}{\varepsilon} \tag{15.3}$$

smoothly dependent on μ. Here $\lambda_0(0) = -1$. Suppose, furthermore, that $\lambda_0'(0) \neq 0$. Then, using the theorem of an explicit function, we find from the relation $\lambda(\mu, \varepsilon) = -1$ the critical value $\mu_*(\varepsilon)$ of the parameter μ, for which, by virtue of (15.3), there holds a similar asymptotic representation.

We set $\mu = \mu_*(\varepsilon) + \nu$, $0 < \nu \ll 1$, in (15.2). If we apply here Theorem 13.1, we get the mapping

$$\xi \mapsto -\xi + \alpha_1(\varepsilon)\nu\xi + \alpha_2(\varepsilon)\xi^3 + \Phi(\xi, \varepsilon, \nu), \tag{15.4}$$

where the function Φ satisfies the inequality

$$|\Phi| \leq M[\xi^4 + \nu^2|\xi| + \nu\xi^2] \tag{15.5}$$

with a constant M independent of ε, and the coefficients α_1, α_2, for $\varepsilon = 0$, turn into the corresponding coefficients of the normal form of the mapping along the trajectories of the relay system (1.1).

THEOREM 15.1. *Suppose we have conditions* $\alpha_1(0) < 0$, $\alpha_2(0) > 0$ *in* (15.4). *Then there exist positive constants* ε_0 *and* ν_0 *such that for* $0 < \varepsilon \leq \varepsilon_0$,

$0 < \nu \leq \nu_0$ mapping (15.4) has an exponentially stable cycle with period 2 in the neighborhood of the unstable zero fixed point.

Theorem 15.1 can be proved by means of a direct verification with inequality (15.5) taken into account.

15.2. Bifurcation of Spatially Nonhomogeneous Relaxation Cycles

Let us consider the two-component parabolic problem

$$\varepsilon \frac{\partial x}{\partial t} = D_1 \frac{\partial^2 x}{\partial s^2} + f(x,y), \qquad \frac{\partial x}{\partial s}\bigg|_{s=0} = \frac{\partial x}{\partial s}\bigg|_{s=1} = 0, \qquad (15.6)$$

$$\frac{\partial y}{\partial t} = (D_2(\varepsilon) - \nu)\frac{\partial^2 y}{\partial s^2} + g(x,y), \qquad \frac{\partial y}{\partial s}\bigg|_{s=0} = \frac{\partial y}{\partial s}\bigg|_{s=1} = 0, \qquad (15.7)$$

where the right-hand sides f and g are the same as in 11.2, $0 < \nu \ll 1$, and the parameters $D_1, D_2 > 0$ are such that (see 11.2)

$$\pi^2 D_2(\varepsilon) = F(\varepsilon, \pi^2 D_1), \qquad F_0(\pi^2 D_1) > 0,$$

$$k^2 F_0(\pi^2 D_1) > F_0(k^2 \pi^2 D_1), \qquad k = 2, 3, \ldots.$$

Considering the Poincaré operator acting along the trajectories of the relaxation system (15.6), (15.7) (for which the analog of Lemma 15.1 is also valid) in the neighborhood of its homogeneous cycle and applying Theorem 13.1, we get the mapping

$$\xi \mapsto \xi + \alpha_1(\varepsilon)\nu\xi + \alpha_2(\varepsilon)\xi^3 + \Phi(\xi, \varepsilon, \nu), \qquad (15.8)$$

where the function Φ satisfies inequality (15.5) and $\alpha_1(0) = \pi^2$, $\alpha_2(0)$ are the coefficients of the mapping for the relaxation system (9.3) similar to mapping (15.8).

The following statement is a direct corollary of Theorem 13.1.

THEOREM 15.2. *Suppose that $\alpha_2(0) < 0$ in (15.8). Then there exist positive constants ε_0 and ν_0 such that for $0 < \varepsilon \leq \varepsilon_0$, $0 < \nu \leq \nu_0$ the boundary-value problem (15.6), (15.7) has two stable spatially nonhomogeneous cycles in a certain (independent of ε) neighborhood of the unstable homogeneous cycle (system (15.6), (15.7) has no other attractors in this neighborhood).*

Chapter 4

Duck-Trajectories of Relaxation Systems

In Chapter 1, for constraints having the character of generical conditions, we proved the theorem on the C^1-proximity of solutions of a multidimensional relaxation system and a relay system, i.e., considered the principal case. The results obtained allow us to judge the dynamics of a relaxation system from the properties of a limiting relay system. In this chapter, we first consider the situation when one of the most important constraints of Sec. 1, the normal switching condition, is generally violated. The peculiarity of this case is that the indicated singularity can appear in relaxation system so that it cannot be removed. It turns out that the so-called duck-trajectories appear in the system in this case. The typical representative of systems of this kind is the well-known van der Pol system with a harmonic external effect on the slow variable.

In the second part of this chapter, we consider the mechanism of generation of an ordinary relaxation cycle from an equilibrium position. This problem was considered before only for systems in the plane in which (as is shown in a number of works with the use of the technique of nonstandard analysis [16, 96]) duck-trajectories preceding a relaxation cycle appear when the equilibrium position passes through a junction point.

The explanation of this phenomenon is connected with the delay of the loss of stability in relaxation systems with one slow variable, discovered by Pontryagin, which is demonstrated, by way of a specific example, in [79]. It follows from this general theoretical proposition that the mechanism of generation of relaxation oscillations developed for two-dimensional systems must also be operative in the case of arbitrary relaxation systems with one slow variable. We show this in what follows. Here the analysis is based on the technique developed in the preceding chapters, the general technique of asymptotic integration of singularly perturbed differential equations [7], and the Bogolyubov–Mitropol'skii reduction principle [71].

In addition, by combining the reduction principle with the classical Pontryagin theorem on the cycles of weakly perturbed Hamiltonian systems [5], we solve the problem of bifurcation of duck-cycles, which makes it possible to get a more complete idea of the dynamics of a relaxation system when its equilibrium position passes through a junction point.

Then we show (and this is important in principle) that in the case of multidimensional systems with an arbitrary number of slow variables the origination of relaxation periodic oscillations is more complicated and the duck-trajectories that appear are unstable in the general case.

The presentation that follows is based on [31, 32, 43].

16. Origination of Duck-Trajectories upon the Violation of the Normal Switching Condition

16.1. Statement of the Problem and Description of the Result

For the sake of simplicity, we shall only consider a three-dimensional case in which our arguments hold true, i.e., for $x \in R$, $y \in R^2$, we shall consider the relaxation system

$$\varepsilon \dot{x} = f(x, y), \qquad \dot{y} = g(x, y), \qquad 0 < \varepsilon \ll 1, \tag{16.1}$$

in the neighborhood of a zero junction point. The generic conditions that are usually assumed mean (see Sec. 1) that the smooth right-hand sides f and g satisfy the constraints

$$f(0,0) = 0, \qquad f'_x(0,0) = 0, \qquad f''_{xx}(0,0) \neq 0, \tag{16.2}$$

$$l(g(0,0)) \neq 0, \qquad l(*) = (\mathrm{grad}_y\, f(0,0), *). \tag{16.3}$$

But we suppose here that inequality (16.3), which is Condition 1.1 of the normal switching, is generically violated, i.e.,

$$l(g(0,0)) = 0, \quad g(0,0) \neq 0, \quad \frac{d}{ds}l(g(0, sg(0,0)))|_{s=0} \neq 0. \tag{16.4}$$

Under conditions (16.2), (16.4) we reduce system (16.1) in the neighborhood of the junction point to the normal form of which we have spoken above. With this aim in view, we first make a linear change of variables in (16.1), i.e., replace x by the variable ξ, changing the direction of the x-axis in accordance with the motion of the phase point in the equation $\dot{x} = f(x, 0)$, and replace y by the variables η_1, η_2, directing the η_1-axis along the vector e_1 such that $l(e_1) = 1$ and the η_2 along the vector $e_2 = g(0,0)$. Then, performing in (16.1) the changes described in 1.4 (see [77]) that transform the surface $f = 0$ into $\eta_1 = -\xi^2$, we arrive at the normal form

$$\varepsilon \dot{\xi} = \xi^2 + \eta_1, \qquad \dot{\eta}_1 = g_1(\xi, \eta_1, \eta_2), \qquad \dot{\eta}_2 = g_2(\xi, \eta_1, \eta_2), \tag{16.5}$$

where, by virtue of condition (16.4) and the choice of coordinates η_1, η_2,

$$g_1 = \pm\eta_2 + a\xi + b\eta_1 + \ldots, \qquad g_2(0,0,\eta_2) \equiv 1, \qquad (16.6)$$

and the dots denote terms of a higher order of smallness. In what follows we suppose that in the first relation of (16.6), η_2 is taken with the minus sign. This corresponds to the negativity of the third term in (16.4). The other case can be analyzed by analogy, and we shall consider it separately.

Let us analyze a degenerate system corresponding to system (16.5). For our purpose we set $\varepsilon = 0$ in (16.5) and take ξ to be the new time. As a result we arrive at the relations

$$\eta_{10}(\xi) = -\xi^2, \qquad \frac{d\eta_{20}}{d\xi} = -\frac{2\xi g_2(\xi, -\xi^2, \eta_{20})}{g_1(\xi, -\xi^2, \eta_{20})}. \qquad (16.7)$$

Note that by virtue of the first relation in (16.6) the equation defining η_{20} for $\xi = \eta_{20} = 0$ is singular. We are interested in its solutions, smooth in the neighborhood of the point $\xi = 0$, which satisfy the condition $\eta_{20}(0) = 0$. Substituting representation $\eta_{20}(\xi) = \varkappa\xi + \ldots$ into (16.7) and taking into account properties (16.6) of the functions g_1 and g_2, we get the ramification equation

$$\varkappa(\varkappa - a) = 2 \qquad (16.8)$$

for $\varkappa$.

It turns out that the roots of Eq. (16.8) are associated with the solutions $\eta_{20}^1(\xi)$ and $\eta_{20}^2(\xi)$, smooth in the neighborhood of the point $\xi = 0$, of the differential equation for η_{20}. To prove this equation, we consider the auxiliary solution

$$\dot\xi = g_1(\xi, -\xi^2, \eta_{20}), \qquad \dot\eta_{20} = -2\xi g_2(\xi, -\xi^2, \eta_{20}). \qquad (16.9)$$

By virtue of (16.6), where the first relation is taken with the minus sign, and (16.8), its zero equilibrium position is of a saddle type. Therefore, there exist smooth functions $\eta_{20} = \eta_{20}^1(\xi)$ and $\eta_{20} = \eta_{20}^2(\xi)$ that define its stable and unstable manifolds.

Note that in the degenerate system (16.7) the phase point passes, in a finite time period, along the trajectory $\eta_{10} = -\xi^2$, $\eta_{20} = \eta_{20}^2(\xi)$, from the unstable slow-motion surface $\eta_1 = -\xi^2$, $\xi \geq 0$, to the stable one $\eta_1 = -\xi^2$, $\xi \leq 0$, and along the trajectories $\eta_{10} = -\xi^2$, $\eta_{20} = \eta_{20}^1(\xi)$ it passes from a stable surface to an unstable one, i.e., we have *duck-trajectories*. It is natural to expect that system (16.5) has trajectories of this kind.

In what follows, we shall consider system (16.5) for $|\xi|, |\eta_1|, |\eta_2| \leq q$, where q is sufficiently small. According to the results from [71], in sufficiently small neighborhoods of the points $\xi = -q$ and $\xi = q$, it has invariant integral manifolds $\eta_1 = H_1(\xi, \eta_2, \varepsilon)$ and $\eta_1 = H_2(\xi, \eta_2, \varepsilon)$, respectively, where the functions H_1, H_2, smooth with respect to the set of variables, are such that

$$H_1(\xi, \eta_2, 0) \equiv H_2(\xi, \eta_2, 0) \equiv -\xi^2.$$

In this case, the manifold $\eta_1 = H_1$ is locally exponentially stable with the coefficient of the exponent of order ε^{-1} and the second manifold is unstable with the same coefficient. It turns out that these manifolds are closely associated with the duck-trajectories, namely, with the trajectories with zero approximations $(-\xi^2, \eta_{20}^1(\xi))$ or $(-\xi^2, \eta_{20}^2(\xi))$, of system (16.5). In particular, any solution of the nonlinear boundary-value problem

$$\frac{d\eta_1}{d\xi} = \frac{\varepsilon g_1(\xi, \eta_1, \eta_2)}{\xi^2 + \eta_1}, \qquad \frac{d\eta_2}{d\xi} = \frac{\varepsilon g_2(\xi, \eta_1, \eta_2)}{\xi^2 + \eta_1}, \tag{16.10}$$

$$\eta_1|_{\xi=-q} = H_1|_{\xi=-q}, \qquad \eta_1|_{\xi=q} = H_2|_{\xi=q} \tag{16.11}$$

is a duck-trajectory.

Moreover, since the choice of the manifolds H_1 and H_2 is not unique, we can use the boundary-value problem (16.10), (16.11) to construct the duck-trajectories of system (16.5). Indeed, if there is a duck-trajectory, then one of its ends necessarily lies on a certain manifold $\eta_1 = H_1$ (since a "drop" occurs onto one of them). The other end lies on a certain manifold $\eta_1 = H_2$ (otherwise there would be a "breakoff" to one of the stable manifolds).

THEOREM 16.1. *The boundary-value problem* (16.10), (16.11) *has exactly two solutions, smoothly dependent on ε, with zero approximations*

$$\eta_{10}(\xi) = -\xi^2, \quad \eta_{20}(\xi) = \eta_{20}^1(\xi) \quad \text{and} \quad \eta_{10}(\xi) = -\xi^2, \quad \eta_{20}(\xi) = \eta_{20}^2(\xi)$$

respectively.

It follows from the proof of the theorem given below that the duck-trajectories are presented for small (in smooth metrics) perturbations of g_1 and g_2. For the sake of comparison, recall [16] that in a relaxation system in the plane duck-trajectories appearing because of the merging of integral manifolds, similar to H_1 and H_2, into a single manifold, are destroyed for arbitrarily smooth perturbations of order $\exp(-c/\varepsilon)$, $c > 0$.

16.2. Formal Asymptotics of Duck-Trajectories

Here we shall give the algorithm of asymptotic calculation of solutions of the boundary-value problem (16.10), (16.11). For definiteness, we shall always take $\eta_{20}^1(\xi)$ as $\eta_{20} = \eta_{20}(\xi)$, since all our arguments remain valid for $\eta_{20}^2(\xi)$.

Setting

$$\eta_1(\xi) = -\xi^2 + \sum_{k=1}^{\infty} \varepsilon^k a_k(\xi), \quad \eta_2(\xi) = \sum_{k=0}^{\infty} \varepsilon^k b_k(\xi), \quad b_0 = \eta_{20}(\xi), \quad (16.12)$$

in (16.10) and successively equating the coefficients in like powers of ε, we arrive at the relations

$$\frac{g_1(\xi, -\xi^2, \eta_{20})}{a_1^2} a_{k+1} = \frac{g'_{1\eta_2}(\xi, -\xi^2, \eta_{20})}{a_1} b_k + \Phi_k(\xi), \quad (16.13)$$

$$\frac{db_k}{d\xi} = \frac{g'_{2\eta_2}(\xi, -\xi^2, \eta_{20})}{a_1} b_k - \frac{g_2(\xi, -\xi^2, \eta_{20})}{a_1} a_{k+1} + \Psi_k(\xi), \quad (16.14)$$

where $a_1 = -g_1(\xi, -\xi^2, \eta_{20})/2\xi$ and the functions Φ_k, Ψ_k depend only on a_r, b_m with subscripts $r \leq k$, $m \leq k-1$. Suppose that a_r and b_m are defined for all $r \leq k$, $m \leq k-1$ and are smooth in the neighborhood of $\xi = 0$. Then the functions Φ_k, Ψ_k are also smooth. Calculating the quantity a_{k+1} from (16.13) and substituting the result into (16.14), we get the differential equation

$$\frac{db_k}{d\xi} = w(\xi) b_k + \varphi_k(\xi) \quad (16.15)$$

for b_k, where, by virtue of (16.6),

$$w(\xi) \sim -\frac{\varkappa^2}{2\xi}, \quad \varphi_k(\xi) = O\left(\frac{1}{\xi}\right) \quad \text{for} \quad \xi \to 0. \quad (16.16)$$

It follows from the asymptotic representations (16.16) that

$$\exp\left\{(H)\int_0^\xi w(s)\,ds\right\} = \gamma(\xi)/|\xi|^{\varkappa^2/2}, \quad \gamma(0) = 1,$$

where, as we know, $(H)\int$ is the Hadamard regularization and γ is a certain smooth function. Therefore, the relation

$$b_k(\xi) = \int_0^\xi \exp\left\{(H)\int_0^\sigma w(s)\,ds\right\} \varphi_k(\sigma)\,d\sigma$$

gives a unique (smooth for $\xi = 0$) solution of (16.15), and then we can use (16.13) to find the smooth function $a_{k+1}(\xi)$, and so on.

16.3. The General Scheme of the Proof

The central feature of the **proof** of Theorem 16.1 is the substantiation of the estimate
$$\sup_{-q \leq \xi, s \leq q} \|G(\xi, s, \varepsilon)\| \leq M \qquad (16.17)$$
for the Green matrix function of the linear boundary-value problem
$$\varepsilon \frac{dh_1}{d\xi} = a(\xi, \varepsilon)h_1 + b(\xi, \varepsilon)h_2,$$
$$\frac{dh_2}{d\xi} = c(\xi, \varepsilon)h_1 + d(\xi, \varepsilon)h_2, \qquad (16.18)$$
$$h_1(-q, \varepsilon) = \delta_1(\varepsilon)h_2(-q, \varepsilon), \qquad h_1(q, \varepsilon) = \delta_2(\varepsilon)h_2(q, \varepsilon), \qquad (16.19)$$
that results from the linearization of the boundary-value problem (16.10), (16.11) on the intervals of series (16.12) of lengths $n+1$ and n, $n \geq 3$, and the subsequent substitution $h_1/\varepsilon \to h_1$. In (16.17), the symbol $\| * \|$ is a matrix norm in R^2 and the same letter M here and in the sequel denotes certain positive constants, independent of ε, whose exact values are of no interest to us.

Suppose we have found estimate (16.17). Then, setting
$$\eta_1 = -\xi^2 + \sum_{k=1}^{n+1} \varepsilon^k a_k(\xi) + \varepsilon h_1, \qquad \eta_2 = \eta_{20}(\xi) + \sum_{k=1}^{n} \varepsilon^k b_k(\xi) + h_2,$$
in (16.10), (16.11), we arrive at the nonlinear boundary-value problem
$$\varepsilon \frac{dh_1}{d\xi} = a(\xi, \varepsilon)h_1 + b(\xi, \varepsilon)h_2 + \Delta_1(\xi, \varepsilon, h_1, h_2) + \Delta_2(\xi, \varepsilon),$$
$$\frac{dh_2}{d\xi} = c(\xi, \varepsilon)h_1 + d(\xi, \varepsilon)h_2 + \Delta_3(\xi, \varepsilon, h_1, h_2) + \Delta_4(\xi, \varepsilon), \qquad (16.20)$$
$$h_1(-q, \varepsilon) = \delta_1(\varepsilon)h_2(-q, \varepsilon) + \Omega_1(h_2(-q, \varepsilon), \varepsilon) + \Omega_2(\varepsilon),$$
$$h_1(q, \varepsilon) = \delta_2(\varepsilon)h_2(q, \varepsilon) + \Omega_3(h_2(q, \varepsilon), \varepsilon) + \Omega_4(\varepsilon), \qquad (16.21)$$
where the functions Δ_j, Ω_k, $j, k = 1, 2, 3, 4$, smooth with respect to the collection of all their variables, are such that
$$\max_{-q \leq \xi \leq q} |\Delta_2(\xi, \varepsilon)|, \quad \max_{-q \leq \xi \leq q} |\Delta_4(\xi, \varepsilon)|, \quad |\Omega_2|, |\Omega_4| \leq M\varepsilon^n, \qquad (16.22)$$
and Δ_1, Δ_3, Ω_1, Ω_3 vanish together with their first derivatives with respect to h_1, h_2 when $h_1 = h_2 = 0$. Next, using the substitution
$$h_2(\xi, \varepsilon) + \frac{1}{\delta_1(\varepsilon)}(\Omega_1(h_2(\xi, \varepsilon), \varepsilon) + \Omega_2(\varepsilon))\frac{q - \xi}{2q} +$$
$$+ \frac{1}{\delta_2(\varepsilon)}(\Omega_3(h_2(\xi, \varepsilon), \varepsilon) + \Omega_4(\varepsilon))\frac{q + \xi}{2q} \to h_2$$

to transfer the nonhomogeneity from the boundary conditions (16.21) to the right-hand sides of (16.20), we pass, with the aid of the Green function of boundary-value problem (16.18), (16.19), from a system of differential equations to an integral equation. The estimate

$$\max_{-q \le \xi \le q} \frac{1}{\varepsilon} \int_{-q}^{q} \|G(\xi, s, \varepsilon)\| \, ds \le M/\varepsilon$$

that follows from (16.22) and properties (16.22) allow us to infer that the right-hand side of the resulting integral equation maps a certain ball of the space $C[-q, q]$ with radius of order ε^{n-1} into itself and is a contraction mapping.

Thus, the proof of the theorem reduces to the derivation of inequality (16.17). In this connection, we shall discuss in greater detail the properties of the boundary-value problem (16.18), (16.19). Its coefficients are obviously smooth with respect to the set of variables and, by virtue of properties (16.6) of the functions g_1, g_2 and the asymptotic formula

$$\frac{\partial H_1}{\partial \eta_2} = -\varepsilon g'_{1\eta_2}(\xi, -\xi^2, \eta_2)/2\xi + O(\varepsilon^2),$$

$$\frac{\partial H_2}{\partial \eta_2} = -\varepsilon g'_{1\eta_2}(\xi, -\xi^2, \eta_2)/2\xi + O(\varepsilon^2),$$

the equalities

$$a(\xi, 0) = 2\varkappa\xi + \ldots, \quad b(0, 0) = -\varkappa, \quad c(0, 0) = -\varkappa^2, \quad d(0, 0) = 0, \quad (16.23)$$

$$\delta_1(0) = g'_{1\eta_2}(-q, -q^2, \eta_{20}(-q))/2q, \quad (16.24)$$

$$\delta_2(0) = -g'_{1\eta_2}(q, -q^2, \eta_2(q))/2q \quad (16.25)$$

hold true. Suppose, furthermore, that $h_{11}(\xi, \varepsilon)$, $h_{21}(\xi, \varepsilon)$ and $h_{12}(\xi, \varepsilon)$, $h_{22}(\xi, \varepsilon)$ are two solutions of (16.18) with the initial conditions

$$\begin{aligned} h_{11}(-q, \varepsilon) &= \delta_1(\varepsilon), & h_{21}(-q, \varepsilon) &= 1; \\ h_{12}(q, \varepsilon) &= \delta_2(\varepsilon), & h_{22}(q, \varepsilon) &= 1 \end{aligned} \quad (16.26)$$

respectively. It will be shown later that they are linearly independent, and, therefore, for the Green function of the boundary-value problem (16.18), (16.19), we have the formula

$$G(\xi, s, \varepsilon) = \begin{cases} U(\xi, \varepsilon) P_1 C(s, \varepsilon), & -q \le \xi \le s, \\ -U(\xi, \varepsilon) P_2 C(s, \varepsilon), & s < \xi \le q, \end{cases} \quad (16.27)$$

where the indicated solutions of (16.18) serve as columns of the matrix U, $C(s,\varepsilon) = U^{-1}(s,\varepsilon)$, and

$$P_1 = \begin{pmatrix} 1 & 0 \\ 0 & 0 \end{pmatrix}, \quad P_2 = \begin{pmatrix} 0 & 0 \\ 0 & 1 \end{pmatrix}.$$

It follows from (16.27) that, in order to substantiate estimate (16.17), we must consider the asymptotic properties of the solutions of system (16.18) with the initial conditions (16.26). Below we construct the required asymptotics separately on the segments

$$-q \leq \xi \leq -\varepsilon^\lambda, \quad -\varepsilon^\lambda \leq \xi \leq \varepsilon^\lambda, \quad \varepsilon^\lambda \leq \xi \leq q,$$

where $0 < \lambda < 1/2$, and then sew them together at the points $\xi = -\varepsilon^\lambda$ and $\xi = \varepsilon^\lambda$. It should be immediately noted that it is sufficient to construct the asymptotics of the first of the indicated solutions since, by virtue of (16.23)–(16.25), we can get the asymptotics of the second solution by means of the substitution of $-\xi$ for ξ.

16.4. Asymptotic Integration of a System of Variational Equations on a Duck-Trajectory

It follows from [71] that for $-q \leq \xi \leq -q_1 < 0$ system (16.18) has an exponentially stable (with the coefficient of the exponent of order ε^{-1}) invariant integral manifold

$$h_1 = \omega(\xi,\varepsilon)h_2. \tag{16.28}$$

Here ω is a function smooth with respect to the collection of variables. Next, since system (16.18) is the linearization of system (16.7), it follows that

$$\omega = \varepsilon^{-1}\frac{\partial H_1}{\partial \eta_2} + O(\varepsilon^n), \tag{16.29}$$

where the derivative is taken for $\eta_2 = \eta_{20}(\xi) + \sum_{k=1}^n \varepsilon^k b_k(\xi)$. Relation (16.29) implies that

$$\omega(-q,\varepsilon) - \delta_1(\varepsilon) = O(\varepsilon^n). \tag{16.30}$$

It follows from (16.26) that with the accuracy indicated in (16.30) the initial condition of the solution $h_{11}(\xi,\varepsilon)$, $h_{21}(\xi,\varepsilon)$ lies on the integral (16.28). This circumstance allows us to reduce the problem to the integration of the scalar linear equation which we obtain from (16.18) if we take into account (16.28). Therefore, our immediate aim is to construct the asymptotics of the function

ω. At the same time we shall show that manifold (16.28) can be extended up to the point $\xi = q$.

Substituting (16.28) into (16.18), we get the equation

$$\varepsilon \frac{d\omega}{d\xi} - a(\xi,\varepsilon)\omega = -\varepsilon c(\xi,\varepsilon)\omega^2 - \varepsilon d(\xi,\varepsilon)\omega + b(\xi,\varepsilon) \tag{16.31}$$

for determining the function ω. Setting in (16.31)

$$\omega = \sum_{k=0}^{\infty} \varepsilon^k \omega_k(\xi), \tag{16.32}$$

where

$$\omega_0(\xi) = -b(\xi,0)/a(\xi,0) = \frac{1}{2\xi} + \ldots, \tag{16.33}$$

and equating the coefficients of like powers of ε, we can find the functions ω_k with the use of the recurrent sequence of algebraic equations of the form

$$- a(\xi,0)\omega_k = R_k(\xi), \tag{16.34}$$

where R_k depends only on ω_m with subscripts $m \leq k-1$. A more detailed analysis of the structure of the right-hand side of (16.34) and the use of induction lead to the asymptotic expansions

$$\omega_k(\xi) = \frac{1}{\xi^{2k+1}} \sum_{m=0}^{\infty} \varphi_{km}\xi^m, \qquad \xi \to -0. \tag{16.35}$$

We omit the simple calculations here.

By virtue of the generic properties of manifold (16.28), the asymptotic representation (16.32) is valid for $-q \leq \xi \leq -q_1 < 0$. To prove its validity up to the point $\xi = -\varepsilon^\lambda$, $0 < \lambda < 1/2$, we set

$$\omega = \sum_{k=0}^{N} \varepsilon^k \omega_k + z, \qquad N \leq n-1,$$

in (16.31) and pass to the integral equation

$$z = \omega(-q,\varepsilon) - \sum_{k=0}^{N} \varepsilon^k \omega_k(-q) +$$

$$+ \frac{1}{\varepsilon} \int_{-q}^{\xi} \exp\left\{\frac{1}{\varepsilon}\int_s^{\xi} a(\sigma,\varepsilon)\,d\sigma\right\} (\varepsilon F_1(s,\varepsilon)z +$$

$$+ \varepsilon F_2(s,\varepsilon)z^2 + F_3(s,\varepsilon))\,ds. \tag{16.36}$$

Here the functions F_1, F_2 are bounded uniformly with respect to the set of variables, and for F_3, by virtue of (16.35), we have the estimate

$$\max_{-q \leq \xi \leq -\varepsilon^\lambda} |F_3(\xi, \varepsilon)| \leq M\varepsilon^{(1-2\lambda)(N+1)}.$$

This inequality and the estimate

$$\max_{-q \leq \xi \leq -\varepsilon^\lambda} \frac{1}{\varepsilon} \int_{-q}^{\xi} \exp\left\{\frac{1}{\varepsilon} \int_s^\xi a(\sigma, \varepsilon)\, d\sigma\right\} ds \leq M\varepsilon^{-\lambda}$$

being verified allow us to infer that the integral operator generated by the right-hand side of (16.36) maps a certain ball of the space $C[-q, -\varepsilon^\lambda]$ with radius of order $\varepsilon^{(1-2\lambda)(N+1)-\lambda}$ into itself and is a contraction operator there. It remains to use contraction mappings that lead to the following statement.

LEMMA 16.1. *Uniformly with respect to* $-q \leq \xi \leq -\varepsilon^\lambda$, $0 < \lambda < 1/2$,

$$\omega(\xi, \varepsilon) = \sum_{k=0}^{N} \varepsilon^k \omega_k(\xi) + O(\varepsilon^{(1-2\lambda)(N+1)-\lambda}). \tag{16.37}$$

Let us show that the integral manifold (16.28) can be extended to the point $\xi = \varepsilon^\lambda$. For $-\varepsilon^\lambda \leq \xi \leq \varepsilon^\lambda$, we make the substitutions $\xi = \mu u$, $\omega = \mu^{-1} v$, $\mu = \varepsilon^{1/2}$ in (16.31), which reduces it to the form

$$\frac{dv}{du} = \mu^{-1} a(\mu u, \mu^2) v - c(\mu u, \mu^2) v^2 - \mu d(\mu u, \mu^2) v + b(\mu u, \mu^2). \tag{16.38}$$

Then we set

$$v = \sum_{k=0}^{\infty} \mu^k v_k \tag{16.39}$$

in (16.38) and equate the coefficients of like powers of μ. As a result, we obtain the recurrent sequence of differential equations

$$\frac{dv_0}{du} = \varkappa^2 v_0^2 + 2\varkappa u v_0 - \varkappa, \tag{16.40}$$

$$\frac{dv_k}{du} = 2\varkappa[\varkappa v_0 + u]v_k + V_k(u), \quad k \geq 1, \tag{16.41}$$

where the functions V_k depend only on v_m with subscripts $m \leq k-1$.

We make the substitutions $\xi = \mu u$, $\varepsilon = \mu^2$ in (16.32), take into account expansion (16.35) of the functions ω_k, and group terms with like powers of μ. It is clear that the coefficients resulting for μ^{k-1} must serve as asymptotic expansions of the functions v_k as $u \to -\infty$. Therefore, we must take as v_0

the unique solution of (16.40), which tends to zero as $u \to -\infty$, and define v_k by the relation

$$v_k = \int_{-\infty}^{u} \exp\left\{\int_s^u 2\varkappa(\varkappa v_0(\sigma) + \sigma)\,d\sigma\right\} V_k(s)\,ds. \qquad (16.42)$$

Note that, just as in Sec. 4, where we constructed the asymptotics of the trajectories of the relaxation system in the neighborhood of a junction point, the meaning of the choice of a particular solution (16.42) of Eq. (16.41) is the same here, namely, we take the solution which is closest to the bounded solution as $u \to -\infty$.

We find from (16.40) that

$$v_0 = u^{-1} \sum_{m=0}^{\infty} \frac{\alpha_{m0}}{u^{2m}} \quad \text{if} \quad u \to -\infty; \qquad v_0 = u \sum_{m=0}^{\infty} \frac{\beta_{m0}}{u^{2m}} \quad \text{if} \quad u \to +\infty, \qquad (16.43)$$

where $\alpha_{00} = 1/2$, $\beta_{00} = -2/\varkappa$. Then we establish by induction that

$$v_k = u^{k-1} \sum_{m=0}^{\infty} \frac{\alpha_{mk}}{u^{2m}} \quad \text{if} \quad u \to -\infty, \qquad (16.44)$$

$$v_k = u^{k+1} \sum_{m=0}^{\infty} \frac{\beta_{mk}}{u^{2m}} \quad \text{if} \quad u \to +\infty, \qquad (16.45)$$

for $k \geq 1$.

We shall only prove the second of these relations, since the proof of the first one is similar. We use the same letter P to denote expressions of the form

$$\sum_{m=0}^{\infty} \frac{P_m}{u^{2m}}.$$

Then we can write (16.45) as $v_k = u^{k+1} P$. In addition, we have the easily verifiable relations

$$P^2 = P, \qquad \int s^m P\,ds = u^{m+1} P, \qquad (16.46)$$

from which, with the use of integration by parts, we find that

$$\int_{-\infty}^{u} \exp\left\{\int_s^u 2\varkappa(\varkappa v_0(\sigma) + \sigma)\,d\sigma\right\} s^m P\,ds = u^{m-1} P. \qquad (16.47)$$

Suppose that relation (16.45) is valid for all $m \le k-1$. Then, substituting the expansions

$$a, b, c, d = \sum_{r=0}^{\infty} \mu^r u^r P$$

into (16.38), we get the representation $V_k = u^{k+2} P$ for the inhomogeneity in (16.41). It remains to use (16.47).

We can assign an exact meaning to the asymptotic expansion (16.39) in the same way as to expansion (16.32), namely, setting

$$v = \sum_{k=0}^{N} \mu^k v_k + z$$

in (16.38), we pass from the resulting differential equation to an integral one, inverting the corresponding linear differential operator by means of a formula similar to (16.42), and then apply the contraction mapping principle. In this way we arrive at the following statement.

LEMMA 16.2. *Uniformly with respect to u, where $-\varepsilon^{\lambda-1/2} \le u \le \varepsilon^{\lambda-1/2}$, $0 < \lambda < 1/2$,*

$$v(u, \mu) = \sum_{k=0}^{N} \mu^k v_k(u) + O(\mu^{2\lambda(N+2)-1}), \qquad \mu = \varepsilon^{1/2}, \qquad (16.48)$$

and for $\xi = -\varepsilon^\lambda$ the functions $w(\xi, \varepsilon)$ and $\mu^{-1} v(\xi/\mu, \mu)$ coincide with an accuracy to within ε^α, where $\alpha > 0$ is arbitrary.

Let us restrict the proof to a new approach, compared to the preceding lemma, namely, the sewing together of asymptotic relations (16.32) and (16.39) for the value $\xi = -\varepsilon^\lambda$. Let us consider the interval $-2\varepsilon^\lambda \le \xi \le -\varepsilon^\lambda$ on which the functions $w(\xi, \varepsilon)$ and $\mu^{-1} v(\xi/\mu, \mu)$ are defined. Suppose that

$$w(-\varepsilon^\lambda, \varepsilon) - \mu^{-1} v(-\varepsilon^{\lambda-1/2}, \mu) \sim \varepsilon^\beta, \qquad \beta > 0.$$

Then we take $\mu^{-1} v(-\varepsilon^{\lambda-1/2}, \mu)$ as the initial condition for (16.31) for $\xi = -\varepsilon^\lambda$ and integrate it in the direction of the decrease of ξ. By virtue of the linear approximation, the deviation of this solution from $w = w(\xi, \varepsilon)$ increases as $\exp(\varkappa \xi^2/\varepsilon)$, and, therefore, it will be of order $\varepsilon^\beta \exp(4\varkappa \varepsilon^{2\lambda-1})$ for $\xi = -2\varepsilon^\lambda$, i.e., it will become exponentially large. On the other hand, by virtue of (16.32), (16.39) and properties (16.35), (16.44) of the functions w_k and v_k, this deviation remains a magnitude of order ε^{β_1} for a certain β_1 throughout the interval $-2\varepsilon^\lambda \le \xi \le -\varepsilon^\lambda$. We have arrived at a contradiction.

Thus, we have extended the integral manifold (6.28) to the point $\xi = \varepsilon^\lambda$, $0 < \lambda < 1/2$. For $\varepsilon^\lambda \leq \xi \leq q$, using the substitution $\omega = \varepsilon^{-1}w$, we pass from Eq. (16.31) to the equation

$$\varepsilon \frac{dw}{d\xi} = a(\xi,\varepsilon)w - c(\xi,\varepsilon)w^2 - \varepsilon d(\xi,\varepsilon)w + \varepsilon b(\xi,\varepsilon), \qquad (16.49)$$

whose solution we seek in the form

$$w = \sum_{k=0}^{\infty} \varepsilon^k w_k(\xi), \qquad (16.50)$$

where

$$w_0 = a(\xi,0)/c(\xi,0). \qquad (16.51)$$

Substituting expansion (16.50) into (16.49) and equating the coefficients of like powers of ε, we get a recurrent sequence of algebraic equations, from which we find the unknown functions w_k. Here the algorithmic part and the substantiation of expansion (16.50) are completely analogous to the case of $-q \leq \xi \leq -\varepsilon^\lambda$, and relations (16.50) and (16.39) for $\xi = \varepsilon^\lambda$ can be sewn together in the same way as in Lemma 16.2. Therefore, we only give the final result.

LEMMA 16.3. *Uniformly with respect to* $\varepsilon^\lambda \leq \xi \leq q$, $0 < \lambda < 1/2$,

$$w = \sum_{k=0}^{N} \varepsilon^k w_k(\xi) + O(\varepsilon^{(1-2\lambda)N+1-\lambda}), \qquad (16.52)$$

where the asymptotic expansions

$$w_k(\xi) = \frac{1}{\xi^{2k-1}} \sum_{m=0}^{\infty} \psi_{km} \xi^m$$

are valid for the functions w_k *as* $\xi \to +0$.

Lemmas 16.1–16.3 make it possible to construct the asymptotics of the solution $h_{11}(\xi,\varepsilon)$, $h_{21}(\xi,\varepsilon)$ of system (16.18) with any degree of accuracy. However, we need, for our purposes, only the leading terms of this asymptotics which have
the form

$$h_{11} = [\omega_0(\xi) + O(\varepsilon^{1-3\lambda})]h_{21}, \qquad (16.53)$$

$$h_{21} = |\xi|^{-x^2/2}[h_{21,0}(\xi) + O(\varepsilon^{1-3\lambda})], \qquad (16.54)$$

where
$$h_{21,0}(\xi) = q^{\varkappa^2/2} \exp\left\{\int_{-q}^{\xi}[c(s,0)\omega_0(s) + \frac{\varkappa^2}{2s} + d(s,0)]\,ds\right\},$$

for $-q \leq \xi \leq -\varepsilon^\lambda$, $0 < \lambda < 1/3$,
the form
$$h_{11} = \mu^{-1}[v_0(u) + O(\mu^{4\lambda-1})]h_{21}, \tag{16.55}$$

$$h_{21} = c_0 \exp\left\{-(H)\int_{-\infty}^{u}\varkappa^2 v_0(s)\,ds\right\}(1 + O(\mu^{4\lambda-1})), \tag{16.56}$$

where
$$c_0 = \mu^{-\varkappa^2/2}h_{21,0}(0), \qquad u = \xi/\mu,$$

for $-\varepsilon^\lambda \leq \xi \leq \varepsilon^\lambda$, $1/4 < \lambda < 1/2$,
the form
$$h_{11} = \varepsilon^{-1}[a(\xi,0)/c(\xi,0) + O(\varepsilon^{1-\lambda})]h_{21}, \tag{16.57}$$

$$h_{21} = \exp\left\{\frac{1}{\varepsilon}\int_{0}^{\xi}a(s,0)\,ds(1 + O(\varepsilon^{1-\lambda}))\right\} \tag{16.58}$$

for $\varepsilon^\lambda \leq \xi \leq q$, $0 < \lambda < 1/2$.

As was pointed out above, the structure of the asymptotics of the solution $h_{12}(\xi,\varepsilon)$, $h_{22}(\xi,\varepsilon)$ of (16.18) results from (16.53)–(16.58) upon the substitution of $-\xi$ for ξ. This means that in this case we must consider (16.53), (16.54) for $\varepsilon^\lambda \leq \xi \leq q$ and take the function

$$h_{22,0}(\xi) = q^{\varkappa^2/2} \exp\left\{-\int_{\xi}^{q}[c(s,0)\omega_0(s) + \frac{\varkappa^2}{2s} + d(s,0)]\,ds\right\} \tag{16.59}$$

instead of $h_{21,0}(\xi)$. Furthermore, for $-\varepsilon^\lambda \leq \xi \leq \varepsilon^\lambda$ we have the relations

$$h_{12} = \mu^{-1}[-v_0(-u) + O(\mu^{4\lambda-1})]h_{22}, \tag{16.60}$$

$$h_{22} = c_0 \exp\left\{-(H)\int_{u}^{+\infty}\varkappa^2 v_0(-s)\,ds\right\}(1 + O(\mu^{4\lambda-1})), \tag{16.61}$$

where
$$c_0 = \mu^{-\varkappa^2/2}h_{22,0}(0),$$

and the formulas

$$h_{12} = \varepsilon^{-1}[a(\xi,0)/c(\xi,0) + O(\varepsilon^{1-\lambda})]h_{22}, \qquad (16.62)$$

$$h_{22} = \exp\left\{-\frac{1}{\varepsilon}\int_\xi^0 a(s,0)\,ds(1+O(\varepsilon^{1-\lambda}))\right\} \qquad (16.63)$$

are valid for $-q \leq \xi \leq -\varepsilon^\lambda$. Substituting (16.53)–(16.63) into (16.27), we make sure of the validity of estimate (16.17). We have completed the proof of Theorem 16.1.

16.5. The Case $\frac{d}{ds}l(g(0,sg(0,0)))|_{s=0} > 0$

Suppose now that η_2 is taken with the minus sign in the first relation of (16.6). Then, instead of the ramification equation (16.8), we shall have

$$\varkappa(\varkappa + a) = -2. \qquad (16.64)$$

For $|a| < 2\sqrt{2}$, the roots of Eq. (16.64) are complex and, consequently, system (16.5) has no duck-trajectories for these a.

Now if $|a| > 2\sqrt{2}$, then Eq. (16.64) has two real roots, $\varkappa_1$ and $\varkappa_2$, of the same sign, with $0 < |\varkappa_1| < \sqrt{2}$ and $|\varkappa_2| > \sqrt{2}$, $\varkappa_1 \varkappa_2 = 2$. According to the results given in [87], the root $\varkappa_1$ is always associated with the smooth solution $\eta_{20} = \eta_{20}^1(\xi)$ of (16.7), i.e., the relation $\eta_{20} = \eta_{20}^1(\xi)$ defines the integer manifold of system (16.9) in the neighborhood of the origin which corresponds to the eigenvalue $\lambda_1 = -\varkappa_2$ of the zero equilibrium position. And the eigenvalue $\lambda_2 = -\varkappa_1$ is associated with the unique smooth integer manifold $\eta_{20} = \eta_{20}^2(\xi)$ of system (16.9) when the inequalities

$$\varkappa_2/\varkappa_1 \neq n, \qquad n = 2, 3, \ldots, \qquad (16.65)$$

are satisfied, i.e., when there are no resonances. As a result, using the technique developed in the preceding subsections, we arrive at the following statement.

THEOREM 16.2. *Suppose that $|a| > 2\sqrt{2}$ and inequalities (16.65) are satisfied. Then the boundary-value problem (16.10), (16.11) has exactly two solutions. These solutions smoothly depend on ε, and their zero approximations have the forms*

$$\eta_{10}(\xi) = -\xi^2, \qquad \eta_{20}(\xi) = \eta_{20}^1(\xi) \quad \text{and} \quad \eta_{10}(\xi) = -\xi^2, \qquad \eta_{20}(\xi) = \eta_{20}^2(\xi),$$

where η_{20}^1 and η_{20}^2 are the solutions of (16.7) corresponding to the roots $\varkappa_1$ and $\varkappa_2$ of the ramification equation (16.64).

Let us discuss the topological sense of resonances (16.65). Suppose that the two-sided inequality $n + 1 < \varkappa_2/\varkappa_1 < n + 2$ is satisfied under the conditions of Theorem 16.2. In the plane $\xi = -q$ on the curve $\eta_1 = H_1(-q, \eta_2, \varepsilon)$ we consider an interval $|\eta_2 - \eta_2(\varepsilon)| \leq \delta$, where $\eta_2(\varepsilon)$ is the initial condition of the solution of the boundary-value problem (16.10), (16.11) with the zero approximation $(-\xi^2, \eta_{20}^2(\xi))$. For $\delta = \exp(-c/\varepsilon)$ and a sufficiently large $c > 0$ the solutions of system (16.5) with the initial conditions from this interval are extended up to the point $\xi = q$, with the image of this interval in the plane $\xi = q$ versally cutting the curve $\eta_1 = H_2(q, \eta_2, \varepsilon)$. In addition, the band formed by the trajectories under consideration is twisted n times, like the Möbius band, in the neighborhood of the point $\xi = 0$. This occurs because in the neighborhood of $\xi = 0$, after normalizations and removal of the terms of a higher order of smallness described in 16.4, the system of variational equations on the duck-trajectory with the initial condition $\eta_2(\varepsilon)$ reduces to Weber's equation

$$\frac{d^2v}{du^2} - u\frac{dv}{du} + \left(\frac{\varkappa_2^2}{2} - 1\right)v = 0. \qquad (16.66)$$

Every one of its nontrivial solutions has exactly n zeros. Therefore, by virtue of the linear approximation, the deviations from the duck-trajectory along the coordinates η_1, η_2, will have n alternating zeros each, and this corresponds to the twisting of the strip described above.

Note that for $\varkappa_2^2/2 = \varkappa_2/\varkappa_1 = n + 1$, Eq. (16.66) has a solution $v = H_n(u)$, where $H_n(u)$ is a Chebyshev–Hermite polynomial. Therefore, when passing through the resonance, the number of zeros of the solutions of (16.66) changes by unity, and this leads to one more twisting of the band. It follows, in particular, that for $\varkappa_2/\varkappa_1 = n$ the versatility of the intersection of the band with the manifold $\eta_1 = H_2$ is violated when $\xi = q$. Therefore, in the general case, for $\varkappa_2/\varkappa_1 = n$, system (16.5) has no duck-trajectories that would correspond to the root $\varkappa_2$ of the ramification equation (16.64) (duck-trajectories with the zero approximation $(-\xi^2, \eta_{20}^1(\xi))$ are preserved). It should be pointed out in addition that, first, for $\varkappa_2/\varkappa_1 = n$ and certain generic conditions, all solutions of (16.7), that satisfy the conditions $\eta_{20}(0) = 0$, $\eta'_{20}(0) = \varkappa_2$, have only a finite smoothness [87] and that, second, if $\varkappa_2/\varkappa_1 \to n$ because of the variation of the parameters of the system, then the coefficient of ξ^n in the Taylor expansion of the function $\eta_{20}^2(\xi)$ at the point $\xi = 0$ tends to infinity and, therefore, the curve $(-\xi^2, \eta_{20}^2(\xi))$ and the curve $(-\xi^2, \eta_{20}^1(\xi))$ merge.

17. Destruction of an Invariant Torus of van der Pol's System with Harmonic Input

17.1. The Existence and Stability of an Invariant Torus

In this section, we shall analyze van der Pol's relaxation system

$$\varepsilon \dot{x} = y - \frac{1}{3}x^3 + x, \qquad \dot{y} = -x + A\cos\varphi, \qquad \dot{\varphi} = \omega, \qquad (17.1)$$

which it is natural to consider on the cylindrical surface $(x, y, \varphi) \in R^2 \times S^1$, where S^1 is a unit circle. Here the parameters A, $\omega > 0$, $0 < \varepsilon \ll 1$. As was shown in Ch. 1, the properties of system (17.1) are closely connected with the relay system

$$\dot{y} = -x + A\cos\varphi, \qquad \dot{\varphi} = \omega, \qquad x = \begin{cases} \Phi_1(y), \\ \Phi_2(y), \end{cases} \qquad (17.2)$$

where $\Phi_1(y)$, $-\infty < y < 2/3$, $\Phi_2(y)$, $-2/3 < y < \infty$, are the roots of the equation $y = \frac{1}{3}x^3 - x$ for $|x| > 1$.

Recall the method of determining the solutions of system (17.2). For instance, for $y(0) = y_0 < 2/3$, $\varphi(0) = \varphi_0$ we take the solution $y(t)$, $\varphi(t)$ of the Cauchy problem

$$\dot{y} = -\Phi_1(y) + A\cos\varphi, \qquad \dot{\varphi} = \omega, \qquad y|_{t=0} = y_0, \qquad \varphi|_{t=0} = \varphi_0$$

as the solution of system (17.2). Since $\Phi_1(y) < -1$, there exists a first moment of time t_0 such that $y(t_0) = 2/3$. At this moment a switching occurs, i.e., in the sequel $y(t)$, $\varphi(t)$ is a solution of the Cauchy problem

$$\dot{y} = -\Phi_2(y) + A\cos\varphi, \qquad \dot{\varphi} = \omega, \qquad y|_{t=t_0} = 2/3, \qquad \varphi|_{t=t_0} = \varphi(t_0).$$

From similar consideration, among $t > t_0$ there exists a first moment of time $t_1 > t_0$ at which $y(t_1) = -2/3$. Then a switching occurs again, and so on.

Suppose that the conditions of the normal switching

$$1 + A\cos\varphi(t_{2s}) \neq 0, \qquad -1 + A\cos\varphi(t_{2s+1}) \neq 0, \qquad s = 0, 1, \ldots, \qquad (17.3)$$

are fulfilled at the switching moments t_p, $p = 0, 1, \ldots$. Then, by virtue of Theorem 1.1, the solution $x(t, \varepsilon), y(t, \varepsilon), \varphi(t, \varepsilon)$ of (17.1) with the initial conditions $x(0, \varepsilon) = x_0 \leq -1$, $y(0, \varepsilon) = y_0$, $\varphi(0, \varepsilon) = \varphi_0$ converges, as $\varepsilon \to 0$, to the solution $x(t), y(t), \varphi(t)$ of the degenerate system in the C^1-metric, except for certain small neighborhoods of the moments of time $0, t_0, t_1, \ldots$.

THEOREM 17.1. *If $A < 1$, then there exists $\varepsilon_0 > 0$, such that for all $0 < \varepsilon \leq \varepsilon_0$ system (17.1) has a globally exponentially stable two-dimensional invariant torus.*

To **prove** this theorem, we consider, in the plane $y = 0$, the Poincaré operator along the trajectories of the relay system (17.2). By virtue of conditions (17.3) and the inequality $\dot{y}|_{y=0} = -\Phi_1(0) + A\cos\varphi > 0$, it induces the smooth mapping $\varphi \mapsto f_0(\varphi)$, $f_0(\varphi + 2\pi) = f_0(\varphi) + 2\pi$, of a circle into itself which has the property

$$f_0'(\varphi) > 0. \qquad (17.4)$$

Indeed, it follows from the symmetry of system (17.2) that the mapping $\varphi \mapsto f_0(\varphi)$ is the square of the mapping $\varphi \mapsto f_1(\varphi)$, $f_1 = f_3 \circ f_2$, where f_2 is the Poincaré operator acting along the trajectories of the system under consideration from the plane $y = 0$ into the plane $y = 2/3$ and f_3 is the Poincaré operator acting from the plane $y = 2/3$ into the plane $y = 0$ after the switching. Differentiating the relation $y(f_2(\varphi), \varphi) = 2/3$, where $y(s, \varphi)$ is the solution of the Cauchy problem

$$\frac{dy}{ds} = \frac{1}{\omega}(-\Phi_1(y) + A\cos s), \qquad y|_{s=\varphi} = 0,$$

we arrive at the relation

$$f_2'(\varphi) = -\omega h(f_2(\varphi))/[1 + A\cos(f_2(\varphi))] > 0,$$

where $h = h(s)$ is the solution of the Cauchy problem

$$\frac{dh}{ds} = -\frac{1}{\omega}\Phi_1'(y(s,\varphi))h, \qquad h|_{s=\varphi} = -\frac{1}{\omega}(-\Phi_1(0) + A\cos\varphi) < 0.$$

The monotonicity of the mapping f_3 can be verified by analogy.

In the same plane $y = 0$, we shall consider the Poincaré operator $\Pi(\varepsilon)$ acting along the trajectories of the original relaxation system (17.1). By virtue of what was said above, the limiting relation

$$\lim_{\varepsilon \to 0} \Pi(\varepsilon) = \Pi_0, \qquad \Pi_0 = \begin{cases} x \mapsto \Phi_1(0), \\ \varphi \mapsto f_0(\varphi) \end{cases} \qquad (17.5)$$

holds true in the C^1-metric. This relation, together with (17.4), ensures [6] the existence of an exponentially stable invariant curve of the mapping $\Pi(\varepsilon)$, in which the mapping has the form $\varphi \to f(\varphi, \varepsilon)$, where the function $f(\varphi, \varepsilon)$ converges to $f_0(\varphi)$ in the C^1-metric as $\varepsilon \to 0$. We have proved the theorem.

Below we shall consider the mechanism of destruction of the invariant torus of system (17.1) when the parameter A passes through unity.

17.2. Duck-Trajectories of van der Pol's System

As was shown in Sec. 16, when the condition of a normal switching is violated, a relaxation system acquires duck-trajectories. We apply the technique developed there to system (17.1) for $A > 1$ and, since system (17.1) is symmetric, we restrict the consideration to singular points whose coordinate φ can be found from the relation

$$1 + A \cos \varphi = 0. \tag{17.6}$$

From (17.6) we have

$$\varphi_1 = \pi - \arccos(1/A), \qquad \varphi_2 = \pi + \arccos(1/A).$$

We shall first consider the singular point $x = -1$, $y = 2/3$, $\varphi = \varphi_1$. In its neighborhood we reduce system (17.1) to normal form by means of the substitutions $x = -1 + \psi(\xi)$, $y = 2/3 + \eta_1$, $\varphi = \varphi_1 + \eta_2$, where the function $\psi(\xi)$, $\psi(0) = 0$, can be found from the equation $\psi\sqrt{1 - \frac{1}{3}\psi} = \xi$; $\alpha\eta_2 \to \eta_2$, where $\alpha = (\sqrt{A^2 - 1}/\omega)^{1/2}$; $\beta t \to t$, where $\beta = (\omega\sqrt{A^2 - 1})^{1/2}$. We multiply the right-hand sides by $\psi'(\xi)$ and as a result obtain the system

$$\varepsilon \dot{\xi} = \xi^2 + \eta_1, \qquad \dot{\eta}_1 = g_1(\xi, \eta_1, \eta_2), \qquad \dot{\eta}_2 = g_2(\xi, \eta_1, \eta_2), \tag{17.7}$$

where the functions g_1, g_2, analytic with respect to the set of variables, are such that

$$g_1(\xi, \eta_1, \eta_2) = -\eta_2 + a\xi + b\eta_1 + \ldots, \qquad g_2(0, 0, 0) = 1. \tag{17.8}$$

Here $a = -(\omega\sqrt{A^2 - 1})^{-1/2}$ in the first relation, and the dots denote terms of a higher order of smallness.

Properties (17.8) of the functions g_1, g_2 show that we can apply Theorem 16.1, i.e., in the neighborhood of the singular point being considered, system (17.7) has two types of duck-trajectories, along some of which the motion is from the stable slow-motion manifolds to the unstable ones and along the others the motion is converse. It is natural to call a singular point of this kind a saddle point.

Let us now consider the second singular point $(-1, 2/3, \varphi_2)$ of system (17.1). In its neighborhood the normal form has form (17.7) but the first relation of (17.8) must be replaced by

$$g_1(\xi, \eta_1, \eta_2) = \eta_2 + a\xi + b\eta_1 + \ldots \tag{17.9}$$

It follows from (17.9) that for $|a| < 2\sqrt{2}$, i.e., for $A > \sqrt{1 + 1/64\omega^2}$, the roots of the ramification equation (16.64) are complex and, hence, system (17.1) has no duck-trajectories in the neighborhood of the singular point being considered.

Let $A < \sqrt{1 + 1/64\omega^2}$. Then Eq. (16.64) has two positive roots $0 < \varkappa_1 < \sqrt{2}$ and $\varkappa_2 > \sqrt{2}$, and when there are no resonances (16.65), the statement of Theorem 16.2 is valid (by analogy with what was said above, we shall call this singular point a nodal point, or node). It should also be pointed out that the relation $\varkappa_2/\varkappa_1 = n$ is associated with the value

$$A_n = \sqrt{1 + 4n/64\omega^2(n+1)^2}, \qquad n = 2, 3, \ldots, \qquad (17.10)$$

i.e., as $A \to 1 + 0$ all the resonances are succesfully run through.

17.3. Properties of an Unstable Cycle

We denote by $\Phi_+(y)$, $-2/3 < y < 2/3$, the root of the equation $y = \frac{1}{3}x^3 - x$ which is smaller than unity in absolute value. Then, on the unstable slow-motion manifold, system (17.1) for $\varepsilon = 0$ can be written as

$$\frac{dy}{d\varphi} = \frac{1}{\omega}[-\Phi_+(y) + A\cos\varphi]. \qquad (17.11)$$

A simple verification shows that for $A > 0$ the zero state of Eq. (17.11) generates an unstable cycle $y_0(\varphi, A)$ with period 2π, which is preserved for all $0 < A < 1$. Indeed, it cannot disappear since its multiplier, defined by the relation

$$\mu_0(A) = \exp\left\{ \int_0^{2\pi} \frac{d\varphi}{\omega[1 - x_0^2(\varphi, A)]} \right\}, \qquad (17.12)$$

where $x_0(\varphi, A) = \Phi_+(y_0(\varphi, A))$, exceeds unity for all A being considered.

In the coordinates (x, φ) the phase pattern of the slow-motion system

$$\dot{x} = \frac{-x + A\cos\varphi}{x^2 - 1}, \qquad \dot{\varphi} = \omega, \qquad (17.13)$$

for $A < 1$ is shown in Fig. 17.1. Note that for all $0 < A < 1$ the cycle $x_0(\varphi, A)$ remains in the domain $|x| < 1$. However, as A increases, its amplitude grows, and for $A = 1$ it has the phase pattern shown in Fig. 17.2.

Let us return to (17.1). According to the results of [3], for $0 < A < 1$ it has an unstable cycle with zero approximation $(x_0(\varphi, A), y_0(\varphi, A), \varphi)$, one

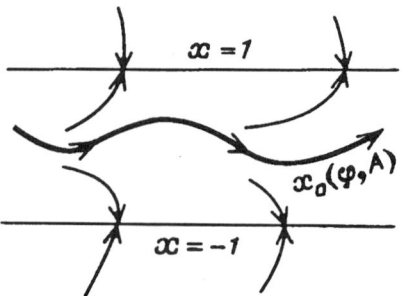

Fig. 17.1

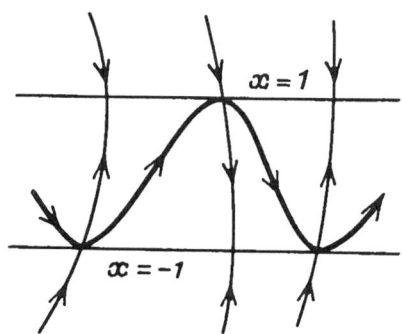

Fig. 17.2

of whose multipliers $\mu_0(A,\varepsilon)$ tends to the finite limit (17.12) as $\varepsilon \to 0$, and for the other we get the asymptotic representation

$$\mu_1(A,\varepsilon) = \mu_0^{-1}(A)\exp\left\{\frac{1}{\varepsilon}\int_0^{2\pi}\frac{1}{\omega}(1-x_0^2(\varphi,A))\,d\varphi\right\}(1+o(1)) \qquad (17.14)$$

from Liouville's formula.

We can sum up everything that was said above as the following theorem.

THEOREM 17.2. *Let $A < 1$. Then there exists $\varepsilon_0 > 0$ such that for all $0 < \varepsilon \leq \varepsilon_0$ system (17.1) has an unstable cycle with the zero approximation $(x_0(\varphi,A), y_0(\varphi,A), \varphi)$. One of the multipliers of this cycle tends to the limit (17.12) as $\varepsilon \to 0$, and asymptotic formula (17.14) holds true for the other.*

Let $1 < A < \sqrt{1+1/64\omega^2}$. For these A the phase pattern of system

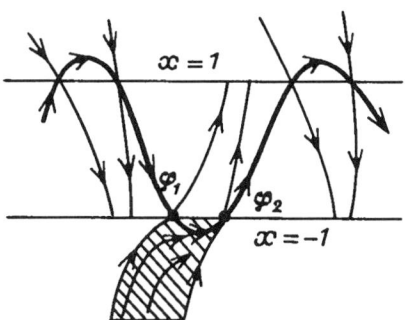

Fig. 17.3

(17.13) is shown in Fig. 17.3. However, this time the closed trajectory of the degenerate system that passes through the points $\varphi = \varphi_1$ and $\varphi = \varphi_2$ is no longer associated, in the general case, with the cycle of the original system (17.1). In fact, for the A being considered, two points of violation of the normal switching condition appear on each of the breakoff curves $x = -1$ and $x = 1$, one of which (denoted by φ_1 in Fig. 17.3) is of a saddle type and the other is of a nodal type.

At the nodal singular point $\varphi = \varphi_2$, for $A \neq A_n$ (see (17.10)), the hypothesis of Theorem 16.2 is satisfied, i.e., we have two duck-trajectories, and for $A = A_n$, in the general case, system (17.1) has no duck-trajectories that would correspond to the leading direction of the zero equilibrium position of system (16.9). As the parameter A increases from A_{n+1} to A_n, the duck-trajectory with the zero approximation $(-\xi^2, \eta_{20}^2(\xi))$ (see 16.5) randomly runs in the sector hatched in Fig. 17.3. If the cycle of the degenerate system was associated with that of system (17.1), it would necessarily be a duck-cycle including the duck-trajectory with the zero approximation $(-\xi^2, \eta_{20}^2(\xi))$. But the latter situation is impossible because of the dynamics of this duck-trajectory described above.

Note that the dissipativeness of system (17.1) and the Brouwer's fixed-point theorem imply that for all A and ω it has a cycle with period $2\pi/\omega$. From this fact and from what we have said above, we infer that for $A > 1$ this cycle is a duck-cycle with an intricately constructed zero approximation.

According to the results of 17.2, for $A > \sqrt{1 + 1/64\omega^2}$, system (17.1) has no duck-trajectories at the singular point $\varphi = \varphi_2$, and the phase pattern of the degenerate system (17.13) corresponding to it is shown in Fig. 17.4.

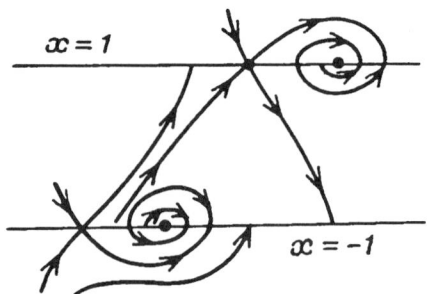

Fig. 17.4

17.4. Destruction of an Invariant Torus

Let us show that for $A > 1$ system (17.1) does not have the invariant torus of which we spoke in Theorem 17.1. Assuming the contrary, we shall consider the neighborhood of the singular point $(-1, 2/3, \varphi_1)$ and take a piece of the torus surface as the manifold $\eta_1 = H_1$, appearing in Theorem 16.1, and the manifold which contains a duck-trajectory that passes from the chosen stable manifold through all the unstable manifolds as $\eta_1 = H_2$ (the zero approximation of this duck-trajectory is shown in Fig. 17.2). But then, the Poincaré operator $\Pi(\varepsilon)$, defined in the plane $\xi = -q$, along the trajectories of system (17.1) on the curve $\eta_1 = H_1(-q, \eta_2, \varepsilon)$ has a point of discontinuity $\eta_2(\varepsilon)$, where $\eta_2(\varepsilon)$ differs from the initial condition of the duck-trajectory described above by a value of order $\exp(-\delta/\varepsilon)$, $\delta > 0$. This is due to the fact that for the trajectory of system (17.1) with initial conditions exponentially close to $\eta_2(\varepsilon)$, taken on the curve $\eta_1 = H_1(-q, \eta_2, \varepsilon)$, all three possibilities are realized, namely, having passed a certain distance in the neighborhood of the indicated duck-trajectory, the trajectory can jump to one of the two stable manifolds or continue its motion along the duck-trajectory.

Thus, the mapping $\Pi(\varepsilon)$ has a point of discontinuity on the assumed invariant curve. We have arrived at a contradiction, which proves the validity of the following statement.

THEOREM 17.3. *Let $A > 1$. Then there exists $\varepsilon_0 > 0$ such that for all $0 < \varepsilon \leq \varepsilon_0$ system (17.1) has no invariant torus, of which we spoke in Theorem 17.1.*

This theorem, in conjunction with what was presented in 17.2 and 17.3, allows us to give the following visual mechanism of the destruction of the torus: as $A \to 1-0$, the invariant torus is approached by an unstable cycle the motion along which is perpendicular to the motion on the torus, and, therefore, the torus loses smoothness and becomes discontinuous, with the discontinuities occurring on the duck-trajectories. The question concerning a more detailed mechanism of the destruction of the torus and the corresponding dynamic effects remains open.

Also open is the question concerning the nature of the attractor that appears in its place. In particular, the following problem is of a certain interest: is it possible for system (17.1) to have a strange attractor containing duck-trajectories?

18. Pontryagin Delay Phenomenon and Stable Duck-Cycles

18.1. The Main Assumptions

For $x \in R^n$, $y \in R$, we consider the singularly perturbed system

$$\varepsilon \dot{x} = f(x, y), \qquad \dot{y} = g(x, y, \mu), \qquad 0 < \varepsilon \ll 1, \qquad (18.1)$$

depending on the auxiliary small parameter μ, whose right-hand sides are assumed to be smooth with respect to the set of variables. We suppose that the equation $f = 0$ defines in R^{n+1} a smooth curve Γ which disintegrates into nonintersecting parts Γ_{1-}, Γ_{10}, Γ_{+}, Γ_{20}, Γ_{2-}. By definition, $(x, y) \in$ $\in \Gamma_{1-} + \Gamma_{2-}$ if all eigenvalues of the matrix f'_x lie on the left of the imaginary axis; $(x, y) \in \Gamma_{+}$ if exactly one simple eigenvalue of the matrix f'_x is positive and all the other eigenvalues lie in the left complex half-plane as before. At the points Γ_{10} and Γ_{20} the matrix f'_x has a simple zero eigenvalue and the real parts of all the other eigenvalues are negative. We suppose that the segments Γ_{1-}, Γ_{+}, Γ_{2-} of the curve Γ are defined by the relations $x = \Phi_{1-}(y)$, $x = \Phi_{+}(y)$, $x = \Phi_{2-}(y)$ respectively, where the smooth functions Φ_{1-}, Φ_{+}, Φ_{2-} are defined for $(-\infty, 0)$, $(y_0, 0)$, $(y_0, +\infty)$ (see Fig. 18.1) and the surface $g(x, y, 0) = 0$ has a unit intersection point Γ_{10} with the curve Γ, with $g > 0$ on Γ_{1-} and $g < 0$ at the other points of the curve Γ.

The further restrictions are mainly connected with the system of fast-motion equations

$$\dot{\varepsilon x} = f(x, y). \qquad (18.2)$$

By virtue of the results [71], for $y = 0$, in the neighborhood of the multiple zero equilibrium position it has an exponentially stable one-dimensional

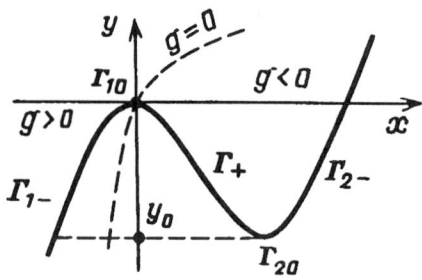

Fig. 18.1

Fig. 18.2

invariant integral manifold on which the motion is described by a certain scalar equation.

CONDITION 18.1. The right-hand side of this scalar equation begins with a quadratic term and the unstable "whisker" emanating from the point $x = 0$ tends to the exponentially stable equilibrium position $x = \Phi_{2-}(0)$ as $t \to \infty$. The same situation is observed at the point Γ_{20}.

CONDITION 18.2. At the points Γ_+, as $t \to \infty$, the phase point tends from the neighborhood of the stable equilibrium position $x = \Phi_+(y)$ of system (18.2) to the equilibrium position $\Phi_{1-}(y)$ along one unstable direction and to equilibrium position $x = \Phi_{2-}(y)$ along the other (for $n = 2$ the situation we have described is shown in Fig. 18.2 when Φ_{1-} is a focal point and Φ_{2-} is a nodal point).

Let us denote by e, h, $(e, h) = 1$ the eigenvectors corresponding to the zero eigenvalues of the matrices $A = f'_x(0, 0)$ and A^*, supposing, respectively, that the direction of the vector e coincides with that of the unstable "whisker" of system (18.2) for $y = 0$ indicated above.

CONDITION 18.3. We suppose that $(f'_y(0, 0), h) \neq 0$ and assume that a similar inequality is satisfied at the point Γ_{20}.

CONDITION 18.4. We suppose that $g'_x(0,0,0)e \neq 0$ (here a row vector is applied to a column vector).

CONDITION 18.5. We suppose that $g'_\mu(0,0,0) \neq 0$.

The first three Conditions 18.1–18.3 are similar to the conditions formulated in Sec. 1. Condition 18.4 ensures the generality of intersection of the surface $g(x,y,0) = 0$ with the curve Γ and Condition 18.5 ensures that the equilibrium position of system (18.1) passes through the point Γ_{10} with nonzero velocity with the variation of μ.

18.2. An Auxiliary Lemma

According to the results of [71], outside of the small neighborhoods of the points Γ_{10} and Γ_{20}, marked in Fig. 18.1, system (18.1) has an exponentially stable invariant manifold $x = M_{1-}(y,\varepsilon)$ and a dichotomous (with one unstable direction) manifold $x = M_+(y,\varepsilon)$, where the smooth functions M_{1-}, M_+ are such that $M_{1-}(y,0) = \Phi_{1-}(y)$, $M_+(y,0) = \Phi_+(y)$. Note that these manifolds are not uniquely defined. This is inessential for the sequel. We shall only suppose that M_{1-} is defined for the values of y which are somewhat smaller than y_0 and later we will have to extend the manifold M_+ to the asymptotically small neighborhood of Γ_{20}. Here we shall prove the existence of the value $\mu_*(\varepsilon)$ of the parameter μ for which the manifolds M_{1-} and M_+ merge in the neighborhood of the origin.

We first reduce system (18.1) in a certain small neighborhood of the origin to the normal form that we have repeatedly mentioned. For this purpose, instead of the variable x we introduce variables (ξ_1, ξ_2), $\xi_1 \in R$, $\xi_2 \in R^{n-1}$, where the coordinate ξ_1 corresponds to the direction of the vector e and the group of coordinates ξ_2 corresponds to the directions orthogonal to the vector h. In accordance with Lemma 3.1, system (18.1) has an exponentially stable (with the coefficient of the exponent of order ε^{-1}) invariant integer manifold

$$\xi_2 = V(\xi_1, y, \varepsilon, \mu), \qquad |\xi_1|, |y| \leq q, \qquad (18.3)$$

where the function V smoothly depends on all the arguments and $q > 0$ is sufficiently small. Substituting (18.3) into (18.1), we get a system of differential equations on the integer manifold

$$\varepsilon\dot{\xi} = f(\xi, \eta, \varepsilon, \mu), \qquad \dot{\eta} = g(\xi, \eta, \varepsilon, \mu), \qquad \xi = \xi_1, \qquad \eta = y.$$

Then, by means of the substitutions described in 4.1, we reduce the last system to normal form (cf. (4.4))

$$\varepsilon\dot{\xi} = \xi^2 + \eta, \qquad \dot{\eta} = \gamma(\xi, \eta, \varepsilon, \mu), \tag{18.4}$$

where

$$\gamma(\xi, \eta, 0, 0) = -\xi + b\eta + \ldots, \qquad \gamma'_\mu(0, 0, 0, 0) \neq 0, \tag{18.5}$$

and denote by dots the terms of a higher order of smallness.

LEMMA 18.1. *There exists a unique value $\mu_*(\varepsilon)$ of the parameter μ for which the integral manifolds $x = M_{1-}(y, \varepsilon)$ and $x = M_+(y, \varepsilon)$ of system (18.1) merge in the neighborhood of the point Γ_{10}, the asymptotic expansion*

$$\mu_*(\varepsilon) = \sum_{k=1}^\infty \mu_k \varepsilon^k \tag{18.6}$$

being valid.

Obviously, the **proof** reduces to the substantiation of the statement formulated for the normal form (18.4). We shall begin by describing the formal scheme of seeking the coefficients of series (18.6). From system (18.4) we pass to the equation

$$\frac{d\eta}{d\xi} = \varepsilon\gamma(\xi, \eta, \varepsilon, \mu)/(\xi^2 + \eta). \tag{18.7}$$

Then we take into account expansion (18.6) and seek solutions of Eq. (18.7) in the form

$$\eta = -\xi^2 + \sum_{k=1}^\infty \varepsilon^k a_k(\xi). \tag{18.8}$$

As a result, equating the coefficients in like powers of ε, we get a recurrent sequence of algebraic equations for the functions $a_k(\xi)$. The constants μ_k can be found from the condition of boundedness of the functions $a_{k+1}(\xi)$ as $\xi \to 0$. Note that the function

$$a_1(\xi) = -\gamma(\xi, -\xi^2, 0, 0)/2\xi$$

is bounded as $\xi \to 0$ by virtue of the properties of the function γ. For the coefficient $a_2(\xi)$ we have the relation

$$a_2(\xi) = -\frac{\gamma^2(\xi, -\xi^2, 0, 0) - \xi\gamma'_\xi(\xi, -\xi^2, 0, 0)\gamma(\xi, -\xi^2, 0, 0)}{8\xi^4} -$$
$$-\frac{\gamma'_\varepsilon(\xi, -\xi^2, 0, 0) + \gamma'_\mu(\xi, -\xi^2, 0, 0)\mu_1}{2\xi},$$

by means of which we find that

$$\mu_1 = -\frac{1}{\gamma'_\mu(0,0,0,0)}(\gamma'_\varepsilon(0,0,0,0) + \frac{1}{4}\gamma'_\eta(0,0,0,0) + \frac{1}{8}\gamma''_{\xi\xi}(0,0,0,0)).$$

The other coefficients of series (18.6) and (18.8) are determined by analogy.
In (18.7) we set $\mu = \mu_{k_0}(\varepsilon) + \varepsilon^{k_0-1}\delta$, where

$$\mu_{k_0}(\varepsilon) = \sum_{k=1}^{k_0-2} \mu_k \varepsilon^k, \qquad k_0 \geq 3,$$

and denote by $\eta = F_1(\xi,\varepsilon,\delta)$, $\eta = F_2(\xi,\varepsilon,\delta)$ its solutions whose initial conditions belong to the stable and unstable manifolds of system (18.4) and are defined for $\xi = -q$ and $\xi = q$ respectively. We shall show that these solutions can be extended up to the point $\xi = 0$ and the function $\delta = \delta(\varepsilon)$ can be uniquely determined from the equation

$$\varphi(\delta,\varepsilon) \equiv F_1(0,\varepsilon,\delta) - F_2(0,\varepsilon,\delta) = 0. \tag{18.9}$$

We make the substitution

$$\eta = -\xi^2 + \sum_{k=1}^{k_0-1} \varepsilon^k a_k(\xi) + \varepsilon z$$

in (18.7), which reduces it to the form

$$\varepsilon\frac{dz}{d\xi} = b(\xi,\varepsilon,\delta)z + \Phi(z,\xi,\varepsilon,\delta) + \varepsilon^{k_0-1}\Psi(\xi,\varepsilon,\delta), \tag{18.10}$$

where the functions b, Φ, Ψ, smooth with respect to the set of variables, satisfy the conditions

$$b(\xi,\varepsilon,\delta) = b_0(\xi) + O(\varepsilon), \qquad \Phi(0,\xi,\varepsilon,\delta) = \frac{\partial\Phi}{\partial z}(0,\xi,\varepsilon,\delta) \equiv 0,$$

$$b_0(\xi) = 4\xi + O(\xi^2). \tag{18.11}$$

From Eq. (18.10) we pass to the integer equation

$$z(\xi,\varepsilon,\delta) = z(-q,\varepsilon,\delta)\exp\left\{\frac{1}{\varepsilon}\int_{-q}^{\xi} b(s,\varepsilon,\delta)\,ds\right\} +$$

$$+ \frac{1}{\varepsilon}\int_{-q}^{\xi}\exp\left\{\frac{1}{\varepsilon}\int_{\tau}^{\xi} b(s,\varepsilon,\delta)\,ds\right\}[\Phi(z,\tau,\varepsilon,\delta) + \varepsilon^{k_0-1}\Psi(\tau,\varepsilon,\delta)]\,d\tau, \tag{18.12}$$

where

$$z(-q, \varepsilon, \delta) = \varepsilon^{-1}\left(F_1(-q, \varepsilon, \delta) + q^2 - \sum_{k=1}^{k_0-1} \varepsilon^k a_k(-q)\right) = O(\varepsilon^{k_0-1}). \quad (18.13)$$

Using properties (18.11) and (18.13) and the inequality

$$\max_{-q \leq \xi \leq 0} \frac{1}{\varepsilon} \int_{-q}^{\xi} \exp\left\{\frac{1}{\varepsilon} \int_{\tau}^{\xi} b(s, \varepsilon, \delta) \, ds\right\} d\tau \leq M/\sqrt{\varepsilon}, \quad M > 0, \quad (18.14)$$

we easily find out that the operator generated by the right-hand side of (18.12) maps the ball of the space $C[-q, 0]$ with radius of order $\varepsilon^{k_0-3/2}$ with center at zero into itself and is a contraction operator. Therefore, the function $z_1(\xi, \varepsilon, \delta)$, $-q \leq \xi \leq 0$, smooth with respect to the set of variables, can be uniquely defined. The function $z_2(\xi, \varepsilon, \delta)$, $0 \leq \xi \leq q$, can be constructed by analogy.

We have thus extended the solutions $\eta = F_1(\xi, \varepsilon, \delta)$ and $\eta = F_2(\xi, \varepsilon, \delta)$ of Eq. (18.7) to the point $\xi = 0$. To investigate Eq. (18.9), we take into account, in greater detail, the structure of the nonhomogeneity of Φ in (18.12):

$$\Phi = \gamma'_\mu(\xi, -\xi^2, 0, 0)\delta/a_1(\xi) + \Psi_0(\xi) + O(\varepsilon), \quad (18.15)$$

where Ψ_0 is a certain smooth function whose explicit expression is of no interest to us. Then we make the substitutions $s/\sqrt{\varepsilon} \to s$, $\tau/\sqrt{\varepsilon} \to \tau$ in the integrals appearing on the right-hand side of (18.12) and use the properties (18.11) and (18.15). As a result we arrive at the asymptotic relations

$$\varphi(\delta, \varepsilon) = \varepsilon^{k_0-1/2}(\varkappa_1 \delta + \varkappa_2) + O(\varepsilon^{k_0}), \quad (18.16)$$

$$\frac{\partial \varphi}{\partial \delta} = \varepsilon^{k_0-1/2}\varkappa_1 + O(\varepsilon^{k_0}), \quad (18.17)$$

where the value of the constants $\varkappa_2$ is inessential and

$$\varkappa_1 = 2\gamma'_\mu(0, 0, 0, 0) \int_{-\infty}^{\infty} \exp(-2s^2) \, ds \neq 0. \quad (18.18)$$

We have proved the lemma.

The proof of a statement analogous to our lemma is fragmentarily given in [93]. We give it here because we shall need the constructions used there in 18.3.

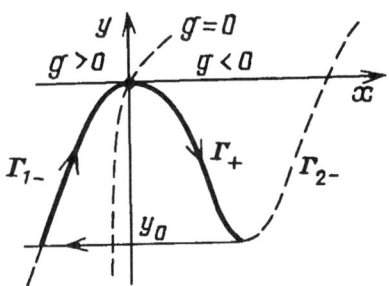

Fig. 18.3

18.3. The Conditions of Existence of Stable Duck-Cycles

Suppose that $\lambda_{1-}(y)$, $\lambda_{2-}(y)$ are the real parts of the eigenvalues of the matrix f'_x at the points of the branches Γ_{1-} and Γ_{2-}, respectively, which are the closest to the imaginary axis and $\lambda_+(y)$ is a simple positive eigenvalue of the matrix f'_x for $(x,y) \in \Gamma_+$. We introduce into consideration the balance functions

$$H_1(y) = \int_y^0 \frac{\lambda_{1-}(s)}{g(\Phi_{1-}(s), s, 0)}\, ds + \int_0^y \frac{\lambda_+(s)}{g(\Phi_+(s), s, 0)}\, ds, \qquad (18.19)$$

$$H_2(y) = \int_{y_0}^0 \frac{\lambda_{1-}(s)\, ds}{g(\Phi_{1-}(s), s, 0)} + \int_0^y \frac{\lambda_+(s)\, ds}{g(\Phi_+(s), s, 0)} + \int_y^{y_0} \frac{\lambda_{2-}(s)\, ds}{g(\Phi_{2-}(s), s, 0)}, \qquad (18.20)$$

where $y \in [y_0, 0]$, which play a principal part in the problem of the existence and stability of the cycles of system (18.1) whose zero approximations are shown in Figs. 18.3 and 18.4. We shall call the cycles shown in Figs. 18.3 and 18.4(a) headless ducks of heights y_0 and y_*, respectively, and the cycle shown in Fig. 18.4(b), a duck with head of height y_*.

In system (18.1) we set

$$\mu = \mu_*(\varepsilon) + \Delta\mu, \qquad \Delta\mu = \pm\exp(-a/\varepsilon), \qquad a > 0.$$

In this case, the manifolds M_{1-} and M_+ of (18.1) are disconnected but, since $\Delta\mu$ is small, the phenomenon of the delay of the loss of stability discovered by Pontryagin is preserved. We denote by $x(t,\varepsilon)$, $y(t,\varepsilon)$ the trajectory of system (18.1) corresponding to the unique integer manifold for $\mu = \mu_*(\varepsilon)$, assuming that the moment $t = 0$ on it is associated with $\xi = 0$ (see 18.2).

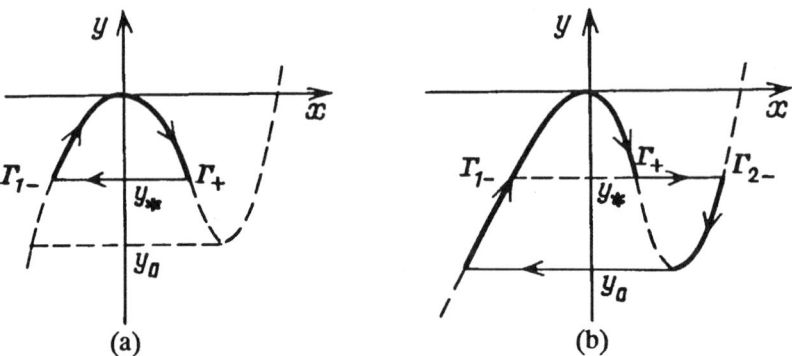

Fig. 18.4

Then we consider the Cauchy problem with zero initial conditions for the system of variational equations

$$\varepsilon \dot{h}_1 = A(t,\varepsilon)h_1 + B(t,\varepsilon)h_2,$$
$$\dot{h}_2 = C(t,\varepsilon)h_1 + D(t,\varepsilon)h_2 + K(t,\varepsilon)\Delta\mu, \quad (18.21)$$

where $A = f'_x$, $B = f'_y$, $C = g'_x$, $D = g'_y$, $K = g'_\mu$ and the derivatives are taken for $x = x(t,\varepsilon)$, $y = y(t,\varepsilon)$, $\mu = \mu_*(\varepsilon)$. It follows from the constraints imposed on system (18.1) that for $t \geq 0$ during which the point $(x(t,\varepsilon), y(t,\varepsilon))$ moves along the unstable branch Γ_+ of the curve Γ, the matrix $A(t,\varepsilon)$ has a simple nonnegative eigenvalue $\lambda(t,\varepsilon)$, $\lambda(0,0) = 0$, with the eigenvector $e(t,\varepsilon)$, $e(0,0) = e$, and its other eigenvalues lie in the left complex half-plane. This circumstance makes it possible to make a linear substitution with respect to h_1 in (18.21), which splits the matrix $A(t,\varepsilon)$, and get the asymptotic formula

$$\theta(t,\varepsilon) = (1 + o(1))\theta_0(\varepsilon)\exp\left\{\frac{1}{\varepsilon}\int_0^t \lambda(\tau,0)\,d\tau\right\}, \quad (18.22)$$

where $\lambda(\tau,0) = \lambda_+(y(\tau,0))$ and

$$\theta_0(\varepsilon) = g'_\mu(0,0,0)\Delta\mu, \quad (18.23)$$

for the variable $\theta(t,\varepsilon)$ corresponding to the direction of $e(t,\varepsilon)$. Formulas (18.22), (18.23) obviously imply the existence of a duck-trajectory of height

y_* (maybe not closed), where $y_* \in [y_0, 0)$ is a root of the equation (provided that it exists)

$$\int_0^y \frac{\lambda_+(s)}{g(\Phi_+(s), s, 0)} ds = a. \tag{18.24}$$

This is a headless duck-trajectory if $\theta_0(\varepsilon) < 0$ and a duck-trajectory with the head if $\theta_0(\varepsilon) > 0$. The following statements answer the natural question concerning the existence of closed duck-trajectories.

THEOREM 18.1. *Suppose that in system (18.1) the value of the parameter μ is such that $\mu = \mu_*(\varepsilon) + \Delta\mu$, where $\Delta\mu = \pm \exp(-a/\varepsilon)$, $a > 0$, and y_* is the root of Eq. (18.24). Then system (18.1) has a stable headless duck-cycle of height y_* for*

$$\theta_0(\varepsilon) < 0, \qquad H_1(y_*) < 0,$$

and a stable duck-cycle with head of height y_* for

$$\theta_0(\varepsilon) > 0, \qquad H_2(y_*) < 0.$$

THEOREM 18.2. *Suppose that*

$$|\mu - \mu_*(\varepsilon)| \leq \exp\left\{\frac{1}{\varepsilon}\int_0^{y_0} \frac{\lambda_+(s)}{g(\Phi_+(s), s, 0)} ds\right\}$$

in system (18.1). Then, for $H_1(y) < 0$, the duck-cycle of maximum height (of the type shown in Fig. 18.3) exists and is stable.

We begin by **proving** Theorem 18.1. We introduce into consideration a hyperplane L that generically cuts Γ_{1-} at a point with coordinate $\tilde{y} \in (y_*, 0)$ such that

$$\int_{\tilde{y}}^0 \frac{\lambda_{1-}(s)}{g(\Phi_{1-}(s), s, 0)} ds + \int_0^{y_*} \frac{\lambda_+(s)}{g(\Phi_+(s), s, 0)} ds < 0 \tag{18.25}$$

in the case of a headless duck, whereas in the case of a duck with a head $\tilde{y} \in (y_0, 0)$ is chosen such that

$$\int_{\tilde{y}}^0 \frac{\lambda_{1-}(s) ds}{g(\Phi_{1-}(s), s, 0)} + \int_0^{y_*} \frac{\lambda_+(s) ds}{g(\Phi_+(s), s, 0)} + \int_{y_*}^{y_0} \frac{\lambda_{2-}(s) ds}{g(\Phi_{2-}(s), s, 0)} < 0. \tag{18.26}$$

Note that, by virtue of the hypothesis of Theorem 18.1, we can always attain the satisfaction of inequality (18.25) or (18.26) by choosing $\tilde{y}$ sufficiently close to y_* in the case of a headless duck and close to y_0 in the case of a duck with a head. The advantage of such a choice of an intersecting hyperplane is that in this case the Poincaré operator Π along the trajectories of system (18.1) maps into itself a certain (independent of ε) neighborhood of the point of intersection of L with the duck-trajectory (otherwise this property does not exist). This follows from the fact that when inequality (18.25) (or (18.26)) is satisfied, all trajectories with the initial condition from a sufficiently small neighborhood of the intersection of L with the duck-trajectory have the same asymptotics.

Indeed, first there occurs a drop of the stable manifold M_{1-} that can be described by means of the technique from [7]. Next, using the construction presented in 18.2, we extend the trajectory up to the point $\xi = 0$. Then we estimate the deviation of the trajectory from the manifold M_+ with the use of a linear approximation. By virtue of the choice of $\tilde{y}$ and inequalities (18.25) and (18.26), the initial condition for the system of variational equations (18.21) is much smaller than $\Delta\mu$ in the norm, and this allows us to consider it to be zero.

We have thus established the existence of a duck-cycle. To prove its uniqueness and stability, we must make note of the fact that the system of variational equations on it possesses the same property as the original nonlinear system, i.e., all of its solutions with initial conditions independent of ε have the same asymptotics coincident with the asymptotics for $\dot{x}(t,\varepsilon)$, $\dot{y}(t,\varepsilon)$, where $x(t,\varepsilon)$, $y(t,\varepsilon)$ are components of the cycle. Therefore, the eigenvalues of the monodromy matrix, different from unity, tend to zero as $\varepsilon \to 0$.

Theorem 18.2 can be substantiated by analogy, the only difference being the possibility of extending the manifold M_+ to the asymptotically small neighborhood of the point Γ_{20} following from our preceding results.

18.4. Bifurcation Theorem

Since, in system (18.4), the equilibrium position loses stability in an oscillatory way, we can apply to it the well-known Andronov bifurcation theorem. However, it is clear a priori that the range of its applications is not large. Therefore, we shall carry out an analysis that will take into consideration the specific properties of singularly perturbed equations.

Carrying out the normalizing substitutions

$$\xi = \nu u, \qquad \eta = \nu^2 v, \qquad \varepsilon = \nu^2, \qquad \mu = \alpha\nu^2, \qquad t \to \nu t$$

in (18.4) we obtain, with an accuracy up to ν^2,
$$\dot{u} = u^2 + v, \qquad \dot{v} = -u + \nu(b_1 v + b_2 u^2 + b_3 + \alpha b_4), \tag{18.27}$$
where $b_1 = \gamma'_\eta$, $b_2 = \frac{1}{2}\gamma''_{\xi\xi}$, $b_3 = \gamma'_\varepsilon$, $b_4 = \gamma'_\mu$ and the derivatives are taken for zero arguments. For $\nu = 0$, we multiply the right-hand sides of system (18.27) by $2\exp(2v)$, first replacing u by $-u$. As a result it becomes Hamiltonian with the Hamilton function
$$H(u,v) = \frac{1}{2} + (u^2 + v - \frac{1}{2})\exp(2v),$$
whose level lines $H(u,v) = \delta$ are closed for $0 < \delta < 1/2$. Below, for the indicated δ we shall need the functions
$$I_1(\delta) = \iint_{\Omega_\delta} \exp(2v)\,dudv, \qquad I_2(\delta) = -\iint_{\Omega_\delta} v\exp(2v)\,dudv,$$
$$I_3(\delta) = \iint_{\Omega_\delta} u^2 \exp(2v)\,dudv, \tag{18.28}$$
where the domain Ω_δ is bounded by the curve $H(u,v) = \delta$.

THEOREM 18.3. *System (18.27) has a stable (unstable) cycle with the zero approximation $H(u,v) = \delta_0$ if*
$$I(\delta_0) = 0, \qquad I'(\delta_0) < 0, \qquad (>0),$$
where
$$I(\delta) = (b_1 + 2b_3 + 2b_4\alpha)I_1 - 2b_1 I_2 + 2b_2 I_3. \tag{18.29}$$

The **proof** follows immediately from the classical Pontryagin theorem [5]. It is useful to bear in mind that
$$I_1(\delta) \sim \pi\delta, \quad I_2(\delta) \sim \frac{\pi}{3}\delta^{3/2}, \quad I_3(\delta) \sim \frac{\pi}{3}\delta^{3/2}, \quad \delta \to +0; \tag{18.30}$$
$$I'_1(\delta) \sim \sqrt{2}\left(\ln\frac{1}{\frac{1}{2}-\delta}\right)^{1/2}, \qquad I'_2(\delta) \sim \frac{1}{3\sqrt{2}}\left(\ln\frac{1}{\frac{1}{2}-\delta}\right)^{3/2},$$
$$I'_3(\delta) \sim \frac{1}{3\sqrt{2}}\left(\ln\frac{1}{\frac{1}{2}-\delta}\right)^{3/2}, \qquad \delta \to \frac{1}{2} - 0. \tag{18.31}$$

It follows from (18.28)–(18.31) and from Theorem 18.1–18.3 that a situation is possible when the relaxation system (18.10) simultaneously has a stable ordinary relaxation cycle and the stable cycle described by Theorem 18.3.

Thus, using the standard methods of the relaxation oscillation theory, we have managed to consider completely enough the problem of stable duck-cycles of relaxation systems with one slow variable.

18.5. Diffusion Instability of Duck-Cycles

For $x \in R$, we take into account the diffusion terms in (18.1) and set Neumann's boundary conditions. As a result we get the relaxation parabolic system ($D_1, D_2 > 0$)

$$\varepsilon \frac{\partial x}{\partial t} = D_1 \Delta x + f(x, y), \qquad \left.\frac{\partial x}{\partial \nu}\right|_{\partial \Omega} = 0, \qquad (18.32)$$

$$\frac{\partial y}{\partial t} = D_2 \Delta y + g(x, y, \mu), \qquad \left.\frac{\partial y}{\partial \nu}\right|_{\partial \Omega} = 0. \qquad (18.33)$$

Suppose that in (18.32), (18.33) the functions f and g are such that we can apply Theorem 18.1 and the parameter $\mu = \mu(\varepsilon)$ is chosen such that the system of ordinary differential equations corresponding to the boundary-value problem (18.32), (18.33) has the exponentially orbitally stable headless duck-cycle shown in Fig 18.5. Let us pose the question concerning its stability in the framework of the boundary-value problem (18.32), (18.33).

Recall (see 11.2) that this is equivalent to investigating the properties of the stability of the linear $T(\varepsilon)$-periodic system ($T(\varepsilon)$ is the period of the duck-cycle $x(t, \varepsilon)$, $y(t, \varepsilon)$)

$$\begin{aligned} \varepsilon \dot{h}_1 &= -z_1 h_1 + a(t, \varepsilon) h_1 + b(t, \varepsilon) h_2, \\ \dot{h}_2 &= -z_2 h_2 + c(t, \varepsilon) h_1 + d(t, \varepsilon) h_2 \end{aligned} \qquad (18.34)$$

with arbitrary parameters $z_1, z_2 \geq 0$, which is derived in the same way as system (11.7). Therefore, in what follows, the object of our analysis will be system (18.34).

Let us suppose that by virtue of the zero approximation $(x_0(t), y_0(t))$ of the duck-cycle we can choose time such that the zero moment is associated with the point A (see Fig. 18.5), the moment $t_0 > 0$ is associated with the point B, and the moment $t = T_0$, where T_0 is the zero approximation of $T(\varepsilon)$, is associated with the point D. As in 11.2, we denote by $a_0(t)$, $b_0(t)$, $c_0(t)$, $d_0(t)$ the zero approximations of the coefficients of system (18.34).

Note that on the section AB of the duck-cycle the coefficient $a_0(t)$ is negative, on the section BD it is positive, and at the point t_0 it has a simple zero. Therefore, if the parameter z_1 in (18.34) is so large that

$$z_1 > \max_{t_0 \leq t \leq T_0} a_0(t), \qquad (18.35)$$

then we can apply to (18.34) the standard technique of asymptotic integration from [7], which leads to the following statement.

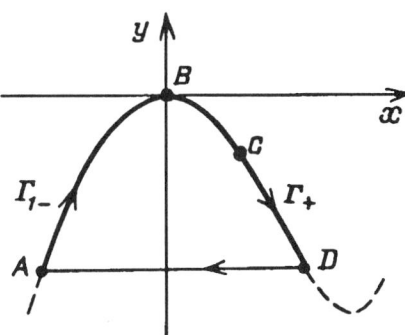

Fig. 18.5

THEOREM 18.4. *Suppose that in system* (18.34) *the parameters* z_1, z_2 *are such that inequality* (18.35) *is satisfied and*

$$z_2 - F_0(z_1) > 0, \quad (< 0),$$

where the function F_0 *is defined by* (11.10). *Then its two multipliers are smaller than unity in absolute value (there exists a multiplier larger than unity in absolute value).*

It follows from Theorem 18.4 that if the diffusion coefficient D_1 in (18.32) is sufficiently large, then the answer to the question concerning the stability of a homogeneous duck-cycle can be formulated just as for an ordinary relaxation cycle (see Theorem 11.3).

Suppose now that inequality (18.35) is violated. We shall assume for simplicity that on the interval (t_0, T_0) the function $a_0(t) - z_1$ has a single simple zero t_1 and the moment t_1 is associated with the point C in Fig. 18.5. Then, under certain generic conditions, the homogeneous duck-cycle of the boundary-value problem (18.32), (18.33) is always unstable.

THEOREM 18.5. *When the conditions formulated in the preceding paragraph are fulfilled and the inequalities*

$$c_0(T_0 - 0) \neq 0, \quad na'_0(t_1) + b_0(t_1)c_0(t_1) \neq 0, \quad n = 0, 1, \ldots, \qquad (18.36)$$

are satisfied, system (18.34) *has a multiplier of order* $\exp(c/\varepsilon)$, *where* $c > 0$.

We can **prove** Theorem 18.5 by means of the asymptotic integration of the linear system (18.34), which is carried out in the same way as the integration of system (16.18). Namely, for $0 \le t \le t_1 - \varepsilon^\lambda$, $0 < \lambda < 1/2$, we generalize the technique of asymptotic integration from [7]. For $t_1 - \varepsilon^\lambda \le$

$t \leq t_1 + \varepsilon^\lambda$, after the normalizations described in Sec. 16, we get Weber's equation (cf. (16.66))

$$\frac{d^2v}{du^2} - u\frac{dv}{du} - \left[1 + \frac{b_0(t_1)c_0(t_1)}{a'_0(t_1)}\right]v = 0,$$

whose solutions grow as $\exp(u^2/2)$, as $|u| \to \infty$, when inequalities (18.36) are satisfied, and this means that at the moment of time $t_1 + \delta$, $\delta > 0$, the functions h_1, h_2 acquire values of order $\exp(c/\varepsilon)$, $c > 0$. For $t_1 + \delta \leq t \leq T(\varepsilon)$, we can again use the technique described in [7], which, in conjunction with the inequality $c_0(T_0 - 0) \neq 0$, leads to the statement of the theorem.

In conclusion, it should be pointed out that all the results presented here are specific enough since they refer only to relaxation systems with one slow variable. Of a more general character is Theorem 18.5, which implies the assumption that stable duck-cycles are not typical of relaxation systems with many slow variables. The next section is devoted to this problem.

19. Instability of Duck-Cycles of Multidimensional Relaxation Systems

19.1. Statement of the Problem

For simplicity, we shall only consider the case of one fast variable $x \in R$, $y \in R^m$, $m \geq 2$. Under this condition we shall consider the relaxation system

$$\varepsilon \dot{x} = f(x,y), \qquad \dot{y} = g(x,y,\mu), \qquad 0 < \varepsilon \ll 1, \qquad \mu \in R^m, \qquad (19.1)$$

with smooth right-hand sides. We assume that $x = 0$, $y = 0$ is a junction point, i.e.,

$$f(0,0) = 0, \qquad f'_x(0,0) = 0, \qquad f''_{xx}(0,0) \neq 0. \qquad (19.2)$$

We suppose additionally that

$$g(0,0,0) = 0, \qquad \det g'_\mu(0,0,0) \neq 0, \qquad (19.3)$$

$$(\text{grad}_y f(0,0), g'_x(0,0,0))\text{sign } f''_{xx}(0,0) < 0. \qquad (19.4)$$

Conditions (19.3) mean that system (19.1) is in the equilibrium position which, upon the generical variation of the parameter μ, passes through the breakoff manifold. Inequality (19.4), which is similar to Condition 18.4 of the preceding section, ensures the existence of a duck-trajectory of the degenerate system corresponding to (19.1) [16], i.e., a trajectory along which the phase point gets, in a finite time interval, from the slow-motion stable surface $f = 0$, $f'_x < 0$ to the unstable one $f = 0$, $f'_x > 0$.

By means of the substitutions described in 4.1, we reduce system (19.1) to the normal form

$$\varepsilon\dot{\xi} = \xi^2 + \eta_1, \quad \dot{\eta}_1 = g_1(\xi, \eta_1, \eta_2, \varepsilon, \mu), \quad \dot{\eta}_2 = g_2(\xi, \eta_1, \eta_2, \varepsilon, \mu). \tag{19.5}$$

In (19.5) we have $\eta_1 \in R$, $\eta_2 \in R^{m-1}$,

$$g_1|_{\varepsilon=0,\mu=0} = -\xi + a_{11}\eta_1 + a_{12}\eta_2 + \ldots,$$

$$g_2|_{\varepsilon=0,\mu=0} = a_{21}\eta_1 + A_{22}\eta_2 + \ldots, \tag{19.6}$$

and the dots denote terms of a higher order of smallness. Setting $\varepsilon = 0$, $\mu = 0$ in (19.5) and taking ξ as the new time, we arrive at the relations

$$\eta_{10}(\xi) = -\xi^2, \quad \frac{d\eta_{20}}{d\xi} = -\frac{2\xi g_2(\xi, -\xi^2, \eta_{20}, 0, 0)}{g_1(\xi, -\xi^2, \eta_{20}, 0, 0)}. \tag{19.7}$$

In the neighborhood of $\xi = 0$, the differential equation for η_{20} has a unique smooth solution $\eta_{20}(\xi) = -\frac{2}{3}a_{21}\xi^3 + \ldots$, for which

$$\dot{\xi} = -g_1(\xi, -\xi^2, \eta_{20}(\xi), 0, 0)/2\xi \sim 1 \quad \text{for} \quad \xi \to 0.$$

This follows from the fact that in the neighborhood of the zero equilibrium position of the system

$$\dot{\xi} = g_1(\xi, -\xi^2, \eta_{20}, 0, 0), \quad \eta_{20} = -2\xi g_2(\xi, -\xi^2, \eta_{20}, 0, 0)$$

there is a unique smooth invariant curve $\eta_{20} = \eta_{20}(\xi)$ during the motion along which the solutions exponentially decay.

In what follows, for purely technical reasons, it is convenient to deal with system (19.5), considering its right-hand sides to be defined in a sufficiently wide neighborhood of the point $(0,0,0)$.

19.2. The Fundamental Lemma

In the neighborhoods of the points $\xi = -q$ and $\xi = q$, $q > 0$, system (19.5) has (see [71]) invariant integer manifolds $\eta_1 = H_1(\xi, \eta_2, \varepsilon, \mu)$ and $\eta_1 = H_2(\xi, \eta_2, \varepsilon, \mu)$, respectively, where $H_1|_{\varepsilon=0} \equiv H_2|_{\varepsilon=0} = -\xi^2$, the former being exponentially stable with the coefficient of the exponent of order ε^{-1}

and the latter being unstable with the same coefficient of the exponent. As was shown in Sec. 16, these manifolds are closely connected with the duck-trajectories of system (19.5), which are trajectories with zero approximation $(-\xi^2, \eta_{20}(\xi))$, since, obviously, the solution of the boundary-value problem

$$\frac{d\eta_1}{d\xi} = \frac{\varepsilon g_1(\xi, \eta_1, \eta_2, \varepsilon, \mu)}{\xi^2 + \eta_1}, \qquad \frac{d\eta_2}{d\xi} = \frac{\varepsilon g_2(\xi, \eta_1, \eta_2, \varepsilon, \mu)}{\xi^2 + \eta_1}, \tag{19.8}$$

$$\eta_1|_{\xi=-q} = H_1|_{\xi=-q}, \qquad \eta_1|_{\xi=q} = H_2|_{\xi=q} \tag{19.9}$$

is a duck-trajectory.

LEMMA 19.1. *There exists a smooth function $\mu_*(\alpha, \varepsilon)$ such that for $\mu = \mu_*(\alpha, \varepsilon)$ the solution of system (19.8) with the initial conditions*

$$\eta_1|_{\xi=-q} = H_1|_{\xi=-q}, \qquad \eta_2|_{\xi=-q} = \eta_{20}(-q) + \varepsilon\alpha \tag{19.10}$$

satisfies the boundary conditions (19.9). Moreover, the asymptotic expansions

$$\mu_*(\alpha, \varepsilon) = \sum_{k=1}^{\infty} \mu_k(\alpha)\varepsilon^k, \tag{19.11}$$

$$\eta_1 = -\xi^2 + \sum_{k=1}^{\infty} a_k(\xi, \alpha)\varepsilon^k, \qquad \eta_2 = \eta_{20}(\xi) + \sum_{k=1}^{\infty} b_k(\xi, \alpha)\varepsilon^k \tag{19.12}$$

hold true.

In order to **prove** the lemma, we shall first describe the formalism of seeking the coefficients of series (19.11), (19.12). Substituting them into (19.8) and equating the expressions for like powers of ε, we get a recurrent sequence of equations

$$\frac{g'_{1\mu}(\xi, -\xi^2, \eta_{20}, 0, 0)}{a_1}\mu_k - \frac{g_1(\xi, -\xi^2, \eta_{20}, 0, 0)}{a_1^2}a_{k+1}+$$
$$+\frac{g'_{1\eta_2}(\xi, -\xi^2, \eta_{20}, 0, 0)}{a_1}b_k = \varphi_k(\xi, \alpha), \tag{19.13}$$

$$\frac{db_k}{d\xi} = \frac{g'_{2\mu}(\xi, -\xi^2, \eta_{20}, 0, 0)}{a_1}\mu_k + \frac{g'_{2\eta_2}(\xi, -\xi^2, \eta_{20}, 0, 0)}{a_1}b_k-$$
$$-\frac{g_2(\xi, -\xi^2, \eta_{20}, 0, 0)}{a_1^2}a_{k+1} + \psi_k(\xi, \alpha), \tag{19.14}$$

where $k = 1, 2, \ldots$, $a_1 = -g_1(\xi, -\xi^2, \eta_{20}, 0, 0)/2\xi = \frac{1}{2} + \ldots$, and the functions φ_k, ψ_k depend only on a_r, b_s, μ_j with subscripts $r \le k$, $s, j \le k-1$.

Finding the function a_{k+1} from (19.13) and substituting it into (19.14), we find for the function b_k a differential equation of the form

$$\frac{db_k}{d\xi} = \varkappa(\xi)b_k + F_k(\xi), \qquad b_k(-q) = \begin{cases} \alpha & \text{for } k=1, \\ 0 & \text{for } k>1, \end{cases}$$

where the matrix $\varkappa(\xi)$ and the vector-function F_k are smooth in the neighborhood of $\xi = 0$. Returning to Eq. (19.13), we infer that in order to find the smooth function $a_{k+1}(\xi, \alpha)$ from it, it is necessary to satisfy the relations

$$2g'_{1\mu}(0,0,0,0)\mu_k = \varphi_k(0,\alpha), \qquad b_k|_{\xi=0} = 0, \qquad (19.15)$$

from which we find the unknown vector μ_k (the system of linear equations (19.15) can be uniquely solved by virtue of inequality (19.3)).

We have thus described the formalism of finding series (19.11), (19.12) and now we can complete the proof of the lemma with the use of the standard scheme, setting

$$\mu = \sum_{k=1}^{k_0} \mu_k(\alpha)\varepsilon^k + \delta_{k_0}\varepsilon^{k_0+1}$$

in (19.8), (19.9), and show that the solutions of (19.8) with the initial condition (19.10) and the initial condition

$$\eta_1|_{\xi=q} = H_2|_{\xi=q}, \qquad \eta_2|_{\xi=q} = \eta_{20}(q) + \sum_{k=1}^{k_0} \varepsilon^k b_k(q, \alpha)$$

respectively, can be extended up to the point $\xi = 0$. From the condition of coincidence of these solutions for $\xi = 0$ we find the unknown function $\delta_{k_0}(\alpha, \varepsilon)$.

The only difficulty that is encountered when this scheme is realized is the analysis of the linear system

$$\varepsilon\frac{dh_1}{d\xi} = a(\xi,\varepsilon)h_1 + b(\xi,\varepsilon)h_2, \qquad \frac{dh_2}{d\xi} = c(\xi,\varepsilon)h_1 + d(\xi,\varepsilon)h_2, \qquad (19.16)$$

which is the linearization of (19.8) on finite intervals of series (19.12). In (19.16) all coefficients smoothly depend on their arguments and

$$a(\xi, 0) = 4\xi + \ldots, \qquad b(0,0) = 2g'_{1\eta_2}(0,0,0,0,0),$$
$$c(0,0) = 0, \qquad d(0,0) = 2g'_{2\eta_2}(0,0,0,0,0). \qquad (19.17)$$

Therefore, the properties of system (19.16) are in many respects similar to those of system (16.18) studied in Sec. 16. In particular, here we also have to construct, for $-q \leq \xi \leq -\varepsilon^\lambda$ and $-\varepsilon^\lambda \leq \xi \leq 0$, $0 < \lambda < 1/2$, the asymptotics of the integer manifold $h_1 = \omega(\xi,\varepsilon)h_2$, where the row matrix $\omega(\xi,\varepsilon)$ can be found from the equation

$$\varepsilon \frac{d\omega}{d\xi} + \varepsilon\omega c(\xi,\varepsilon)\omega = a\omega + b.$$

We shall not give the corresponding calculations but only point out that the solutions of system (19.16) with initial conditions independent of ε behave, for $\xi = -q$, as follows: the component $h_2(\xi,\varepsilon)$ remains bounded uniformly with respect to ε throughout the interval $[-q, 0]$ and the component h_1 is bounded on any interval $[-q, -q_1]$, $q_1 > 0$, but acquires the order $\varepsilon^{-1/2}$ at the point $\xi = 0$.

19.3. Instability of Duck-Cycles

Let us suppose that the degenerate system (19.7) has a duck-trajectory such that $\eta_{20}(-q_0) = \eta_{20}(q_0)$, where $q_0 < q$, and

$$-\int_{-q}^{q_0} \xi^2 \, d\xi / g_1(\xi, -\xi^2, \eta_{20}(\xi), 0, 0) < 0. \tag{19.18}$$

Under similar conditions a relaxation system in the plane has a stable duck-cycle, and when the inequality opposite to (19.18) is satisfied for $q = q_0$, any cycle with the zero approximation being considered is unstable.

Suppose that in (19.5) we have the value of the parameter $\mu = \mu_*(\alpha,\varepsilon) + \Delta\mu$, $\Delta\mu = \mu_0 \exp(-\delta_0/\varepsilon)$, where

$$g'_{1\mu}(0,0,0,0,0)\mu_0 < 0,$$

$$\delta_0 = -\int_0^{q_0} 4\xi^2 \, d\xi / g_1(\xi, -\xi^2, \eta_{20}(\xi), 0, 0) > 0. \tag{19.19}$$

It follows from Lemma 19.1 and conditions (19.18), (19.19) that the trajectories of system (19.5) with the initial conditions $\eta_1|_{\xi=-q} = H_1|_{\xi=-q} + \varepsilon\beta$, $\eta_2|_{\xi=-q} = \eta_{20}(-q) + \varepsilon\alpha$ are at a distance of order ε from the closed duck-trajectory, introduced above, of a degenerate system. Therefore, the operator

$$\Pi : \begin{array}{l} \beta \longrightarrow \Delta(\alpha,\beta,\varepsilon)\exp(-\gamma/\varepsilon), \\ \alpha \longrightarrow \Phi(\alpha,\beta,\varepsilon), \end{array} \tag{19.20}$$

where $\gamma > 0$, Δ and Φ are smooth functions, and

$$\Phi(\alpha, 0, \varepsilon) = B\alpha + b + O(\varepsilon), \qquad \Phi'_\alpha(\alpha, 0, \varepsilon) = B + O(\varepsilon), \qquad (19.21)$$

is defined in the plane $\xi = -q$. Taking into account expansions (19.11), (19.12) in (19.8), we can obtain an explicit formula for the matrix B which implies that unity is not its eigenvalue for

$$\det[b'_{1\alpha}(-q_0, \alpha) - b'_{1\alpha}(q_0, \alpha)] \neq 0. \qquad (19.22)$$

Since $b_1(\xi, \alpha)$ linearly depends on α, the verification of condition (19.22) reduces to the calculation of the determinant of the constant matrix.

THEOREM 19.1. *Let inequality (19.22) be satisfied. Then, for $\mu = \mu_*(\alpha(\varepsilon), \varepsilon) + \Delta\mu$, where $\alpha(\varepsilon) = (I - B)^{-1}b + O(\varepsilon)$ is a component of the fixed point of operator (19.20), system (19.5) has a unique duck-cycle with the zero approximation described above, which is unstable if $a_{12} \cdot a_{21} \neq 0$.*

The statement of the theorem on unstability follows from the properties of system (19.16), namely, its solutions with initial conditions, independent of ε, defined for $\xi = -q$ become of order $\exp(\delta/\varepsilon)$, $\delta > 0$, for $\xi = q$.

We can show that for certain conditions every a priori existing duck-cycle of an arbitrary structure is unstable in the general case. Thus, in contrast to the systems with one slow variable, the generation of an arbitrary relaxation cycle in system (19.1) is not connected with duck-cycles.

Chapter 5

Nonclassical Relaxation Auto-Oscillations

We begin this chapter by elucidating the conditions under which singularly perturbed systems of the Lotka–Volterra type, which are widely used in biophysics, have specific relaxation auto-oscillations: the fast component is δ-like and the slow component is close to a discontinuous one. We show that relaxation systems of this kind have a limit object, namely, a specific one-dimensional monotonic mapping, and this means that in some respects the situation is close to the classical one, where, as we know, a suitable relay system is regarded as the limit object. At the end of the chapter we introduce a new class of singularity perturbed systems of the generic position whose relaxation cycle is close to the classical one by some of its properties and close to the specific relaxation cycles in systems of the Lotka–Volterra type by other properties.

Note that the results presented below can find application in the problem of simulation of complicated biophysical processes. This assurance is based, in particular, on the fact that the source of these results is [56], where the principal asymptotics of the nonclassical relaxation cycle of the mathematical model of the well-known Belousov reaction was constructed.

The results given in this chapter were published in [34].

20. Systems of Lotka–Volterra Type

20.1. Statement of the Problem

Changing slightly the notations used above, we shall discuss a two-dimensional problem of Lotka–Volterra type,

$$\dot{x} = g(x,y), \qquad \varepsilon \dot{y} = y f(x,y), \qquad 0 < \varepsilon \ll 1, \qquad (20.1)$$

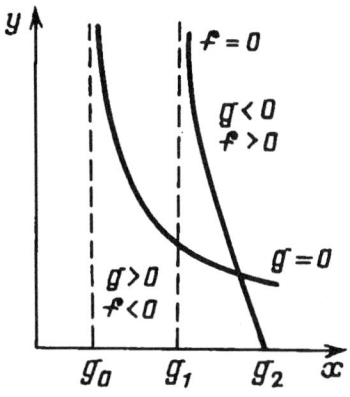

Fig. 20.1

where the right-hand sides $g, f \in C^\infty(\Omega)$,

$$\Omega = \{(x, y) : x > -\delta_0, \ y > -\delta_0\}, \quad \delta_0 > 0.$$

We shall consider its cycles belonging to the first quadrant K when a number of sufficiently natural restrictions are satisfied.

CONDITION 20.1. Suppose that $g > 0$ for $y < \varphi_1(x)$ and $g < 0$ for $y > \varphi_1(x)$, where $\varphi_1(x)$ is a certain function, defined for $0 \le q_0 < x < \infty$ and belonging to $C^\infty(q_0, \infty)$, such that $\varphi_1(x) > 0$ and $\varphi_1(x) \to \infty$ as $x \to q_0 + 0$.

CONDITION 20.2. For $q_1 < x \le q_2$, where $q_1 > q_0$, the equation $f = 0$ defines in K the curve $y = \varphi_2(x)$ with the following properties: $\varphi_2 \in C^\infty(q_1, q_2]$,

$$\varphi_2(x) \sim \alpha/(x - q_1) \quad \text{for} \quad x \to q_1 + 0,$$

where α is a positive constant, $\varphi_2(q_2) = 0$ and $\varphi_2(x) > 0$ for $x < q_2$ and the curves $y = \varphi_1(x)$ and $y = \varphi_2(x)$ meet at a single point. In addition to what we have said, we suppose that $f(x, 0) < 0$ for $0 \le x < q_2$ and $f(x, 0) > 0$ for $x > q_2$.

The formulated conditions are classed with the properties and position of the isoclines of system (20.1). Figure 20.1 makes them visual.

CONDITION 20.3. As $y \to +\infty$ uniformly with respect to x from any closed and bounded subset of the semiaxis $x \geq 0$, the asymptotic expansions

$$g(x,y) = y^{n_1}\left[a_0(x) + \frac{a_1(x)}{y} + \frac{a_2(x)}{y^2} + \cdots\right],$$
$$f(x,y) = y^{n_2-1}\left[b_0(x) + \frac{b_1(x)}{y} + \frac{b_2(x)}{y^2} + \cdots\right],$$
(20.2)

where the integers $n_1, n_2 \geq 1$, with $p = n_1 - n_2 \geq -1$, hold true. We suppose that expansions (20.2) can be differentiated with respect to x, y in any order and any number of times.

For the asymptotic expansions (20.2) to correspond to Conditions 20.1 and 20.2, we also assume the following:

$a_0(x) < 0$ for $q_0 < x < \infty$ and $a_0(x) > 0$ for $0 \leq x < q_0$,

$b_0(x) > 0$ for $q_1 < x < \infty$ and $b_0(x) < 0$ for $0 \leq x < q_1$,

with $b_0'(q_1) > 0$, $b_1(q_1) < 0$.

The second relation of (20.2) implies the relation

$$\alpha = -b_1(q_1)/b_0'(q_1),$$

where α is the constant from Condition 20.2.

CONDITION 20.4. The equation

$$-\int_{q_0}^{q_2} \frac{f(\sigma,0)}{g(\sigma,0)} d\sigma = \int_{q_2}^{x} \frac{f(\sigma,0)}{g(\sigma,0)} d\sigma \qquad (20.3)$$

has a root $q_3 \in (q_2, \infty)$, which is unique since the right-hand side is monotonic.

The system

$$\dot{x} = 1 - x^3 y, \qquad \varepsilon \dot{y} = y\left[a(x-1) - \frac{b+1}{b+y}\right], \qquad a, b > 0, \qquad (20.4)$$

can be given as an example for which the enumerated restrictions are fulfilled. Indeed, here

$$\varphi_1(x) = 1/x^3, \qquad \varphi_2(x) = -b + (b+1)/a(x-1),$$

$q_0 = 0, \qquad q_1 = 1, \qquad q_2 = 1 + (b+1)/ab, \qquad q_3 = 2(1 + (b+1)/ab),$

$n_1 = n_2 = 1, \qquad a_0(x) = -x^3, \qquad b_0(x) = a(x-1).$

20.2. Heuristic Arguments

Before formulating rigorous statements we shall give some heuristic arguments.

We denote by $x(t,\varepsilon)$, $y(t,\varepsilon)$ the solutions of (20.1) with the initial conditions $x(0,\varepsilon) = x_0 \in [q_0, q_1]$, $y(0,\varepsilon) = 1$. According to the results given in [7], first, in an asymptotically short time, a "drop" occurs approximately along the straight line $x = x_0$ onto the stable integral manifold $y = 0$ and the motion along it obeys the law $\dot{x} = g(x,0)$, where, as we know, $g(x,0) > 0$. For $x > q_2$ the integral manifold $y = 0$ becomes unstable. However, by virtue of the delay effect discovered by Pontryagin, of which we have spoken above, the "breakoff" from it does not occur immediately but approximately for $x = x_1$, where x_1 is the root of the equation

$$-\int_{x_0}^{q_2} \frac{f(\sigma,0)}{g(\sigma,0)}d\sigma = \int_{q_2}^{x} \frac{f(\sigma,0)}{g(\sigma,0)}d\sigma, \tag{20.5}$$

whose existence is ensured by Condition 20.4.

In order to follow the motion of the phase point of (20.1) after the "breakoff," we make the following constructions, whose meaning will be explained later.

In the first place, we divide the first equation of (20.1) by the second and make the substitution $u = \varepsilon^{1/(p+1)} y$ for $p \geq 0$ and $u = \varepsilon \ln y$ for $p = -1$. Then we use expansion (20.2) in order to delete terms of a higher order of smallness. As a result, for $p \geq 0$, we arrive at the equation

$$\frac{dx}{du} = u^p \frac{a_0(x)}{b_0(x)}, \tag{20.6}$$

and, for $p = -1$, at the equation

$$\frac{dx}{du} = \frac{a_0(x)}{b_0(x)}. \tag{20.7}$$

In the second place, we describe certain properties of the solution $x = v_{01}(u)$, $v_{01}(0) = x_1$, of (20.6) or (20.7). Since by Condition 20.3 we have $a_0(x)/b_0(x) < 0$ for $q_1 < x \leq x_1$ and

$$a_0(x)/b_0(x) = O(1/(x - q_1)), \qquad x \to q_1, \tag{20.8}$$

this solution is defined only for $0 \leq u \leq u_*$, where

$$u_* = \left[-(p+1)\int_{q_1}^{x_1} \frac{b_0(s)}{a_0(s)}ds\right]^{1/(p+1)} \qquad \text{if} \qquad p \geq 0, \tag{20.9}$$

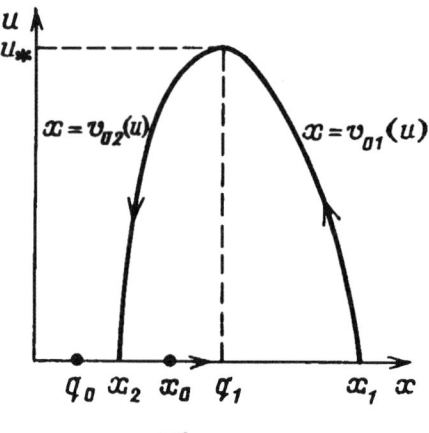

Fig. 20.2

or
$$u_* = -\int_{q_1}^{x_1} \frac{b_0(s)}{a_0(s)} ds \quad \text{if} \quad p = -1, \tag{20.10}$$

and the asymptotic relation
$$v_{01}(u) = q_1 + O(\sqrt{u_* - u}), \quad u \to u_* - 0$$

is satisfied at the point u_* itself.

In the third place, suppose that $x = v_{02}(u)$ is the solution of (20.6) or (20.7) which passes through the singular point (q_1, u_*) and lies in the half-plane $x \leq q_1$. By virtue of (20.8), this solution exists and is unique. In accordance with Condition 20.3, as u diminishes from u_* to $-\infty$, the solution $v_{02}(u)$ decreases monotonically and tends to the equilibrium position $x = q_0$. Therefore, for the constant $x_2 = v_{02}(0)$, which can be found from the equation
$$\int_{x}^{q_1} \frac{b_0(s)}{a_0(s)} ds = -\int_{q_1}^{x_1} \frac{b_0(s)}{a_0(s)} ds, \tag{20.11}$$

we have (see Fig. 20.2)
$$q_0 < x_2 < q_1. \tag{20.12}$$

From the constructions that we have carried out, it follows that after the "breakoff" from the manifold $y = 0$ for $x = x_1$ the phase point of (20.1) moves approximately along the curve, given in Fig. 20.2, defined by the

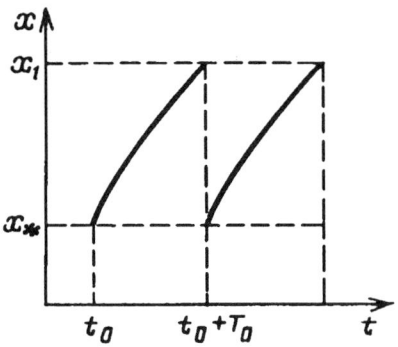

Fig. 20.3

relations $x = v_{01}(u)$ and $x = v_{02}(u)$, where $0 \leq u \leq u_*$. Therefore, the Poincaré operator $\Pi(x_0, \varepsilon)$ along the trajectories of (20.1) is defined on the segment $y = 1$, $q_0 \leq x_0 \leq q_1$, and its zero approximation has the form $\Pi_0(x_0) = x_2$.

Let us consider in greater detail the properties of the function $\Pi_0(x)$. Differentiating (20.5) with respect to x_0 and (20.11) with respect to x_1, we make sure that

$$\Pi_0'(x) = \frac{a_0(x_2)b_0(x_1)f(x,0)g(x_1,0)}{a_0(x_1)b_0(x_2)f(x_1,0)g(x,0)} > 0$$

for all $q_0 \leq x \leq q_1$. Furthermore, it follows from (20.12) that $\Pi_0(q_0) > q_0$, $\Pi_0(q_1) < q_1$, i.e., $\Pi_0(x)$ maps the interval $[q_0, q_1]$ strictly into itself. Therefore, the equation $\Pi_0(x) = x$ has at least one solution x_* and, in the general case, $\Pi_0'(x_*) \neq 1$. It turns out (this will be substantiated below) that every solution of this kind is associated with a relaxation cycle which is exponentially orbitally stable (unstable) for $\Pi_0'(x_*) < 1$ (> 1). In the variables x, u its zero approximation is shown in Fig. 20.2, where we must assume that $x_0 = x_2 = x_*$.

From the analysis carried out below, it follows that the time of motion along the trajectories of (20.1) from the neighborhood of the point $(x_1, 1)$ to the straight line $y = 1$ after the "turn" is of order $O(\varepsilon)$ for $n_2 > 1$ and $O(\varepsilon \ln \frac{1}{\varepsilon})$ for $n_2 = 1$. This means that the change of the component $y(t, \varepsilon)$ of the relaxation cycle is δ-like and the graph of the zero approximation of the component $x(t, \varepsilon)$ has the form shown in Fig. 20.3 (here x_1 is the root of Eq. (20.5) for $x_0 = x_*$ and T_0 is the zero approximation of the period). Note that by virtue of the results given in [7] the component $y(t, \varepsilon)$ of the relaxation cycle remains a quantity of order $\exp(-c/\varepsilon)$, $c > 0$, as long as the phase point of (20.1) remains close to the integer manifold $y = 0$ for $x_* + \delta \leq x \leq x_1 - \delta$, where $\delta > 0$ is arbitrary. Finally, by the change that we have carried out when deriving equation (20.6) or (20.7), we have

$$\max_t y(t, \varepsilon) \sim u_* \varepsilon^{-1/(p+1)}$$

for $p \geq 0$ and

$$\max_t y(t, \varepsilon) \sim \exp(u_*/\varepsilon)$$

for $p = -1$.

20.3. Analysis of an Example

We shall illustrate what we have said above by the example of system (20.4), for which Eq. (20.5) has the form

$$\int_{x_0}^{x} \left[a(s-1) - \frac{b+1}{b} \right] ds = 0.$$

From this we obtain

$$x_1(x_0) = 2(1+z) - x_0, \qquad 0 \leq x_0 \leq 1, \qquad (20.13)$$

where $z = (b+1)/ab$. From the equation for x_2,

$$\frac{1}{2x_2^2} - \frac{1}{x_2} = \frac{1}{2x_1^2} - \frac{1}{x_1},$$

similar to (20.11), we find that

$$x_2 = x_1/(2x_1 - 1). \qquad (20.14)$$

Substituting (20.13) into (20.14), we get

$$\Pi_0(x_0) = \frac{2(1+z) - x_0}{4z + 3 - x_0}, \qquad 0 \leq x_0 \leq 1.$$

A simple verification shows that on the interval $[0, 1]$ the equation $\Pi_0(x_0) = x_0$ has the unique root

$$x_* = 1 + z - \sqrt{z(z+1)}$$

and

$$\Pi_0'(x_*) = 1/(1 + 2z + 2\sqrt{z(z+1)})^2 < 1.$$

Therefore, for any $a, b > 0$, system (20.4) has a unique orbitally exponentially stable relaxation cycle.

Also, note the important relation

$$\lim_{\varepsilon \to 0} \int_0^{T(\varepsilon)} y(t,\varepsilon)dt = \lim_{\varepsilon \to 0} \oint_{\Gamma(\varepsilon)} \frac{y\,dx}{1-x^3 y} = \int_{x_*}^{x_1(x_*)} \frac{dx}{x^3} =$$

$$= 2\sqrt{z/(z+1)}, \qquad (20.15)$$

where $y(t,\varepsilon)$ is a component of the relaxation cycle $\Gamma(\varepsilon)$ and $T(\varepsilon)$ is its period. Relation (20.15) shows that as $\varepsilon \to 0$, the component $y(t,\varepsilon)$ converges to the δ-function multiplied by a constant, since its integral tends to a finite positive limit. In the general case, passing to the variable u in the integral

$$\varepsilon \int \frac{dy}{f(x,y)}$$

and setting $x = v_{01}(u)$ or $x = v_{02}(u)$, we see that the same statement is valid for $n_1 = n_2 = 1$ and $n_1 = 1, n_2 = 2$ (see Condition 20.3), and the corresponding limit is

$$\int_0^{u_*} \frac{du}{b_0(v_{01}(u))} - \int_0^{u_*} \frac{du}{b_0(v_{02}(u))}.$$

In other situations this approach leads to the conclusion that limit (20.15) is zero, i.e., that the oscillations $y(t,\varepsilon)$ are δ-like only superficially.

20.4. Applications to Lotka–Volterra Systems

With a natural variation of Conditions 20.1–20.4 the map $\Pi_0(x)$ is constructed for the Lotka–Volterra system

$$\dot{x} = xg(x,y), \qquad \varepsilon\dot{y} = yf(x,y). \qquad (20.16)$$

Namely, we suppose that in Condition 20.1 the function $\varphi_1(x) \in C^\infty[0,\infty)$, i.e., the curve $g = 0$ does not have a vertical asymptote. Condition 20.2 remains valid and Condition 20.3 is complemented by the requirement that $a_0(x) < 0$ for all $x \geq 0$. We replace Condition 20.4 by the limiting relation

$$\lim_{x \to \infty} \int_{q_2}^{x} \frac{f(\sigma,0)}{\sigma g(\sigma,0)} d\sigma = \infty, \qquad (20.17)$$

whose necessity is caused by the fact that as $x_0 \to +0$, the left-hand side of the relation

$$-\int_{x_0}^{q_2} \frac{f(\sigma,0)}{\sigma g(\sigma,0)} d\sigma = \int_{q_2}^{x} \frac{f(\sigma,0)}{\sigma g(\sigma,0)} d\sigma \qquad (20.18)$$

increases indefinitely. Therefore, condition (20.17) ensures the unique solvability of (20.18) for every $x_0 \in (0, q_1]$ and the corresponding root $x_1 \in (q_2, \infty)$. Acting now as in 20.2, we first introduce an equation

$$\int_{x}^{q_1} \frac{b_0(s)}{s a_0(s)} ds = -\int_{q_1}^{x_1} \frac{b_0(s)}{s a_0(s)} ds \qquad (20.19)$$

for x_2, similar to (20.11), and then set $\Pi_0(x_0) = x_2$. Here, as before, $\Pi_0'(x) > 0$ for $x \in (0, q_1]$, $\Pi_0(q_1) < q_1$.

In order to elucidate the behavior of the function $\Pi_0(x)$ as $x \to +0$, we first suppose that

$$-\int_{q_1}^{\infty} \frac{b_0(s)}{s a_0(s)} ds < \infty. \qquad (20.20)$$

We shall use inequality (20.20) somewhat later, but first we must note the asymptotic relation

$$x_1 = x_1(x_0) \sim \Phi(\alpha_1 \ln \frac{1}{x_0}), \quad x_0 \to +0; \quad \alpha_1 = -\frac{f(0,0)}{g(0,0)} \qquad (20.21)$$

that follows from (20.18), where Φ is the inverse of the function

$$\Psi(x) = \int_{q_1}^{x} \frac{f(\sigma,0)}{\sigma g(\sigma,0)} d\sigma.$$

Also recall (see Conditions 20.1, 20.2) that $f(0,0)/g(0,0) < 0$. Therefore, by virtue of the limiting relation (20.17), it follows from (20.21) that $x_1 \to \infty$ as $x_0 \to +0$. Now we can use inequality (20.20), from which we deduce that, as $x_0 \to +0$, the root $x_2 = x_2(x_0)$ of (20.19) tends to the finite limit $x_{2,\infty} > 0$. Here $x_{2,\infty}$ is the root of Eq. (20.19), whose right-hand side includes integral (20.20). Extending the definition of $\Pi_0(x)$ by continuity at zero, i.e., setting $\Pi_0(0) = x_{2,\infty}$, we get a map whose properties are similar to those enumerated in 20.2.

If integral (20.20) diverges, then it follows from (20.19), (20.21) that

$$x_2(x_0) \sim \exp\{-\alpha_2 R(\Phi(\alpha_2 \ln \frac{1}{x_0}))\}, \quad x_0 \to +0, \qquad (20.22)$$

where
$$\alpha_2 = \frac{a_0(0)}{b_0(0)}, \qquad R(x) = -\int_{q_1}^{x} \frac{b_0(s)}{s a_0(s)} ds.$$

Since $a_0(0)/b_0(0) > 0$ (see Condition 20.3), $\Pi_0(x) \to 0$ as $x \to +0$. Note that the asymptotic relation (20.22) can be differentiated with respect to x_0 (this can be verified by a successive differentiation of (20.18) and (20.19)). From this we find, in turn, that

$$\lim_{x \to +0} \Pi_0'(x) = 0 \quad \text{or} \quad \infty \qquad (20.23)$$

if

$$\lim_{x \to \infty} \left[\frac{x}{\alpha_1} - \alpha_2 R(\Phi(x)) + \ln R'(\Phi(x)) + \ln \Phi'(x) \right] = -\infty \quad \text{or} \quad \infty. \qquad (20.24)$$

Recall that, by virtue of (20.17) and the supposed divergence of integral (20.20), we have $R(x), \Phi(x) \to \infty$ as $x \to \infty$. Therefore, the limiting relations (20.24) (and, hence, (20.23)) characterize a certain general position.

Thus, if integral (20.20) converges and also if it diverges but limit (20.23) is equal to infinity, the map $\Pi_0(x)$ is responsible for the dynamics of system (20.16). Otherwise, i.e., when integral (20.20) diverges and the limit (20.23) is zero, the question concerning the correspondence of the stable zero fixed point of the map $\Pi_0(x)$ to some cycle of system (20.16) remains open.

As an example illustrating the constructions that we have carried out, we can give the system

$$\dot{x} = x[1 - (x+1)^{1-\gamma}y], \qquad \varepsilon \dot{y} = y\left[a(x-1) - \frac{b+1}{b+y}\right], \qquad a, b > 0,$$

for which

$$\Psi(x) = \int_1^x \frac{a(\sigma-1) - (b+1)/b}{\sigma} d\sigma, \qquad R(x) = \int_1^x \frac{a(s-1)}{s(s+1)^{1-\gamma}} ds. \qquad (20.25)$$

We can infer from (20.25) that

$$\Psi(x) \sim ax; \qquad R(x) \sim \frac{a}{\gamma} x^\gamma \quad \text{for} \quad \gamma > 0,$$
$$R(x) \sim a \ln x \qquad \qquad \text{for} \quad \gamma = 0, \qquad (20.26)$$

as $x \to \infty$, and for negative γ integral (20.20) converges. Taking the asymptotics (20.26) into account in (20.24), we see that limit (20.23) is equal to infinity for $0 \le \gamma < 1$ and to zero for $\gamma \ge 1$.

21. Relaxation Cycles of Systems of Lotka–Volterra Type

21.1. Description of the Results

The following statement is the main result.

THEOREM 21.1. *Suppose that Conditions 20.1–20.4 are satisfied. Then there exists $\varepsilon_0 > 0$ such that for $0 < \varepsilon \leq \varepsilon_0$ on the interval $y = 1$, $q_0 \leq x \leq q_1$, the Poincaré operator $\Pi(x,\varepsilon)$ along the trajectories of system (20.1) is defined and the asymptotic expansion*

$$\Pi(x,\varepsilon) = \Pi_0(x) + \sum_{n=1}^{\infty} \varepsilon^{n/(p+1)} \sum_{m=0}^{[\frac{n}{p+1}]} \Pi_{n,m}(x) \ln^m \frac{1}{\varepsilon} \quad \text{for } p \geq 0; \quad (21.1)$$

$$\Pi(x,\varepsilon) = \Pi_0(x) + \sum_{n=1}^{\infty} \varepsilon^n \Pi_n(x) \quad \text{for } p = -1 \quad (21.2)$$

is valid for it in the metric of the space $C^1[q_0, q_1]$.

Let us now discuss the problem of the correspondence between the fixed points of the map $\Pi_0(x)$ and the relaxation cycles of system (20.1).

THEOREM 21.2. *Suppose that under the Conditions of Theorem 21.1 the map $\Pi_0(x)$ contains a fixed point x_* such that $\Pi'_0(x_*) \neq 1$. Then system (20.1) of Lotka–Volterra type has a relaxation cycle which is exponentially orbitally stable (unstable) for $\Pi'_0(x_*) < 1 \ (> 1)$. For its period $T(\varepsilon)$ and the initial condition $x(\varepsilon)$, defined on the straight line $y = 1$ for $t = 0$, the asymptotic expansions*

$$T(\varepsilon) = T_0 + \sum_{n=1}^{\infty} \varepsilon^{n/(p+1)} \sum_{m=0}^{[\frac{n}{p+1}]} T_{n,m} \ln^m \frac{1}{\varepsilon}, \quad (21.3)$$

$$x(\varepsilon) = x_* + \sum_{n=1}^{\infty} \varepsilon^{n/(p+1)} \sum_{m=0}^{[\frac{n}{p+1}]} x_{n,m} \ln^m \frac{1}{\varepsilon} \quad (21.4)$$

are valid for $p \geq 0$ and

$$T(\varepsilon) = T_0 + \sum_{n=1}^{\infty} \varepsilon^n T_n, \quad x(\varepsilon) = x_* + \sum_{n=1}^{\infty} \varepsilon^n x_n \quad (21.5)$$

for $p = -1$. In (21.3) and (21.5) the leading term

$$T_0 = \int_{x_*}^{x_1} \frac{dx}{g(x,0)},$$

where x_1 is the root of Eq. (20.5) for $x_0 = x_*$.

The statements of Theorem 21.2 regarding the existence, stability, and the asymptotics of the initial condition of the relaxation cycle are direct corollaries of Theorem 12.1, and the relation for $T(\varepsilon)$ follows from the asymptotic properties of the trajectories of system (20.1) given in the proof of this theorem.

Note that the statements of Theorems 21.1 and 21.2 remain valid for system (20.16) under similar constraints (see 20.4) with the only difference that in Theorem 21.1 we take as q_0 any number from the interval $(0, q_1)$. The substantiation of the corresponding theorems also remains the same.

Recall that in Condition 20.3 we have $p = n_1 - n_2 \geq -1$, i.e., as $y \to \infty$, the function $f(x, y)$ grows not very rapidly relative to $g(x, y)$. This is precisely the reason why the trajectory of system (20.1) in the neighborhood of the point (q_1, u_*) (see Fig. 20.2) makes a "turn" and returns to the neighborhood of the straight line $y = 0$. Now if $p \leq -2$, then all trajectories of system (20.1) with the initial conditions indicated in Theorem 21.1 do not return to the interval $\{y = 1, q_0 \leq x \leq q_1\}$.

THEOREM 21.3. *Suppose that all restrictions from 20.1 are satisfied but in Condition 20.3 we have $p \leq -2$. Then, for any $x_0 > q_2$, there exists $\varepsilon_0 = \varepsilon_0(x_0) > 0$ such that for $0 < \varepsilon \leq \varepsilon_0$ the solution $x(y, \varepsilon)$ of the equation*

$$\frac{dx}{dy} = \varepsilon \frac{g(x, y)}{y f(x, y)}$$

with the initial condition $x(1, \varepsilon) = x_0$ is defined for all $y \geq 1$ and has a finite limit $x_\infty(\varepsilon)$ as $y \to \infty$, for which the asymptotic expansion

$$x_\infty(\varepsilon) = x_0 + \sum_{n=1}^{\infty} \varepsilon^n x_{\infty, n} \tag{21.6}$$

holds true.

21.2. Conclusion

Thus, in contrast to classical relaxation oscillations in the plane, where the relaxation cycle is unique, the global dynamics of system (20.1) is defined by the map

$$x \mapsto \Pi_0(x),$$

which can have several fixed points. Moreover, here the multipliers of the corresponding relaxation cycles tend to finite limits as $\varepsilon \to 0$ (recall that the multiplier of an ordinary relaxation cycle is of order $\exp(-c/\varepsilon)$, $c > 0$). It should also be pointed out that by virtue of the results given earlier, under the constraints of the type of general conditions the classical relaxation auto-oscillations are structurally stable, i.e., vary only slightly under a small smooth perturbation (independent of ε) of the right-hand sides of the differential equations.

Note that the concept of structural stability treated in this way, which is natural for relaxation systems, differs from the ordinary one, since after the division by ε we find that the speed of one of the variables formally changes strongly, whereas the relaxation auto-oscillations that we have constructed do not behave in this way, namely, in the general case, the global integer manifold $y = 0$ is destroyed already under perturbations of order $\exp(-c/\varepsilon)$, $c > 0$. However, in the class of systems of form (20.1) or (20.16) they are structurally stable if the perturbations remain small and as $y \to \infty$. This means that we have discovered relaxation oscillations specific for the classes of systems under consideration.

22. Construction of Complete Asymptotics of Trajectories

22.1. Asymptotics of Trajectories on the Segment $y \leq 1$

We denote by L the trajectory of system (20.1) with the initial condition $(x_0, 1)$, $x_0 \in [q_0, q_1]$, defined for $t = 0$. Below we consider its asymptotic properties for $p \geq 0$ (see Condition 20.3). We shall discuss the case $p = -1$ separately.

Here we shall only construct the asymptotics for the segment of the trajectory L that lies not higher than the straight line $y = 1$ (see Fig. 22.1; recall that x_1 is the root of Eq. (20.5)). Note that we could have constructed it proceeding from the instructions given in [7], but it is more convenient to use the technique suggested in [77], which is as follows: the trajectory is divided into segments on each of which one of the variables is taken as the new time. Therefore, we always deal with a scalar equation whose integration is relatively simple.

Let us divide the part of the trajectory of (20.1) being considered into five segments (see Fig. 22.1), namely, the "drop" segment

$$\exp(-1/\varepsilon^{1-\lambda}) \leq y \leq 1, \qquad 0 < \lambda < 1,$$

the beginning of the slow-motion segment, from the secant $y = \exp(-1/\varepsilon^{1-\lambda})$ to the straight line $x = r_1$, $r_1 \in (x_0, q_2)$, the middle part of the slow-motion

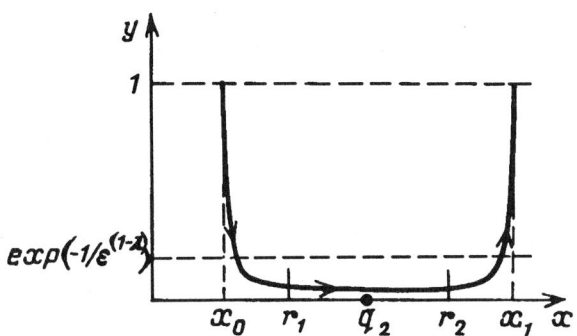

Fig. 22.1

segment, from $x = r_1$ to $x = r_2$, where $r_2 \in (q_2, x_1)$, the end of the slow-motion segment, from $x = r_2$ to the secant $y = \exp(-1/\varepsilon^{1-\lambda})$, $0 < \lambda < 1$, and the "breakoff" segment, from $y = \exp(-1/\varepsilon^{1-\lambda})$ to $y = 1$.

We shall first study the asymptotic behavior of the trajectory of (20.1) on the "drop" segment

$$\exp(-1/\varepsilon^{1-\lambda}) \leq y \leq 1, \qquad 0 < \lambda < 1.$$

With this aim in view, we divide the first equation of (20.1) by the second and denote by $x(y, \varepsilon)$ the solution of the resulting equation

$$\frac{dx}{dy} = \varepsilon \frac{g(x, y)}{y f(x, y)} \tag{22.1}$$

that satisfies the initial condition $x(1, \varepsilon) = x_0$. We seek its asymptotics as the series

$$x(y, \varepsilon) = x_0 + \sum_{k=1}^{\infty} \varepsilon^k x_{1k}(y). \tag{22.2}$$

Substituting (22.2) into (22.1) and equating the coefficients of like powers of ε, we arrive at the equation

$$\frac{dx_{11}}{dy} = \frac{G(x_0, y)}{y}, \qquad \frac{dx_{1k}}{dy} = \frac{1}{y} F_k(y), \qquad k \geq 2, \tag{22.3}$$

where $G(x, y) = g(x, y)/f(x, y)$ and

$$F_k(y) = \sum_{r=1}^{k-1} \frac{1}{r!} \frac{\partial^r G}{\partial x^r}(x_0, y) \sum_{\substack{i_1 + \ldots + i_r = k-1 \\ i_j \geq 1}} x_{1 i_1} \ldots x_{1 i_r}, \tag{22.4}$$

that defines the unknown functions $x_{1k}(y)$. It follows from (22.3), (22.4) that the right-hand sides of Eqs. (22.3) depend only on x_{1m} with subscripts $m \leq k-1$. Therefore,

$$x_{11}(y) = \int_1^y \frac{1}{\sigma} G(x_0, \sigma) d\sigma, \quad x_{1k}(y) = \int_1^y \frac{1}{\sigma} F_k(\sigma) d\sigma, \quad k \geq 2. \tag{22.5}$$

In order to give a precise meaning to the asymptotic expansion (22.2) on the segment $\exp(-1/\varepsilon^{1-\lambda}) \leq y \leq 1$, $0 < \lambda < 1$, we elucidate the behavior of the functions $x_{1k}(y)$ as $y \to +0$. It turns out that the asymptotic representations

$$x_{1k}(y) = \sum_{r=0}^{k-1} \ln^r y \sum_{m=0}^{\infty} \delta_{mr}^k y^m + \delta_k \ln^k y, \quad y \to +0, \quad k \geq 1, \tag{22.6}$$

hold true.

We carry out the proof by induction. For $k = 1$, relation (22.6) can be immediately verified (see the first formula in (22.5)). Suppose that expansions (22.6) are valid up to the subscript $k-1$. We denote by P_n an arbitrary polynomial of degree n of the variable $\ln y$ with coefficients that can be expanded in series in integer powers of y. Then we set $\tilde{P}_n = P_n + \delta_{n+1} \ln^{n+1} y$, where δ_{n+1} is a constant. In these notations (22.6) assumes the form

$$x_{1k} = \tilde{P}_{k-1}. \tag{22.7}$$

We can easily verify that

$$\tilde{P}_n \tilde{P}_m = \tilde{P}_{n+m+2}, \quad \int \frac{1}{y} P_n dy = \tilde{P}_n, \tag{22.8}$$

where the same letters P_n and $\tilde{P}_n$ denote different representatives of the classes of functions that we have introduced. Using the indicated properties of P_n and $\tilde{P}_n$ from (22.4) and (22.5), we successively derive

$$F_k(y) = P_{k-1}, \quad x_{1k} = \int \frac{1}{\sigma} P_{k-1}(\sigma) d\sigma = \tilde{P}_{k-1},$$

and this completes the proof.

LEMMA 22.1. *The asymptotic representation*

$$x(y, \varepsilon) = x_0 + \sum_{k=1}^N \varepsilon^k x_{1k}(y) + O\left(\varepsilon^{\lambda(N+1)}\right), \quad N \geq 1, \tag{22.9}$$

is valid uniformly on the interval $\exp(-1/\varepsilon^{1-\lambda}) \leq y \leq 1$, where $0 < \lambda < 1$.

The proof is based on the fact that after the substitution $\zeta = -\varepsilon \ln y$ the right-hand side of (22.1) becomes regular, i.e., it can have any number of derivatives with respect to x which are bounded uniformly with respect to ε and $0 \leq \zeta \leq \varepsilon^\lambda$. Therefore, passing from the variable x to the variable h in (22.1) with the use of the formula $x = x_{N+1}(y, \varepsilon) + h$, where

$$x_{N+1} = x_0 + \sum_{k=1}^{N+1} \varepsilon^k x_{1k}(y),$$

and carrying out the indicated substitution, we arrive at the equation

$$\frac{dh}{d\zeta} = a(\zeta, \varepsilon)h + \Phi(\zeta, \varepsilon, h) + \psi(\zeta, \varepsilon), \qquad (22.10)$$

where

$$a = -\frac{\partial G}{\partial x}\bigg|_{x=x_{N+1}, y=\exp(-\zeta/\varepsilon)}, \quad \Phi\bigg|_{h=0} \equiv \frac{\partial \Phi}{\partial h}\bigg|_{h=0} \equiv 0,$$

$$\psi = O(\varepsilon^{\lambda(N+1)}) \qquad (22.11)$$

uniformly on $0 \leq \zeta \leq \varepsilon^\lambda$. Passing from (22.10) to the integer equation

$$h = \int_0^\zeta \exp\left\{\int_s^\zeta a(\sigma, \varepsilon) d\sigma\right\} [\Phi(s, \varepsilon, h) + \psi(s, \varepsilon)] ds$$

and using properties (22.11), we see that its right-hand side maps a ball of the space $C[0, \varepsilon^\lambda]$ with radius of order $O(\varepsilon^{\lambda(N+1)})$ into itself and is a contraction mapping there. We have proved the lemma.

We have thus constructed the asymptotics of the trajectory L of (20.1) on the "drop" segment. Next we make the substitution $\zeta = -\varepsilon \ln y$ in (22.1), as a result of which Eq. (22.1) assumes the form

$$\frac{dx}{d\zeta} = -G(x, \exp(-\zeta/\varepsilon)). \qquad (22.12)$$

We denote by $x(\zeta, \varepsilon)$ the solution of the resulting equation with the initial condition

$$x|_{\zeta=\varepsilon^\lambda} = x(y, \varepsilon)|_{y=\exp(-1/\varepsilon^{1-\lambda})}. \qquad (22.13)$$

Since we are only interested in the power asymptotics, we can replace (22.12) by the equation

$$\frac{dx}{d\zeta} = G_0(x), \quad G_0(x) = -G(x, 0), \qquad (22.14)$$

whose solution with the initial condition (22.13) we also denote by $x(\zeta,\varepsilon)$. Substituting $y = \exp(-\zeta/\varepsilon)$ into series (22.2), taking into account expansions (22.6), and rejecting terms of the form $\zeta^m \exp(-k\zeta/\varepsilon)$, $m \geq 0$, $k \geq 1$, we find that the asymptotic expansion

$$x(\zeta,\varepsilon) = x_{20}(\zeta) + \sum_{k=1}^{\infty} \varepsilon^k x_{2k}(\zeta), \qquad (22.15)$$

where

$$x_{20}(0) = x_0, \qquad x_{2k}(0) = \delta_{00}^k, \qquad k \geq 1, \qquad (22.16)$$

and the constants δ_{00}^k are the same as in (22.6), holds true. It follows from (22.15) and (22.16) that $x_{20}(\zeta)$ is the solution of the Cauchy problem

$$\frac{dx}{d\zeta} = G_0(x), \qquad x|_{\zeta=0} = x_0, \qquad (22.17)$$

and the other coefficients of series (22.15) are determined recurrently.

Recall that by virtue of Conditions 20.1, 20.2, we have

$$G(x,0) = g(x,0)/f(x,0) < 0, \qquad q_0 < x < q_2, \qquad (22.18)$$

$$G(x,0) > 0, \qquad x > q_2, \qquad (22.19)$$

and for $x = q_2$ the denominator in (22.18) vanishes. Therefore, expansion (22.15) can be valid only for $\varepsilon^\lambda \leq \zeta \leq \zeta_0$, where ζ_0 is any number from the interval $(0,\zeta_1)$ and ζ_1 is the root of the equation $x_{20}(\zeta) = q_2$. According to (22.17) and (22.18),

$$\zeta_1 = -\int_{x_0}^{q_2} \frac{f(x,0)}{g(x,0)} dx > 0.$$

LEMMA 22.2. *The asymptotic relation*

$$x(\zeta,\varepsilon) = x_{20}(\zeta) + \sum_{k=1}^{N} \varepsilon^k x_{2k}(\zeta) + O(\varepsilon^{N+1}) \qquad (22.20)$$

holds true uniformly on $\varepsilon^\lambda \leq \zeta \leq \zeta_0$, *where* $\zeta_0 \in (0,\zeta_1)$.

Since the right-hand side of (22.14) is regular (it belongs to the class $C^\infty[0,q_2)$), the substantiation of the asymptotic relation (22.20) is elementary, and we shall not discuss it here.

The next step is the construction of the asymptotics of the trajectory L between the straight lines $x = r_1$ and $x = r_2$, where $r_1 \in (x_0, q_2)$ and $r_2 \in (q_2, x_1)$. First, from the equation

$$x(\zeta,\varepsilon) = r_1 \qquad (22.21)$$

we find the constant $\zeta_2 = \zeta_2(\varepsilon)$, which can be expanded, according to Lemma 22.2, in the asymptotic series

$$\zeta_2(\varepsilon) = \zeta_{20} + \sum_{k=1}^{\infty} \varepsilon^k \zeta_{2k}, \qquad (22.22)$$

where

$$\zeta_{20} = -\int_{x_0}^{r_1} \frac{f(x,0)}{g(x,0)} dx < \zeta_1,$$

and then we consider the equation

$$\frac{d\zeta}{dx} = G_1(x), \qquad G_1(x) = 1/G_0(x), \qquad (22.23)$$

which immediately yields the formula that we need for the trajectory L, namely,

$$\zeta(x,\varepsilon) = \zeta_2(\varepsilon) + \int_{r_1}^{x} G_1(\sigma) d\sigma. \qquad (22.24)$$

We formulate the result as the following lemma.

LEMMA 22.3. *For $r_1 \leq x \leq r_2$ the trajectory L of system (20.1) is defined by (22.24), where the asymptotic expansion (22.22) is valid for the constant $\zeta_2(\varepsilon)$.*

The construction of the asymptotics of the trajectory L on the remaining two segments, namely, at the end of the slow-motion segment and on the "breakoff" segment, is practically similar to that for the beginning of the slow-motion segment and for the "drop" segment, respectively. Therefore, we shall consider only the characteristic features.

When constructing the asymptotics of the trajectory L at the end of the slow-motion segment, we return to Eq. (22.14), which we shall now integrate in the direction of decreasing ζ. We denote by $x(\zeta, \varepsilon)$ its solution with the initial condition

$$x\big|_{\zeta=\zeta_3(\varepsilon)} = r_2,$$

where

$$\zeta_3(\varepsilon) = \zeta_2(\varepsilon) + \int_{r_1}^{r_2} G_1(\sigma) d\sigma. \qquad (22.25)$$

Note that, by virtue of (22.22), $\zeta_3(\varepsilon)$ can be expanded in an asymptotic series in integer powers of ε. Therefore, a similar expansion,

$$x(\zeta,\varepsilon) = x_{30}(\zeta) + \sum_{k=1}^{\infty} \varepsilon^k x_{3k}(\zeta),$$

where $x_{30}(\zeta)$ is the solution of the Cauchy problem

$$\frac{dx}{d\zeta} = G_0(x), \qquad x|_{\zeta=\zeta_{30}} = r_2, \qquad \zeta_{30} = \zeta_{20} + \int_{r_1}^{r_2} G_1(\sigma) d\sigma,$$

holds true for the solution $x(\zeta, \varepsilon)$.

We sum up all that we have said above as the following lemma.

LEMMA 22.4. *The asymptotic representation*

$$x(\zeta, \varepsilon) = x_{30}(\zeta) + \sum_{k=1}^{N} \varepsilon^k x_{3k}(\zeta) + O(\varepsilon^{N+1}), \qquad N \geq 1, \qquad (22.26)$$

is valid uniformly on $\varepsilon^\lambda \leq \zeta \leq \zeta_3(\varepsilon)$, where $0 < \lambda < 1$, for the trajectory L at the end of the slow-motion segment.

To construct the asymptotics of the trajectory L on the "breakoff" segment, we denote by $x(y, \varepsilon)$ the solution of (22.1) with the initial condition

$$x|_{y=\exp(-1/\varepsilon^{1-\lambda})} = x(\zeta, \varepsilon)|_{\zeta=\varepsilon^\lambda}, \qquad (22.27)$$

where $x(\zeta, \varepsilon)$ is function (22.26). It follows from (22.26) and (22.27) that the solution $x(y, \varepsilon)$ can be expanded in the asymptotic series

$$x(y, \varepsilon) = x_1 + \sum_{k=1}^{\infty} \varepsilon^k x_{4k}(y), \qquad (22.28)$$

where, in contrast to (22.5),

$$x_{41}(y) = x_{31}(0) + (H) \int_0^y \frac{1}{\sigma} G(x_1, \sigma) d\sigma, \qquad (22.29)$$

$$x_{4k}(y) = x_{3k}(0) + (H) \int_0^y \frac{1}{\sigma} F_k(\sigma) d\sigma, \qquad k \geq 2, \qquad (22.30)$$

and the symbol (H) in front of the interval (see 4.3) means the Hadamard regularization. The function $F_k(y)$ in (22.30) is defined by (22.4), where x_0 must be replaced by x_1. Also note that as $y \to +0$, asymptotic relations of the form (22.6) are valid for functions (22.29), (22.30) and lead (see the substantiation of Lemma 22.1) to the following statement.

LEMMA 22.5. *The asymptotic expansion*

$$x(y, \varepsilon) = x_1 + \sum_{k=1}^{N} \varepsilon^k x_{4k}(y) + O(\varepsilon^{\lambda(N+1)}), \qquad N \geq 1, \qquad (22.31)$$

holds true uniformly on $\exp(-1/\varepsilon^{1-\lambda}) \leq y \leq 1$, where $0 < \lambda < 1$, for the trajectory L on the "breakoff" segment.

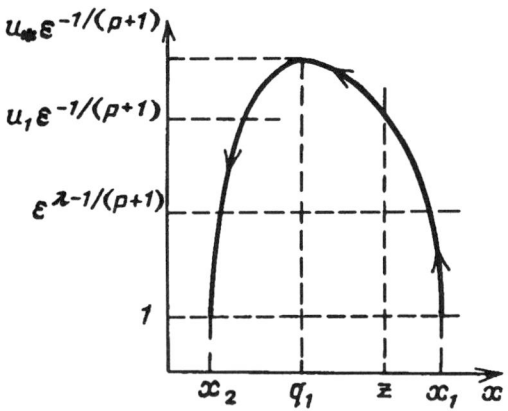

Fig. 22.2

22.2. Asymptotics of Trajectories on the Segment $y \geq 1$

Let us make a concluding step, namely, construct the asymptotics of the trajectory L on the remaining segment, i.e., for $y \geq 1$. It is also convenient to divide this part of the trajectory into five segments (see Fig. 22.2; recall that x_2 is the root of Eq. (20.11)): the beginning of the "take-off" segment, from the straight line $y = 1$ to the secant $y = \varepsilon^{\lambda - 1/(p+1)}$, $0 < \lambda < 1/(p+1)$, the end of the "take-off" segment, from $y = \varepsilon^{\lambda - 1/(p+1)}$ to the straight line $x = z$, $z \in (q_1, x_1)$, the segment of the "turn," from $x = z$ to $y = u_1 \varepsilon^{-1/(p+1)}$, $u_1 \in (0, u_*)$, the beginning of the "return" segment, from the straight line $y = u_1 \varepsilon^{-1/(p+1)}$ to $y = \varepsilon^{\lambda - 1/(p+1)}$, and the end of the "return" segment, from $y = \varepsilon^{\lambda - 1/(p+1)}$ to $y = 1$.

On the first two and the last two segments, which jointly constitute the "take-off" and "return" segments, we shall integrate (22.1). Therefore, it is useful to bear in mind that, by virtue of Condition 20.3, as $y \to \infty$, its right-hand side admits of the asymptotic representation

$$\frac{\varepsilon}{y} G(x, y) = \varepsilon y^p \left(g_0(x) + \frac{g_1(x)}{y} + \frac{g_2(x)}{y^2} + \ldots \right), \qquad (22.32)$$

which can be differentiated with respect to x, y any number of times. It follows from (22.32) that for $1 \leq y \leq y_0 \varepsilon^{-1/(p+1)}$, $0 \leq x \leq q_3$, where y_0 is an arbitrary positive number, the right-hand side of (22.1) can have any number of derivatives with respect to x, bounded uniformly with respect to the x, y under consideration by a constant of order $\varepsilon^{1/(p+1)}$. Thus, on the "take-off" and "return" segments, Eq. (22.1) is regular and, consequently, the substantiation of the asymptotic formulas from Lemmas 22.6, 22.7 and 22.9, 22.10 given below is trivial (Lemma 22.8 will be considered separately).

In order to obtain the asymptotics of the trajectory L at the beginning of the "take-off" segment, we shall again consider series (22.28) and elucidate the behavior of its coefficients as $y \to \infty$. From the asymptotic representation (22.32) we obtain, by induction, the relations

$$x_{4k}(y) = y^{k(p+1)} \sum_{r=0}^{k-1} \frac{\ln^r y}{y^{r(p+1)}} \sum_{s=0}^{\infty} \frac{\alpha_{r,s}^k}{y^s} + \alpha_k \ln^k y, \quad k \geq 1, \quad y \to \infty. \quad (22.33)$$

Just as the proof of (22.6), the substantiation of (22.33) is based on certain algebraic properties of its right-hand side. We denote by Q_n an arbitrary polynomial of degree $n-1$ of the variable $\ln y/y^{p+1}$ with coefficients that can be expanded in series in integer powers of $1/y$. Next we set $\tilde{Q}_n = y^{n(p+1)} Q_n + \alpha_n \ln^n y$, where α_n is a constant. Then we can write (22.33) as

$$x_{4k} = \tilde{Q}_k. \quad (22.34)$$

For x_{4k}, (22.34) immediately follows from (22.29) and (22.32). Let us suppose that representations (22.34) are valid up to the subscript $k - 1$. Then, taking into account the relations

$$Q_n Q_m = Q_{n+m-1}, \quad \tilde{Q}_{i_1} \cdots \tilde{Q}_{i_k} = y^{(i_1+\ldots+i_k)(p+1)} Q_{i_1+\ldots+i_k+1},$$

which are easy to verify, we infer from (22.4) (where x_0 is replaced by x_1) that $F_k = y^{k(p+1)} Q_k$. This implies that

$$x_{4k} = \int y^{k(p+1)-1} Q_k \, dy = y^{k(p+1)} Q_k + \alpha_k \ln^k y = \tilde{Q}_k.$$

We have proved relation (22.33).

The asymptotic representations (22.28) and (22.33) allow us to formulate the following statement.

LEMMA 22.6. *The asymptotic relation*

$$x(y, \varepsilon) = x_1 + \sum_{k=1}^{N} \varepsilon^k x_{4k} + O(\varepsilon^{(N+1)\lambda(p+1)}), \quad (22.35)$$

where x_{4k}, $k \geq 1$, *are functions* (22.29), (22.30), *is valid uniformly on*

$$1 \leq y \leq \varepsilon^{\lambda - 1/(p+1)}, \quad \text{where} \quad 0 < \lambda < 1/(p+1)$$

for the trajectory L at the beginning of the "take-off" segment.

For $y \geq \varepsilon^{\lambda - 1/(p+1)}$, we make the change $y = u/\mu$, $\mu = \varepsilon^{1/(p+1)}$ in (22.1) as a result of which, with due account of expansion (22.32), it assumes the form

$$\frac{dx}{du} = u^p \left(g_0(x) + \mu \frac{g_1(x)}{u} + \mu^2 \frac{g_2(x)}{u^2} + \cdots \right), \qquad (22.36)$$

where (see Condition 20.3) $g_0(x) = a_0(x)/b_0(x)$. We are interested in the solution $x(u, \mu)$ of this equation with the initial condition

$$x\big|_{u = \mu^{\lambda(p+1)}} = x(y, \varepsilon)\big|_{y = \varepsilon^{\lambda - 1/(p+1)}}, \qquad (22.37)$$

where $x(y, \varepsilon)$ is function (22.35). Relations (22.37), (22.35), and (22.33) show that the solution $x(u, \mu)$ should be sought as the asymptotic series

$$x(u, \mu) = v_{01}(u) + \sum_{n=1}^{\infty} \mu^n v_{1n}(u, \mu), \qquad (22.38)$$

where

$$v_{1n}(u, \mu) = \sum_{m=0}^{[\frac{n}{p+1}]} v_{1n}^m(u) \ln^m \frac{1}{\mu} \qquad (22.39)$$

and the function $v_{01}(u)$ was introduced in 20.2.

Substituting series (22.38) into (22.36) and equating the coefficients of μ^n, we arrive at the relation

$$\frac{dv_{1n}}{du} = \sum_{k=0}^{n-1} \sum_{r=1}^{n-k} \frac{g_k^{(r)}(v_{01}(u))}{u^{k-p} r!} \sum_{\substack{i_1 + \cdots + i_r = n-k \\ i_j \geq 1}} v_{1i_1} \cdots v_{1i_r} + \frac{g_n(v_{01}(u))}{u^{n-p}}. \qquad (22.40)$$

It follows from the structure of the right-hand side of (22.40) that this equation has the form

$$\frac{dv_{1n}}{du} = u^p g_0'(v_{01}(u)) v_{1n} + H_n(u, \mu), \qquad (22.41)$$

where the function H_n depends only on v_{1k} with subscripts $k \leq n - 1$. Analyzing (22.40) and taking into account the relation

$$\max_{\substack{i_1 + \cdots + i_r = n-k \\ i_j \geq 1}} \left\{ \left[\frac{i_1}{p+1}\right] + \cdots + \left[\frac{i_r}{p+1}\right] \right\} = \left[\frac{n-k}{p+1}\right],$$

we infer that

$$H_n(u, \mu) = \sum_{m=0}^{[\frac{n}{p+1}]} H_{n,m}(u) \ln^m \frac{1}{\mu}.$$

Thus, Eq. (22.41) disintegrates into the sequence of equations

$$\frac{dv_{1n}^m}{du} = u^p g_0'(v_{01}(u))v_{1n}^m + H_{n,m}, \quad m = 0, 1, \ldots, \left[\frac{n}{p+1}\right]. \quad (22.42)$$

Before choosing particular solutions of (22.42), we shall carry out the following formal operation. We make the substitution $y = u/\mu$ in (22.28), take into account expansions (22.33), and regroup the terms of the resulting series. We see that as $u \to +0$, the asymptotic expansions

$$v_{1n}^m(u) = \sum_{l=0}^{[\frac{n}{p+1}]-m} \ln^l u \sum_{k=m+1+l}^{\infty} \gamma_{nm}^{lk} u^{k(p+1)-n} + \beta(n)\gamma_{\frac{n}{p+1}} \cdot \ln^{\frac{n}{p+1}-m} u, \quad (22.43)$$

where

$$\beta(n) = \begin{cases} 1 & \text{if } n \equiv 0 \mod (p+1), \\ 0 & \text{if } n \not\equiv 0 \mod (p+1) \end{cases} \quad (22.44)$$

must hold for the solution of (22.42). Therefore, the particular solutions of (22.42) should be taken in the form

$$v_{1n}^m = (H) \int_0^u \frac{\varkappa(u)}{\varkappa(\sigma)} H_{n,m}(\sigma) d\sigma + \psi_{n,m} \varkappa(u), \quad (22.45)$$

where

$$\varkappa(u) = \exp\left\{\int_0^u \sigma^p g_0'(v_{01}(\sigma))d\sigma\right\};$$

$$\psi_{n,m} = 0 \quad \text{for} \quad n \not\equiv 0 \mod (p+1);$$

$$\psi_{n,m} = \alpha_{m,n-m(p+1)}^{\frac{n}{p+1}}, \quad m = 0, 1, \ldots, \frac{n}{p+1} - 1, \quad \psi_{n,\frac{n}{p+1}} = \alpha_{\frac{n}{p+1}}$$

$$\text{for} \quad n \equiv 0 \mod (p+1),$$

and the constants $\alpha_{m,n-m(p+1)}^{\frac{n}{p+1}}$, $\alpha_{\frac{n}{p+1}}$ are the same as in expansions (22.33).

We have thus described the method of recurrent determination of the coefficients of series (22.38). Relations (22.43) give the exact meaning to this asymptotic series for $\mu^{\lambda(p+1)} \leq u \leq u_0$, where u_0 is an arbitrary number from the interval $(0, u_*)$, namely, the following statement holds true.

LEMMA 22.7. *The asymptotic representation*

$$x(u,\mu) = v_{01}(u) + \sum_{n=1}^{N} \mu^n v_{1n}(u,\mu) + O(\mu^{(N+1)(1-\lambda(p+1))+\lambda(p+1)^2}) \quad (22.46)$$

holds true uniformly with respect to

$$\mu^{\lambda(p+1)} \le u \le u_0, \quad \text{where} \quad 0 < \lambda < 1/(p+1), \quad u_0 \in (0, u_*),$$

for the trajectory L at the end of the "take-off" segment.

Before constructing the asymptotics of the trajectory L on the "turn" segment, we find from the equation

$$x(u, \mu) = z$$

the constant $u_0 = u_0(\mu)$, which can be expanded, by (22.46), in the asymptotic series

$$u_0(\mu) = u_{00} + \sum_{n=1}^{\infty} \mu^n \sum_{m=0}^{[\frac{n}{p+1}]} u_{nm}^0 \ln^m \frac{1}{\mu}, \qquad (22.47)$$

where

$$u_{00} = \left[-(p+1) \int_z^{x_1} \frac{b_0(s)}{a_0(s)} ds \right]^{1/(p+1)}.$$

Then we consider the Cauchy problem

$$\frac{du}{dx} = 1/u^p \left(g_0(x) + \mu \frac{g_1(x)}{u} + \mu^2 \frac{g_1(x)}{u^2} + \cdots \right), \qquad (22.48)$$

$$u|_{x=z} = u_0(\mu). \qquad (22.49)$$

Note that Eq. (22.48) is regular, i.e., its right-hand side can have any number of derivatives with respect to u, bounded uniformly with respect to μ. This circumstance (together with asymptotics (22.47) of the initial condition) leads to the following statement.

LEMMA 22.8. *The asymptotic representation*

$$u(x, \mu) = w_0(x) + \sum_{n=1}^{\infty} \mu^n \sum_{m=0}^{[\frac{n}{p+1}]} u_{nm}(x) \ln^m \frac{1}{\mu}, \qquad (22.50)$$

where

$$w_0(x) = \left[-(p+1) \int_x^{x_1} \frac{b_0(s)}{a_0(s)} ds \right]^{1/(p+1)},$$

holds true uniformly on $x \in [z_1, z_2]$, *where* $z_1 \in (x_2, q_1)$, $z_2 \in (q_1, x_1)$, *for the trajectory* L.

In many respects the construction of the asymptotics of the trajectory L at the beginning of the "return" segment is similar to that for the end of the "take-off" segment, and, therefore, without considering technical details, we shall only point out the peculiarities appearing here.

Again returning to (22.36), we denote by $x(u, \mu)$ its solution with the initial condition $x(u_1, \mu) = \tilde{x}(\mu)$, where $\tilde{x}(\mu)$ is the root of the equation

$$u(x, \mu) = u_1.$$

According to Lemma 22.8, the asymptotic expansion

$$\tilde{x}(\mu) = z_0 + \sum_{n=1}^{\infty} \mu^n \sum_{m=0}^{[\frac{n}{p+1}]} z_{nm} \ln^m \frac{1}{\mu}, \qquad (22.51)$$

where $z_0 \in (x_2, q_1)$ is the root of the equation

$$\frac{u_1^{p+1}}{p+1} = -\int_x^{x_1} \frac{b_0(s)}{a_0(s)} ds,$$

holds true for this root. Therefore, for the solution $x(u, \mu)$ we have a similar expansion

$$x(u, \mu) = v_{02}(u) + \sum_{n=1}^{\infty} \mu^n \sum_{m=0}^{[\frac{n}{p+1}]} v_{2n}^m(u) \ln^m \frac{1}{\mu}. \qquad (22.52)$$

The function v_{02} appearing here was introduced in 20.2. The algorithm of the recurrent calculation of the coefficients of (22.52) completely coincides with that for determining the functions v_{1n}^m from (22.39) if we replace v_{01} by v_{02} in (22.40)–(22.42). However, in accordance with the initial condition (22.51), instead of (22.45) we have here

$$v_{2n}^m = -\int_u^{u_1} \frac{\varkappa(u)}{\varkappa(\sigma)} H_{n,m}(\sigma) d\sigma + z_{nm} \frac{\varkappa(u)}{\varkappa(u_1)}. \qquad (22.53)$$

As $u \to +0$, the asymptotic expansions of the functions v_{2n}^m also differ from expansion (22.43), namely, in this case the relations

$$v_{2n}^m = \sum_{k=0}^{\infty} \lambda_{n,m}^k u^{k(p+1)} +$$

$$+ u^{(m+1)(p+1)-n} \sum_{l=0}^{[\frac{n}{p+1}]-m} u^{l(p+1)} \ln^l u \sum_{q=0}^{n-(m+l)(p+1)} u^q \sum_{r=0}^{\infty} \omega_{n,m,l}^{q,r} u^{r(p+1)} +$$

$$+ \beta(n) \omega_{n,m} \ln^{\frac{n}{p+1}-m} u, \qquad u \to +0 \qquad (22.54)$$

hold true.

Representations (22.54) can be proved by induction. As an example, we give the substantiation for $m = 0$ (for $m > 0$ the substantiation of relations (22.54) is similar but more cumbersome). For $n = 1$, the validity of expansion (22.54) can be immediately verified with the use of the equation

$$\frac{dv_{21}^0}{du} = u^p g_0'(v_{02}(u))v_{21}^0 + u^{p-1} g_1(v_{02}(u))$$

and relation (22.53). Suppose that expansions (22.54) are valid for the functions v_{2k}^0 with subscripts $k \leq n-1$. Substituting them into the right-hand side of (22.40) (in which v_{01} is replaced by v_{02}), we see that the function $H_{n,0}(u)$ admits of the representation

$$H_{n,0}(u) = \sum_{l=0}^{[\frac{n}{p+1}]} \ln^l u \sum_{s=0}^{n-l(p+1)} u^{p-s} c_{ls}(u) +$$

$$+ \sum_{k=0}^{n-1} \beta(n-k) c_k(u) \ln^{\frac{n-k}{p+1}} u, \qquad (22.55)$$

where $\beta(n)$ is defined by (22.44) and c_{ls}, c_k, are some functions that can be expanded in series in the integer powers of u^{p+1}. Substituting (22.55) into (22.53) and integrating by parts, we arrive at (22.54) for $m = 0$. It must be pointed out in conclusion that (22.43) is a special case of (22.54) when $\lambda_{n,m}^k = 0$ and $q = n - (m+l)(p+1)$.

Relations (22.52)–(22.54) lead to the following statement.

LEMMA 22.9. *The asymptotic relation*

$$x(u,\mu) = v_{02}(u) + \sum_{n=1}^{N} \mu^n \sum_{m=0}^{[\frac{n}{p+1}]} v_{2n}^m(u) \ln^m \frac{1}{\mu} +$$

$$+ O(\mu^{(N+1)(1-\lambda(p+1))+\lambda(p+1)^2}) \qquad (22.56)$$

is valid uniformly on

$$\mu^{\lambda(p+1)} \leq u \leq u_1, \quad \text{where} \quad 0 < \lambda < 1/(p+1), \quad u_1 \in (0, u_*),$$

for the trajectory L at the beginning of the "return" segment.

It remains to consider the last segment of the trajectory L, the end of the "return" segment. We make the substitution $\varepsilon = \mu^{p+1}$ in (22.1) and obtain the relation

$$\frac{dx}{dy} = \mu^{p+1} \frac{1}{y} G(x,y), \qquad (22.57)$$

whose solution with the initial condition

$$x|_{y=\varepsilon^{\lambda-1/(p+1)}} = x(u,\mu)|_{u=\mu^{\lambda(p+1)}}, \qquad (22.58)$$

where $x(u,\mu)$ is function (22.56), is denoted by $x(y,\mu)$. Relations (22.54), (22.56), and (22.58) leads us to the conclusion that this solution can be expanded in the asymptotic series

$$x = x_2 + \sum_{n=1}^{\infty} \mu^n w_n(y,\mu), \qquad (22.59)$$

where

$$w_n(y,\mu) = \sum_{s=0}^{[\frac{n}{p+1}]} w_{n,s}(y) \ln^s \frac{1}{\mu}. \qquad (22.60)$$

Here we give an algorithm of calculating the coefficients $w_{n,s}$. Substituting (22.59) into (22.57) and equating the coefficients in like powers of μ, we successively obtain

$$\frac{dw_n}{dy} \equiv 0, \qquad 1 \le n < p+1, \qquad (22.61)$$

$$\frac{dw_{p+1}}{dy} = \frac{1}{y} G(x_2, y), \qquad (22.62)$$

$$\frac{dw_n}{dy} = \sum_{k=1}^{n-p-1} \frac{G_x^{(k)}(x_2, y)}{y k!} \sum_{\substack{i_1+\ldots+i_k=n-p-1 \\ i_j \ge 1}} w_{i_1} \ldots w_{i_k}, \qquad n > p+1. \qquad (22.63)$$

Note that the right-hand side of (22.63) depends only on w_m with subscripts $m \le n - p - 1$ and is a polynomial of the form

$$\sum_{s=0}^{[\frac{n}{p+1}]-1} V_{n,s}(y) \ln^s \frac{1}{\mu}.$$

Hence, we arrive at the equation with separable variables

$$\frac{dw_{n,s}}{dy} = V_{n,s}, \quad s = 0, 1, \ldots, \left[\frac{n}{p+1}\right] - 1, \quad \frac{dw_{n,[\frac{n}{p+1}]}}{dy} \equiv 0 \qquad (22.64)$$

for $w_{n,s}$.

Here is how the particular solutions of (22.61), (22.62), (22.64) can be chosen. Setting $u = \mu y$ in (22.52), taking into account expansions (22.54), and regrouping terms, we infer that, as $y \to \infty$, the functions $w_{n,s}$ must be expandable in the asymptotic series

$$w_{n,0} = \beta(n)a_n y^n + y^{n-1} \sum_{\varkappa=0}^{[\frac{n}{p+1}]-1} \ln^{\varkappa} y \sum_{k=0}^{\infty} \frac{a_{\varkappa,k}^{n,0}}{y^k} +$$

$$+ \beta(n)a_{n,0} \ln^{\frac{n}{p+1}} y, \qquad (22.65)$$

$$w_{n,s} = y^{n-1} \sum_{\varkappa=0}^{[\frac{n}{p+1}]-s-1} \ln^{\varkappa} y \sum_{k=0}^{\infty} \frac{a_{\varkappa,k}^{n,s}}{y^k} + \beta(n)a_{n,s} \ln^{\frac{n}{p+1}-s} y,$$

$$s = 1,\ldots, \left[\frac{n}{p+1}\right] - 1. \qquad (22.66)$$

Therefore we must choose the functions

$$w_n \equiv w_{n,0} \equiv \lambda_{n,0}^0, \qquad 1 \leq n < p+1,$$

$$w_{p+1,0} = -(H)\int_y^{\infty} \frac{G(x_2,\sigma)}{\sigma} d\sigma + \omega_{p+1,0,0}^{0,0} + \lambda_{p+1,0}^0,$$

$$w_{n,s} = -(H)\int_y^{\infty} V_{n,s}(\sigma) d\sigma + \sum_{r=0}^{[\frac{n}{p+1}]-s-1} \omega_{n,s,0}^{n-(s+r+1)(p+1),r} + \lambda_{n,s}^0,$$

$$s = 0,1,\ldots, \left[\frac{n}{p+1}\right] - 1,$$

$$w_{n,[\frac{n}{p+1}]} \equiv \lambda_{n,[\frac{n}{p+1}]}^0 + \beta(n)\omega_{n,\frac{n}{p+1}},$$

where all the constants are taken from (22.54), as particular solutions of (22.61), (22.62), and (22.64).

Using expansions (22.65), (22.66), we make sure of the validity of the following statement, which concludes the construction of the asymptotics of the trajectory L.

LEMMA 22.10. *The asymptotic representation*

$$x(y,\mu) = x_2 + \sum_{n=1}^{N} \mu^n w_n(y,\mu) + O(\mu^{\lambda(p+1)(N+1)}) \qquad (22.67)$$

holds true uniformly on

$$1 \leq y \leq \varepsilon^{\lambda - 1/(p+1)}, \quad \text{where} \quad 0 < \lambda < 1/(p+1),$$

for the trajectory L at the end of the "return" segment.

22.3. Completion of the Proof of Theorem 21.1

We have thus constructed the complete asymptotics of the trajectory L of system (20.1) for $p \geq 0$. Now if $p = -1$, then all the results given in 22.1 remain valid, and for $y \geq 1$, after the substitution $1/y \to y$, we again obtain an equation similar to (22.1). This reduces the problem of constructing the asymptotics to the situation that we have already studied. We have thus substantiated the asymptotic expansions (21.1) and (21.2) in the metric $C[q_0, q_1]$.

It is sufficient that all the asymptotic relations for the trajectory L given above can be differentiated any number of times with respect to the variable x_0. Indeed, after a formal differentiation with respect to x_0, we get linear nonhomogeneous equations whose solutions can be written in explicit form. Analyzing them with due account of the formulas obtained for the trajectory L, we infer that expansions (21.1), (21.2) are valid in the metric $C^k[q_0, q_1]$ for any $k \geq 0$. This means that we have completed the proof of Theorem 21.1, which is our main statement.

As was pointed out in 21.1, Theorem 21.2 is a simple corollary of Theorem 21.1, and, therefore, we pass to the substantiation of Theorem 21.3 under whose conditions $p = -k_0$, $k_0 \geq 2$.

We denote by $x(y, \varepsilon)$ the solution of (22.1) with the initial condition $x(1, \varepsilon) = x_0 > q_2$ and seek it in the form of the asymptotic series

$$x(y, \varepsilon) = x_0 + \sum_{k=1}^{\infty} \varepsilon^k x_k(y). \qquad (22.68)$$

Omitting simple calculations, we want to point out that here the functions $x_1(y)$, $x_k(y)$, $k \geq 2$, can be determined with the use of relations (22.5) and, as $y \to \infty$, the asymptotic expansions

$$x_k(y) = b_{k0} + \sum_{n=k_0-1}^{\infty} \frac{b_{kn}}{y^n} \qquad (22.69)$$

hold true for them.

The other arguments are standard (see the proof of Lemma 22.1), namely, we set in (22.1) $x = x_N(y, \varepsilon) + h$, where

$$x_N(y, \varepsilon) = x_0 + \sum_{n=1}^{N} \varepsilon^n x_n(y),$$

and pass to an integer equation for determining h. Expansions (22.69) show that the right-hand side of the resulting integer equation maps a ball of the space $C[1,\infty)$ with radius of order ε^{N+1} into itself and is a contraction mapping. Therefore,

$$x(y,\varepsilon) = x_0 + \sum_{k=1}^{N} \varepsilon^k x_k(y) + O(\varepsilon^{N+1}) \qquad (22.70)$$

uniformly on $1 \leq y < \infty$. Moreover, it follows from Conditions 20.1 and 20.2 that for all sufficiently large y the function $x(y,\varepsilon)$ decreases monotonically and, hence, has a finite limit $x_\infty(\varepsilon)$ as $y \to \infty$. Combining relations (22.69) and (22.70), we arrive at (21.6). We have proved Theorem 21.3.

23. Multidimensional Systems of the Lotka–Volterra Type

23.1. Generalization to the Case $x \in R^m$, $y \in R$

Under certain constraints, the constructions given in Secs. 20–22 are preserved if we assume that $x = (x_1, \ldots, x_m)$, $m \geq 2$, in (20.1) and the right-hand sides $f, g \in C^\infty(\Lambda)$,

$$\Lambda = \{(x,y) : x_k > -\delta_0, \quad k = 1, \ldots, m, \quad y > -\delta_0\}, \quad \delta_0 > 0.$$

CONDITION 23.5. The equation $f(x,0) = 0$ defines the smooth surface l_1, which divides the domain $K_0 = \{x : x_k > 0, k = 1, \ldots, m\}$ into two subdomains

$$K_0^+ = K_0 \cap \{x : f(x,0) > 0\} \quad \text{and} \quad K_0^- = K_0 \cap \{x : f(x,0) < 0\}.$$

The formulated condition means that the integer manifold $y = 0$ of (20.1) consists of two parts, a stable part and an unstable part (in Sec. 20 the point $x = q_2$ plays the part of the curve l_1).

We fix a certain domain Ω_0 with boundary Γ_0 which lies entirely on the stable part of the manifold $y = 0$ (i.e., $\bar{\Omega}_0 = \Omega_0 + \Gamma_0 \subset K_0^-$) and analyze the behavior of the trajectory of system (20.1) with the initial conditions $(x_0, 1)$, $x_0 \in \bar{\Omega}_0$, defined for $t = 0$. Our aim is to formulate the constraints under which the Poincaré operator along the trajectories of (20.1), $x \in R^m$, $y \in R$, is defined in the hyperplane $y = 1$ and its zero approximation $\Pi_0 : \bar{\Omega}_0 \to \bar{\Omega}_0$ can be constructed in the same way as the function $\Pi_0(x)$ from Sec. 20.

We denote by $x_0(t)$ the solution of the system $\dot{x} = g(x,0)$ with the arbitrary initial condition $x_0(0) = x_0 \in \bar{\Omega}_0$. Below $[0,T]$ is an interval on which all solutions of this kind are assumed to be defined.

CONDITION 23.6. For any $x_0 \in \bar{\Omega}_0$ there is a moment of time $t_0 = t_0(x_0) \in (0, T)$ such that
$$x_0(t) \in K_0 \quad \text{for} \quad 0 \le t < t_0 \quad \text{and} \quad x_0(t) \in K_0^+ \quad \text{for} \quad t_0 < t \le T,$$
and for $t = t_0$ the curves $x_0(t)$ and l_1 generically intersect, i.e.,
$$(g(x, 0), \operatorname{grad}_x f(x, 0))|_{x=x_0(t_0)} \ne 0. \tag{23.1}$$

CONDITION 23.7. For every $x_0 \in \bar{\Omega}_0$ the equation
$$-\int_0^{t_0} f(x_0(s), 0) ds = \int_{t_0}^{t} f(x_0(s), 0) ds \tag{23.2}$$
has a root $t_1 = t_1(x_0) \in (t_0, T)$, which is unique since the right-hand side is monotonic.

Under the indicated conditions, first a "drop" occurs from the point $(x_0, 1)$ onto the integer manifold $y = 0$ and then a "breakoff" from it approximately at the moment of time $t = t_1$ (see the analogous place in Sec. 20). In order to follow the motion of the phase point of (20.1) after the "breakoff," we must impose constraints upon the behavior of the right-hand sides f and g as $y \to \infty$. With this aim in view, we denote by Π_1 the mapping which puts the point $x_0(t_1)$ into correspondence with the arbitrary point $x_0 \in \bar{\Omega}_0$, and set $\Omega_1 = \Pi_1 \Omega_0$.

CONDITION 23.8. As $y \to \infty$, the asymptotic representations
$$g(x, y) = y^{n_1} \left[a_0(x) + \frac{a_1(x)}{y} + \frac{a_2(x)}{y^2} + \ldots \right],$$
$$f(x, y) = y^{n_2-1} \left[b_0(x) + \frac{b_1(x)}{y} + \frac{b_2(x)}{y^2} + \ldots \right], \tag{23.3}$$
where (just as in Condition 20.3) the natural numbers n_1, n_2 are such that $p = n_1 - n_2 \ge -1$, hold true uniformly with respect to x from any bounded subdomain of the domain K_0. We can differentiate (23.3) with respect to x, y in any order any number of times. We assume additionally that the equation $b_0(x) = 0$ defines the smooth surface l_2 that divides the domain K_0 into the two subdomains
$$F^+ = K_0 \cap \{x : b_0(x) > 0\} \quad \text{and} \quad F^- = K_0 \cap \{x : b_0(x) < 0\},$$
with $\bar{\Omega}_0 \subset F^-$, $\bar{\Omega}_1 \subset F^+$ (in Sec. 20 the part of the curve l_2 is played by the point $x = q_1$). Finally, let
$$\begin{aligned} f(x, y) < 0 & \quad \text{for} \quad (x, y) \in \bar{\Omega}_0 \times \{y : y \ge 0\}, \\ f(x, y) > 0 & \quad \text{for} \quad (x, y) \in \bar{\Omega}_1 \times \{y : y \ge 0\}. \end{aligned} \tag{23.4}$$

As in Sec. 20, relations (23.3) allow us to introduce a system
$$\frac{dx}{du} = u^p \frac{a_0(x)}{b_0(x)} \tag{23.5}$$
for $p \geq 0$ and
$$\frac{dx}{du} = \frac{a_0(x)}{b_0(x)} \tag{23.6}$$
for $p = -1$. We denote by $x = v_{01}(u)$ the solution of (23.5) or (23.6) with the initial condition $v_{01}(0) = x_1$, where x_1 is an arbitrary point from $\bar{\Omega}_1$.

CONDITION 23.9. For any point $x_1 \in \bar{\Omega}_1$ there exists $u_* = u_*(x_1) > 0$ such that $v_{01}(u) \in F^+$ for $0 \leq u < u_*$, $v_{01}(u_*) \in l_2$, and
$$(a_0(x), \operatorname{grad} b_0(x))|_{x=v_{01}(u_*)} \neq 0. \tag{23.7}$$

It follows from inequality (23.7) that another solution, namely, $x = v_{02}(u) \in F^-$, of (23.5) or (23.6), passes through the singular point $(v_{01}(u_*), u_*)$ for $u_* - \delta \leq u < u_*$, where $\delta > 0$ is sufficiently small.

CONDITION 23.10. For any $x_1 \in \bar{\Omega}_1$ the solution $x = v_{02}(u)$ is defined, with the inclusion $v_{02}(u) \in F^-$ preserved up to the point $u = 0$ and $v_{02}(0) \in \bar{\Omega}_0$.

We have formulated all the constraints allowing us to follow the motion of the phase point of (20.1), $x \in R^m$, $y \in R$, after the "breakoff" from the manifold $y = 0$ for $t = t_1$. Indeed, the second inequality of (23.4) shows that first the trajectory of (20.1) freely "takes off" almost along the straight line $x = x_1$, $x_1 = x_0(t_1)$, to an asymptotically great height with respect to y. Then expansion (23.3) becomes valid and the trajectory moves approximately along the curve $(v_{01}(u), u)$, $0 \leq u \leq u_*$. A "turn" occurs in the neighborhood of the point $(v_{01}(u_*), u_*)$ and the trajectory returns, first along the curve $(v_{02}(u), u)$ and then (by virtue of the first inequality of (23.4)) almost along the straight line $x = v_{02}(0)$, to the intersecting hyperplane $y = 1$. Thus, the zero approximation of the Poincaré operator Π_0 along the trajectories of system (20.1) has the form $\Pi_2 \cdot \Pi_1$, where the operator Π_2 puts the value $x_2 = v_{02}(0)$ into correspondence with the arbitrary point $x_1 \in \bar{\Omega}_1$ (see Fig. 23.1, which shows the case $m = 2$).

When Conditions 23.1–23.6 are satisfied, the statement of the base Theorem 21.1 is completely valid for system (20.1), $x \in R^m$, $y \in R$, with the only difference that now the asymptotic relations (21.1) and (21.2) are valid in the metric $C^k(\bar{\Omega}_0)$ for any fixed $k \geq 0$. Its substantiation remains practically the same (in all of the asymptotic expansions encountered in Sec. 22 the scalar coefficients are replaced by vector ones, but this is inessential).

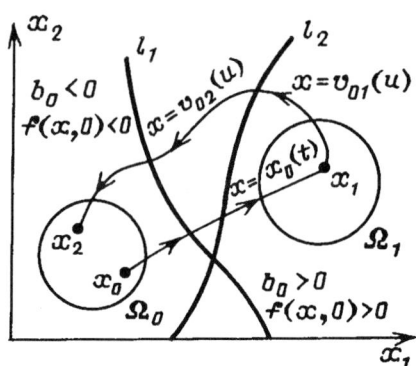

Fig. 23.1

A new circumstance is found in the hypothesis of Theorem 21.2, when the mapping Π_0 has an exponentially stable (or dichotomous) cycle with period $n_0 > 1$. Such a cycle is associated with a cycle of system (20.1) of the same stability which has n_0 fast-motion and slow-motion portions. For $n_0 = 2$ and $m = 2$ in the variables (x_1, x_2, u) the zero approximation of this cycle is shown in Fig. 23.2, where AB, CD and BC, DA are slow-motion and fast-motion portions respectively.

In this case, the statement of Theorem 21.3 is completely preserved if we require that $x_0 \in \bar{\Omega}_1$ instead of $x_0 > q_2$.

In conclusion, it should be pointed out that Theorem 21.1 makes it possible to establish a correspondence not only between the cycles of mapping Π_0 and system (20.1) but also between other structurally stable objects (tori, hyperbolic sets, etc.).

23.2. Diffusion Instability of Cycles

Let us suppose that the mapping $\Pi_0(x)$ introduced in 20.2 has a fixed point $x_* \in (q_0, q_1)$, with

$$\Pi_0'(x_*) = \frac{a_0(x_*)b_0(x_1)f(x_*,0)g(x_1,0)}{a_0(x_1)b_0(x_*)f(x_1,0)g(x_*,0)} < 1, \tag{23.8}$$

where, we recall, x_1 is a root of Eq. (20.5) for $x_0 = x_*$. Then, in accordance with Theorem 21.2, system (20.1), $x \in R^m$, $y \in R$, has a relaxation cycle

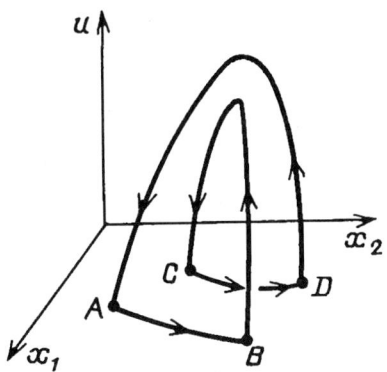

Fig. 23.2

$x(t, \varepsilon)$, $y(t, \varepsilon)$.

Let us add terms $D_1 \Delta x$, $D_2 \Delta y$, $D_1, D_2 > 0$, to the right-hand sides of (20.1) respectively and impose boundary Neumann conditions on the bounded domain $\Omega \subset R^n$, $n \geq 1$. As a result we obtain a parabolic boundary-value problem with the spatially homogeneous cycle $x(t, \varepsilon)$, $y(t, \varepsilon)$. As was shown in 11.3, the special four-dimensional relaxation system

$$\dot{x}_1 = \frac{z_1}{2}(x_2 - x_1) + g(x_1, y_1),$$

$$\dot{x}_2 = \frac{z_1}{2}(x_1 - x_2) + g(x_2, y_2),$$

$$\varepsilon \dot{y}_1 = \frac{z_2}{2}(y_2 - y_1) + y_1 f(x_1, y_1),$$

$$\varepsilon \dot{y}_2 = \frac{z_2}{2}(y_1 - y_2) + y_2 f(x_2, y_2), \qquad z_1, z_2 \geq 0,$$

(23.9)

called a *bilocal model*, is responsible for its stability in the framework of the resulting parabolic system. Thus, we are faced with the problem of extending the results of Secs. 20 and 21 to this special multidimensional case.

Below we give its solution for $z_2 \geq z_{20} > 0$, where z_{20} is a fixed number. We also assume that the functions f, g satisfy Conditions 20.1–20.4 and that

$$f'_x(q_2, 0) \neq 0 \qquad (23.10)$$

which characterizes a certain generality.

We introduce into consideration a Poincaré operator $\Pi(y_1, x_1, x_2, \varepsilon)$ along the trajectories of (23.9), which is defined in some neighborhood of the initial condition of the homogeneous cycle in the intersecting hyperplane $y_1+y_2 = 2$. Our purpose is to construct its zero approximation $\Pi_0(y_1, x_1, x_2)$.

As in 23.1, we shall consider the behavior of the trajectories of (23.9) with the initial conditions (y_1^0, x_1^0, x_2^0) from the hyperplane $y_1 + y_2 = 2$ defined for $t = 0$. We assume that $(x_1^0, x_2^0) \in \Omega_0$, where Ω_0 is a ball of radius r_0 with center at the point (x_*, x_*) and $r_0 > 0$ is sufficiently small.

According to the results of [7], first we observe a "drop" onto the stable integral manifold $y_1 = y_2 = 0$ and a motion along it approximately according to the law

$$\dot{x}_1 = \frac{z_1}{2}(x_2 - x_1) + g(x_1, 0), \qquad x_1(0) = x_1^0,$$
$$\dot{x}_2 = \frac{z_1}{2}(x_1 - x_2) + g(x_2, 0) \qquad x_2(0) = x_2^0. \tag{23.11}$$

In order to follow the further evolution of the phase point of (23.9), we shall consider the equation

$$\lambda(x_1, x_2) = 0, \tag{23.12}$$

where

$$\lambda(x_1, x_2) = \frac{1}{2}[f(x_1, 0) + f(x_2, 0) - z_2] +$$
$$+ \frac{1}{2}\sqrt{z_2^2 + (f(x_1, 0) - f(x_2, 0))^2} \tag{23.13}$$

is the largest eigenvalue of the matrix

$$\begin{pmatrix} f(x_1, 0) - \frac{z_2}{2} & \frac{z_2}{2} \\ \frac{z_2}{2} & f(x_2, 0) - \frac{z_2}{2} \end{pmatrix},$$

which is responsible for the stability of the integer manifold $y_1 = y_2 = 0$. By virtue of the inequality $z_2 \geq z_{20} > 0$, the function (23.13) is smooth in the neighborhood of the straight line $x_1 = x_2$ and, according to (23.10),

$$\text{grad } \lambda|_{x_1=x_2=q_2} = (f'_x(q_2, 0), f'_x(q_2, 0)) \neq 0. \tag{23.14}$$

Therefore, (23.12) defines the smooth curve l in the neighborhood of the point (q_2, q_2) (Fig. 23.3).

Suppose that $x_1(t), x_2(t)$ is the solution of the Cauchy problem (23.11) with the arbitrary initial condition $(x_1^0, x_2^0) \in \Omega_0$. According to inequality (23.10), for a sufficiently small r_0, the curves $(x_1(t), x_2(t))$ and l generically intersect for a certain $t = l_0(x_1^0, x_2^0) > 0$ (see Fig. 23.3) and the equation

$$\int_0^{t_1} \lambda(x_1(t), x_2(t))dt = 0 \tag{23.15}$$

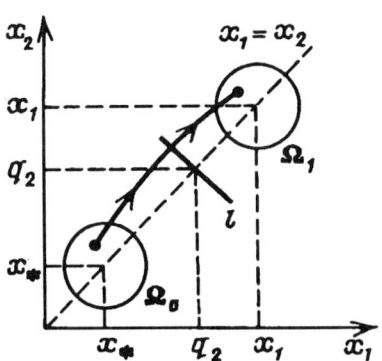

Fig. 23.3

has a unique root $t_1 = t_1(x_1^0, x_2^0) > t_0$ for every $(x_1^0, x_2^0) \in \Omega_0$. Thus, approximately for $t = t_1$, a "breakoff" occurs from the manifold $y_1 = y_2 = 0$ and a "take-off" with respect to the variables y_1, y_2 to an asymptotically great height.

We denote here by Π_1 an operator which puts the point $(x_1^1, x_2^1) = (x_1(t_1), x_2(t_1))$ into correspondence with the arbitrary point $(x_1^0, x_2^0) \in \Omega_0$ and set $\Omega_1 = \Pi_1 \Omega_0$. The constructions given above show that $(0, \Pi_1(x_1^0, x_2^0))$ is the zero approximation of the Poincaré operator along the trajectories of (23.9) from the hyperplane $y_1 + y_2 = 2$ to the same hyperplane after the "breakoff" from the manifold $y_1 = y_2 = 0$. In order to find the zero approximation of the other part of the operator $\Pi(y_1, x_1, x_2, \varepsilon)$, we use the asymptotic representations (20.2), in which we assume that $p \geq 0$ (for $p = -1$ this question remains open). As a result, after the substitution $y_1 = u_1 \varepsilon^{-1/(p+1)}$, $y_2 = u_2 \varepsilon^{-1/(p+1)}$, a suitable normalization of time, and rejection of terms of a higher order of smallness, we get the system

$$\dot{x}_1 = u_1^{n_1} a_0(x_1), \qquad \dot{x}_2 = u_2^{n_1} a_0(x_2),$$
$$\dot{u}_1 = u_1^{n_2} b_0(x_1), \qquad \dot{u}_2 = u_2^{n_2} b_0(x_2) \qquad (23.16)$$

for $n_2 > 1$ and the system

$$\dot{x}_1 = u_1^{n_1} a_0(x_1), \qquad \dot{x}_2 = u_2^{n_1} a_0(x_2),$$
$$\dot{u}_1 = \frac{z_2}{2}(u_2 - u_1) + u_1 b_0(x_1), \qquad (23.17)$$
$$\dot{u}_2 = \frac{z_2}{2}(u_1 - u_2) + u_2 b_0(x_2)$$

for $n_2 = 1$.

First suppose that $n_2 > 1$. Then we find from (23.16) that

$$\frac{dx_1}{du_1} = u_1^p \frac{a_0(x_1)}{b_0(x_1)}, \qquad \frac{dx_2}{du_2} = u_2^p \frac{a_0(x_2)}{b_0(x_2)}.$$

Therefore, the zero approximation of the Poincaré operator from the hyperplane $y_1 + y_2 = 2$, $(x_1^1, x_2^1) \in \Omega_1$, again into the hyperplane $y_1 + y_2 = 2$ has the form $(0, \Pi_2(x_1^1, x_2^1))$, where

$$\Pi_2(x_1^1, x_2^1) = (0, x_1^2, x_2^2), \tag{23.18}$$

and x_1^2, x_2^2 are the roots of the equations

$$\int_{x_1}^{x_1^1} \frac{b_0(s)}{a_0(s)} ds = 0, \qquad \int_{x_2}^{x_2^1} \frac{b_0(s)}{a_0(s)} ds = 0 \tag{23.19}$$

from the interval (q_0, q_1) respectively.

For $n_2 = 1$ we denote by

$$u_2 = \varphi(u_1, x_1^1, x_2^1), \qquad \varphi(0, x_1^1, x_2^1) = 0,$$

$$\varphi'_{u_1}(0, x_1^1, x_2^1) = \left[b_0(x_2^1) - b_0(x_1^1) + \sqrt{(b_0(x_2^1) - b_0(x_1^1))^2 + z_2^2}\right]/z_2$$

the integral manifold of the equilibrium position $u_1 = u_2 = 0$, $x_1 = x_1^1$, $x_2 = x_2^1$ of (23.17) corresponding to the maximum (positive) eigenvalue of the matrix

$$\begin{pmatrix} b_0(x_1^1) - \dfrac{z_2}{2} & \dfrac{z_2}{2} \\ \dfrac{z_2}{2} & b_0(x_2^1) - \dfrac{z_2}{2} \end{pmatrix}.$$

As $t \to \infty$, the trajectory of (23.17), which corresponds to this manifold, tends to one of the equilibrium positions $u_1 = u_2 = 0$, $x_1 = x_1^2$, $x_2 = x_2^2$, where (x_1^2, x_2^2) lies in a certain sufficiently small neighborhood of the point (x_*, x_*) (this follows from the fact that the same occurrence is observed for $x_1^1 = x_2^1$). We have thus determined the operator $\Pi_2 (x_1^1, x_2^1) = (x_1^2, x_2^2)$.

Thus, the zero approximation $\Pi_0(y_1, x_1, x_2)$ of the Poincaré operator along the trajectories of (23.9) has the form

$$\Pi_0 = (0, \Pi_2 \cdot \Pi_1).$$

Using the technique of asymptotic integration developed in Sec. 22, we can make sure that the limiting relation

$$\lim_{\varepsilon \to 0} \Pi(y_1, x_1, x_2, \varepsilon) = \Pi_0$$

is valid in the metric $C^k(\Omega_0)$ for any $k \geq 0$. Therefore, the matrix

$$\Pi'_2(x_1, x_1) \cdot \Pi'_1(x_*, x_*) \tag{23.20}$$

is responsible for the stability of the homogeneous cycle of (23.9). The linearization of system (23.9) on a homogeneous cycle and the projection of the Laplace difference operator onto the eigenvectors $e_1 = (1,1)$ and $e_2 = (1,-1)$ show that these vectors are eigenvectors for matrix (23.20) as well. In this case the vector e_1 is associated with the eigenvalue (23.8). The explicit formula for the other eigenvalue can only be written out for $n_2 > 1$, since in this case only the most complicated operator Π_2 is explicitly defined (see (23.18), (23.19)). Omitting elementary calculations, we give the final formula for the required eigenvalue:

$$\frac{g(x_1, 0) b_0(x_1) a_0(x_*)}{g(x_*, 0) b_0(x_*) a_0(x_1)} \exp\left\{-z_1 \int_{x_*}^{x_1} \frac{dx}{g(x, 0)}\right\}. \tag{23.21}$$

In conclusion, we want to point out the peculiarities encountered here. First, eigenvalue (23.21) is negative by virtue of Conditions 20.1–20.3. Second, if

$$\left|\frac{g(x_1, 0) b_0(x_1) a_0(x_*)}{g(x_*, 0) b_0(x_*) a_0(x_1)}\right| > 1,$$

then, for a suitable decrease of z_1, the homogeneous cycle of (23.9) loses stability when passing through the multiplier equal to -1. For the sake of comparison, recall (see 11.2) that in the general case the classical relaxation cycle loses stability because of the diffusion only divergently, when passing through a unit multiplier. Finally, as in 11.3, for small z_2 it is impossible to investigate a homogeneous cycle for stability by geometric methods since, in this case, as $z_2 \to 0$, function (23.13) and the manifold $u_2 = \varphi_2(u_1, x_1^1, x_2^1)$ lose smoothness.

24. A New Type of Relaxation Oscillations

24.1. Statement of the Problem and Description of the Results

From the presented theory of classical relaxation oscillations in second-order systems

$$\dot{x} = g(x, y), \quad \varepsilon \dot{y} = f(x, y), \quad 0 < \varepsilon \ll 1, \quad f, g \in C^\infty(R^2), \tag{24.1}$$

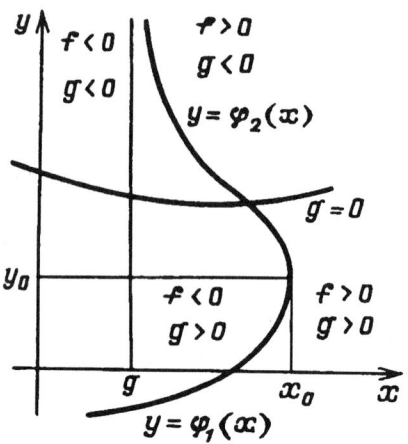

Fig. 24.1

it follows that their cycles are structurally stable, i.e., vary little for small (independent of ε) smooth perturbations of the right-hand sides f and g. Below we describe conditions under which systems of the form (24.1) have structurally stable nonclassical relaxation oscillations. Recall that the concept of structural stability was considered in 21.2. It was also indicated there that the cycles of system (20.1) are structurally stable if the perturbations are of the same class. The results of this subsection allow us to follow their evolution for arbitrary perturbations of the right-hand sides of the equations.

CONDITION 24.11. The equation $f(x, y) = 0$ defines the curves $y = \varphi_1(x)$ and $y = \varphi_2(x)$ with the following properties:

$$\varphi_1(x) \in C^\infty(-\infty, x_0), \qquad \varphi_2(x) \in C^\infty(q, x_0), \qquad q < x_0;$$

$$\varphi_1(x) < \varphi_2(x) \text{ for } q < x < x_0; \qquad \varphi_2 \to \infty \text{ as } x \to q + 0;$$

$\varphi_1(x_0) = \varphi_2(x_0) = y_0$; $f'_y < 0$, $g > 0$ on the curve $y = \varphi_1(x)$ with $f'_y = 0$, $g > 0$ at the point (x_0, y_0); $f'_y > 0$ on the curve $y = \varphi_2(x)$. The isocline $g = 0$ is of the form shown in Fig. 24.1.

CONDITION 24.12. The ordinary nondegeneracy conditions

$$f'_x(x_0, y_0) \neq 0, \qquad f''_{yy}(x_0, y_0) \neq 0$$

are satisfied at the junction point (x_0, y_0).

CONDITION 24.13. As $y \to +\infty$, the asymptotic expansions

$$g(x,y) = y^{n_1}\left[a_0(x) + \frac{a_1(x)}{y} + \frac{a_2(x)}{y^2} + \cdots\right],$$
$$f(x,y) = y^{n_2}\left[b_0(x) + \frac{b_1(x)}{y} + \frac{b_2(x)}{y^2} + \cdots\right],$$
(24.2)

where the integers $n_1, n_2 \geq 1$, $p = n_1 - n_2 \geq -1$, hold true uniformly with respect to x from any bounded set. These expansions can be differentiated with respect to x, y in any order and any number of times. To make Condition 24.3 agree with Condition 24.1, we assume the following:

$$a_0(x) < 0 \quad \text{for all} \quad x;$$

$$b_0(x) > 0 \quad \text{for} \quad q < x < \infty \quad \text{and} \quad b_0(x) < 0 \quad \text{for} \quad -\infty < x < q;$$

$$b_0'(q) > 0, \quad b_1(q) < 0.$$

To formulate the last restriction, we introduce into consideration, as we did in 20.2, the equation

$$\frac{dx}{du} = u^p \frac{a_0(x)}{b_0(x)} \tag{24.3}$$

for $p \geq 0$ and the equation

$$\frac{dx}{du} = \frac{a_0(x)}{b_0(x)} \tag{24.4}$$

for $p = -1$. As before, we denote by $x = v_{01}(u)$ the solution of (24.3) or (24.4) with the initial condition $v_{01}(0) = x_0$. By virtue of Condition 24.3, $a_0(x)/b_0(x) < 0$ for $q < x \leq x_0$ and

$$a_0(x)/b_0(x) = O(1/(x-q)), \quad x \to q, \tag{24.5}$$

and, therefore, the solution $v_{01}(u)$ is only defined for $0 \leq u \leq u_*$, with $v_{01}(u_*) = q$, where u_* is defined by (20.9) or (20.10). It is important that, by virtue of (24.5), Eq. (24.3) or (24.4) has a unique solution $x = v_{02}(u)$, $v_{02}(u_*) = q$, which lies in the half-plane $x \leq q$ and is defined, in general, only locally. Recall that in 20.2 a similar solution was defined for all $u \in (-\infty, u_*]$, since it tends to the equilibrium position as $u \to -\infty$. In our case, Eq. (24.3) or (24.4) has no equilibrium position and, consequently, the necessity of the following restriction arises.

CONDITION 24.14. We suppose that the solution $x = v_{02}(u)$ is defined for $0 \leq u \leq u_*$.

These conditions are defined, for instance, by the system

$$\dot{x} = 2 - y, \qquad \varepsilon \dot{y} = x(y+1)^2 - y,$$

considered in the domain $y > -1$.

At the heuristic level, the behavior of the trajectories of (24.1) with the initial conditions (x, y_0), $x \in [q - \delta, q + \delta]$, $0 < \delta < x_0 - q$, defined for $t = 0$, is clear. Indeed, first, according to the results of [7], the phase point "drops" onto the stable integral manifold $y = \Phi(x, \varepsilon)$, $\Phi(x, 0) = \varphi_1(x)$, moving along it approximately according to the law $\dot{x} = g(x, \varphi_1(x))$. In the neighborhood of the point (x_0, y_0) a "breakoff" occurs from the manifold $y = \Phi(x, \varepsilon)$ and a "take-off" almost along from the straight line $x = x_0$ to an asymptotically great height with respect to y. Then the asymptotic expansions (24.2) are used and, according to (24.3) or (24.4), the phase point begins moving approximately along the curve $x = v_{01}(u)$, where

$$u = \varepsilon^{1/(p+1)} y \quad \text{for} \quad p \geq 0 \quad \text{and} \quad u = \varepsilon \ln y \quad \text{for} \quad p = -1.$$

In the neighborhood of the point (q, u_*) a "turn" occurs and the phase point of (24.1) returns to the segment $l = \{q - \delta \leq x \leq q + \delta, y = y_0\}$, first along the curve $x = v_{02}(u)$ and then, approximately, along the straight line $x = x_1 < x_0$, $x_1 = v_{02}(0)$. Thus, the Poincaré operator $\Pi(x, \varepsilon)$ along the trajectories of (24.1) is defined on this segment, which transforms it into an asymptotically small neighborhood of the point x_1.

THEOREM 24.1. *Under Conditions 24.1–24.4, for all sufficiently small ε, system (24.1) has a unique orbitally exponentially stable (with a multiplier of order $\exp(-c/\varepsilon)$, $c > 0$) relaxation cycle with the leading approximation*

$$\Gamma = \{(x, u) : u = 0, x_1 \leq x \leq x_0\} \cup$$
$$\cup \{(x, u) : x = v_{01}(u), 0 \leq u \leq u_*\} \cup$$
$$\cup \{(x, y) : x = v_{02}(u), 0 \leq u \leq u_*\}.$$

The asymptotic expansions

$$x(\varepsilon) = x_1 + \sum_{n=2, k=1}^{\infty} \varepsilon^{n/3 + k/(p+1)} \sum_{m=0}^{\pi(n-2) + \left[\frac{k}{p+1}\right]} x_{n,k,m} \ln^m \frac{1}{\varepsilon}, \qquad (24.6)$$

$$T(\varepsilon) = T_0 + \sum_{n=2, k=1}^{\infty} \varepsilon^{n/3 + k/(p+1)} \sum_{m=0}^{\pi(n-2) + \left[\frac{k}{p+1}\right]} T_{n,k,m} \ln^m \frac{1}{\varepsilon} \qquad (24.7)$$

are valid for its initial condition $x(\varepsilon)$, defined for $t = 0$ on the segment l, and for the period $T(\varepsilon)$ for $p \geq 0$, and the expansions

$$x(\varepsilon) = x_1 + \sum_{n=2}^{\infty} \varepsilon^{n/3} \sum_{m=0}^{\pi(n-2)} x_{n,m} \ln^m \frac{1}{\varepsilon}, \tag{24.8}$$

$$T(\varepsilon) = T_0 + \sum_{n=2}^{\infty} \varepsilon^{n/3} \sum_{m=0}^{\pi(n-2)} T_{n,m} \ln^m \frac{1}{\varepsilon} \tag{24.9}$$

are valid for $p = -1$. Here

$$T_0 = \int_{x_1}^{x_0} dx/g(x, \varphi_1(x)), \quad \pi(n) = \left[\frac{n}{3}\right] + \begin{cases} 0 & \text{if } n \not\equiv 1 (\mod 3), \\ 1 & \text{if } n \equiv 1 (\mod 3). \end{cases}$$

In order to prove this theorem by means of the technique developed in Secs. 4 and 22, we construct the full asymptotics of the trajectory of (24.1) with the initial conditions for $t = 0$ from the segment l. In this way, we get expansion (24.6) or (24.8) for the operator $\Pi(x, \varepsilon)$ in the C^1-metric, whence the statement of the theorem follows trivially. We also make sure that the relaxation cycle of (24.1) is structurally stable if the perturbations $\Delta g, \Delta f$ are small in the $C^2(\Omega)$-metric, $\Omega = \{(x, y) : |x| \leq \Delta_0, y \geq -\Delta_0\}$ for a suitable $\Delta_0 > 0$.

24.2. The Properties of the Constructed Cycle

Let us consider some peculiarities of the constructed auto-oscillations. We denote by $x_0(t)$ the solution of the Cauchy problem $\dot{x} = g(x, \varphi_1(x))$, $x(0) = x_1$, extended from the interval $[0, T_0]$ to the entire axis by periodicity, and set

$$y_0(t) = \varphi_1(x_0(t)) + \varkappa \sum_{k=-\infty}^{\infty} \delta(t - kT_0),$$

where

$$\varkappa = \int_0^{u_*} \frac{du}{b_0(v_{01}(u))} - \int_0^{u_*} \frac{du}{b_0(v_{02}(u))}$$

for $n_1 = n_2 = 1$ or $n_1 = 1, n_2 = 2$ (see Condition 24.3), $\varkappa = 0$ in the other cases, and $\delta(t)$ is Dirac δ-function.

LEMMA 24.1. For any $t_1 < t_2$, $t_1, t_2 \neq kT_0$, $k = 0, \pm 1, \ldots$, we have

$$\lim_{\varepsilon \to 0} \int_{t_1}^{t_2} \{x(t, \varepsilon) - x_0(t)\} dt = \lim_{\varepsilon \to 0} \int_{t_1}^{t_2} \{y(t, \varepsilon) - y_0(t)\} dt = 0,$$

where $x(t,\varepsilon), y(t,\varepsilon)$, $y(0,\varepsilon) = y_0$, *are components of a relaxation cycle.*

We shall not prove this lemma, referring the reader to a similar place in 20.3.

It must be noted, in addition, that on the interval $[t_0, t_0 + T(\varepsilon)]$, where $t_0 \in (0, T_0)$, outside of an arbitrarily small fixed neighborhood of the point $t = T_0$, the convergence of $x(t,\varepsilon)$, $y(t,\varepsilon)$ to its zero approximations is uniform, and the time the phase point $(x(t,\varepsilon), y(t,\varepsilon))$ remains in the half-plane $y \geq y_0 + 1$ is of order $O(\varepsilon)$ for $n_2 > 1$ and $O(\varepsilon \ln \frac{1}{\varepsilon})$ for $n_2 = 1$. Moreover,

$$\max_t y(t, \varepsilon) \sim \begin{cases} u_* \varepsilon^{-1/(p+1)} & \text{for } p \geq 0, \\ \exp(u_*/\varepsilon) & \text{for } p = -1. \end{cases}$$

Under Condition 24.3, we have $p \geq -1$, i.e., as $y \to \infty$, the function f grows not very rapidly relative to g. Now if $p \leq -2$, then all trajectories of (24.1) with the initial conditions from the interval l recede to infinity almost along the straight line $x = x_0$.

In conclusion, it should be pointed out that for a suitable modification of Conditions 24.1–24.4 the results can be extended to the case $x \in R^m$, $m \geq 2$, $y \in R$ (see Sec. 23). In particular, for $n_2 > 1$, $p \geq 0$, formula (23.21) is responsible for the stability of the relaxation cycle in a medium with diffusion, as before.

24.3. Relaxation Cycle of a Brusselator

Note that Theorem 24.1 remains completely valid if the curve $g = 0$ has a vertical asymptote $x = q_0$, $q_0 < q$. Now if $q_0 = q$, the existence, uniqueness, and stability of a relaxation cycle are preserved but its asymptotics varies. We invite the reader to construct it for this case as an exercise, and shall only consider here the systems of equations

$$\begin{aligned} \dot{x} &= A - (B+1)x + x^2 y, \\ \dot{y} &= Bx - x^2 y, \quad A, B > 0, \end{aligned} \tag{24.10}$$

suggested by Turing in connection with the morphogenesis problem and called a *brusselator*.

We suppose that the parameter B in (24.10) is large. Then, after the normalizations $\varepsilon^2 y \to y$, $x/\varepsilon \to x$, $\varepsilon^2 t \to t$, where $\varepsilon = 1/B$, we arrive at the system

$$\begin{aligned} \varepsilon^3 \dot{x} &= A - (1+\varepsilon)x + x^2 y, \\ \dot{y} &= x - x^2 y, \end{aligned} \tag{24.11}$$

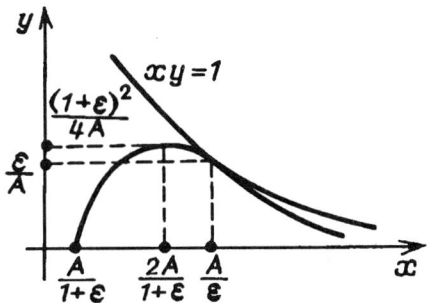

Fig. 24.2

whose null-isoclines are shown in Fig. 24.2. Here the role of the point (x_0, y_0) is played by the point $(2A/(1+\varepsilon), (1+\varepsilon)^2/4A)$, in whose neighborhood an ordinary "breakoff" occurs. After the normalization $x = \varepsilon^{-3}z$, instead of (24.3) or (24.4) we get the equation $dy/dz = -1$, whence we find that $y = \frac{1}{4A} - z$. Thus, in the variables (z, y) the zero approximation of the relaxation cycle of (24.11) has the form

$$\Gamma = \{(z, y) : y = \frac{1}{4A} - z, 0 \le z \le 1/4A\} \cup$$

$$\cup \{(z, y) : y = 0, 0 \le z \le 1/4A\} \cup \{(z, y) : z = 0, 0 \le y \le 1/4A\}.$$

We shall not consider the asymptotic integration of (24.11), also leaving it for the reader to analyze. It should only be pointed out that one of the most important moments, the "turn" of the trajectory in the neighborhood of the point $z = 1/4A$, $y = 0$, is described by the system

$$\varepsilon^3 \dot{v} = -w - \frac{v}{16A^2}, \qquad \dot{w} = \frac{v}{16A^2} + w - \frac{1}{4A}, \qquad (24.12)$$

which results from (24.11) after the substitutions

$$z = 1/4A + \varepsilon w, \qquad y = 4A\varepsilon^3 + \varepsilon^4 v, \qquad \varepsilon^{-3}t \to t$$

and rejection of the terms of a higher order of smallness. Setting $\varepsilon = 0$ in (24.12), we get the equation of "slow" motions $\dot{w} = -1/4A$, from which we see that the variable z begins decreasing in the neighborhood of the value $1/4A$.

Chapter 6

Autowave Processes in Singularly Perturbed Systems of Reaction–Diffusion Type

At the beginning of Ch. 2 we discussed the problem of the origination of spatially nonhomogeneous structures in homogeneous media with diffusion. In Ch. 3 we solved the problem of the structure of the neighborhood of a homogeneous relaxation cycle when it loses stability because of the variation of diffusion coefficients.

The statements given in this chapter refer to parabolic systems with small diffusion on the assumption that the characteristics of the supercritical state have the same order of smallness. Thus we combine here the bifurcation theory in ordinary differential equations and singular perturbations that originate because of the smallness of diffusion coefficients.

This chapter begins with the exposition of the theory of stability of the solutions of linear parabolic systems that smoothly depend on a small parameter. The technique developed is then used to prove two fundamental theorems on the bifurcation of invariant tori. Then we consider some examples which are of interest in themselves. We think that of special interest are the results given in Sec. 30, which throw a new light on the nature of turbulence and, in a certain sense, revive Landau's well-known hypothesis concerning the mechanism of generation of chaos in dynamic systems. In the last section of this chapter we consider a special problem encountered in mathematical ecology.

25. Exponential Dichotomy of Solutions of Linear Parabolic Equations

25.1. Statement of the Problem

In this chapter, we widely use the concept of exponential dichotomy, which is usually associated with the name of Perron. Without substantiating the corresponding statements (which the reader can find in [47]), we only give facts that we shall need later on.

Suppose that E is a real Banach space and H is its ordinary complexification. Let us consider the differential equation

$$\dot{x} + A(t)x = 0 \tag{25.1}$$

in E. We suppose that the domain of definition D of the closed linear operator $A(t)$ does not depend on t and is dense in E. We also suppose that

$$\|[A(t) + \lambda I]^{-1}\|_H \leq M_1[1 + |\lambda|]^{-1}, \qquad \text{Re } \lambda \geq \lambda_0, \tag{25.2}$$

and that

$$\|[A(t_1) - A(t_2)]A_0^{-1}(\tau)\|_E \leq M_2|t_1 - t_2|^\alpha \tag{25.3}$$

for arbitrary t_1, t_2, and τ. In (25.2) and (25.3), M_1, M_2, and λ_0 denote certain constants, $0 < \alpha < 1$,

$$A_0(t) = A(t) + \lambda_0 I, \tag{25.4}$$

and I is an identity operator. There are also two more restrictions of another kind, namely, we suppose that for every $x \in D$ the function $A(t)x$ is almost periodic and that the operator $A_0^{-1}(t)$ is completely continuous for any t.

Inequalities (25.2) and (25.3) imply [85] that Cauchy's initial value problem for the differential equation (25.1) is well posed in the direction of time increase. Without discussing this problem and other well-known problems, we shall go to the heart of the matter.

Let us agree to denote by $C(E)$ the space of continuous functions $x(t)$ which are defined for all $-\infty < t < \infty$ and for which the quantity

$$\|x(t)\|_{C(E)} = \sup_t \|x(t)\|_E$$

is finite. We denote by $C^\alpha(E)$ the subspace of the space $C(E)$ whose elements $x(t)$ satisfy Hölder's condition with exponent α and for which the quantity

$$\|x(t)\|_{C^\alpha(E)} = \|x(t)\|_{C(E)} + \sup_{t_1 \neq t_2} \frac{\|x(t_1) - x(t_2)\|_E}{|t_1 - t_2|^\alpha}$$

is finite.

We introduce the concept of the exponential dichotomy of solutions of the differential equation (25.1). We say that the space E is almost periodically decomposed in the direct sum of two subspaces $E_+(t)$ and $E_-(t)$ if the corresponding families of projectors $P_+(t)$ and $P_-(t)$ almost periodically depend on t in a uniform operator topology and the second family consists of finite-dimensional projectors. The exponential dichotomy of solutions of the differential equation (25.1) is now defined as the union of the following three conditions.

First, there exists an almost periodic splitting of the space E into the direct sum of two subspaces $E_+(t)$ and $E_-(t)$ such that for any τ the solutions $x_+(t)$ and $x_-(t)$ of the differential equation (25.1) with the initial conditions $x_+(\tau) \in E_+(\tau)$ and $x_-(\tau) \in E_-(\tau)$ remain in the subspaces $E_+(t)$ and $E_-(t)$, respectively, as time increases.

Second, each of the solutions $x_-(t)$ can be uniquely extended in the direction of time decrease under the additional condition that $x_-(t) \in E_-(t)$ for all $t \leq \tau$.

Third, there exist two collections of positive constants M_+, γ_+ and M_-, γ_-, respectively, such that

$$\|x_+(t)\|_E \leq M_+ \|x_+(\tau)\|_E \exp[-\gamma_+(t-\tau)], \qquad (25.5)$$

$$\|x_-(t)\|_E \geq M_- \|x_-(\tau)\| \exp[\gamma_-(t-\tau)] \qquad (25.6)$$

for $t \geq \tau$. Note that inequality (25.6) is equivalent to the inequality

$$\|x_-(t)\|_E \leq M_-^{-1} \|x_-(\tau)\|_E \exp[\gamma_-(t-\tau)], \qquad (25.7)$$

which is valid for arbitrary $t \leq \tau$.

Also make note of the fact that the concept of exponential dichotomy that we have introduced differs somewhat from the generally accepted one. The necessity for its modification arises because of its parabolicity and almost periodicity.

We shall describe the concept of regularity of the differential operator

$$Lx(t) = \dot{x}(t) + A(t)x(t), \qquad (25.8)$$

which we assume to be defined on continuously differentiable functions $x(t) \in D \subset C(E)$ such that $x(t) \in D$ and $Lx(t) \in C(E)$ for all t. We say that the differential operator (25.8) is regular if the nonhomogeneous differential equation

$$\dot{x} + A(t)x = f(t) \qquad (25.9)$$

has a unique solution $x(t) \in C(E)$ for an arbitrarily fixed right-hand side $f(t) \in C^\alpha(E)$.

Below we indicate the relationship between the fundamental concepts of exponential dichotomy and regularity and also touch upon some related problems.

25.2. The Results

The first statement refers to the problem of preservation of the regularity property of the perturbed differential operator

$$L_\Delta x(t) = Lx(t) + \Delta(t)x(t) \tag{25.10}$$

under certain constraints imposed on the perturbation operator $\Delta(t)$, namely, we suppose that the linear operator $\Delta(t)$ acts for every t from E_1 into E and is continuous. Here E_1 is a Banach space consisting of elements $x \in D$ and having the norm

$$\|x\|_{E_1} = \sup_t \|A_0(t)x\|_E.$$

We suppose, in addition, that for any $x \in E_1$ the function $\Delta(t)x$ of the variable t with values in E_0 is almost periodic. Finally, we suppose that the quantity

$$\|\Delta(t)\|_{C^\alpha(E)} = \sup_{\|x\|_{E_1} \leq 1} \|\Delta(t)x\|_{C^\alpha(E)}$$

is finite.

THEOREM 25.1. *Suppose that the differential operator* (25.8) *is regular. Then there exists $\delta > 0$ such that under the condition*

$$\|\Delta(t)\|_{C^\alpha(E)} \leq \delta$$

the differential operator (25.10) *is regular.*

We can say that the second statement is fundamental.

THEOREM 25.2. *The regularity of the differential operator* (25.8) *is equivalent to the exponential dichotomy of solutions of the differential equation* (25.1).

The third statement concerns the smoothness of the projectors $P_+(t)$ and $P_-(t)$.

THEOREM 25.3. *The projectors $P_+(t)$ and $P_-(t)$ act in E_1, and for $x \in E_1$ the functions $P_+(t)x$ and $P_-(t)x$ possess almost-periodic derivatives and the relations*

$$\dot{P}_+(t)x = P_+(t)A(t)x - A(t)P_+(t)x, \tag{25.11}$$

$$\dot{P}_-(t)x = P_-(t)A(t)x - A(t)P_-(t)x \qquad (25.12)$$

are satisfied.

The fourth statement is connected with the properties of the family of homogeneous differential equations

$$\dot{x} + A(t,\mu)x = 0 \qquad (25.13)$$

and the corresponding differential operators

$$L(\mu)x(t) = \dot{x}(t) + A(t,\mu)x(t), \qquad (25.14)$$

where $0 \leq \mu \leq 1$. Let us agree that uniformly relative to μ the linear operators $A(t,\mu)$ satisfy the restrictions indicated above. It must be stipulated that the range of validity of the term "uniformly relative to μ satisfy the restrictions" includes, in particular, an assumption concerning a uniform (relative to t) continuity by μ of the function of two variables $A(t,\mu)x$ for every fixed $x \in D$.

THEOREM 25.4. *Suppose that the differential operators (25.14) are regular for all μ. Then the projectors $P_+(t,\mu)$ and $P_-(t,\mu)$, corresponding to the differential equations (25.13) are uniformly (relative to t) continuous by μ in a uniform operator topology.*

26. Estimates of Bounded Solutions of Linear Differential Equations

26.1. Statement of the Problem and Description of the Result

In this section, we give asymptotically exact estimates of the solutions of linear differential equations with constant coefficients which are bounded throughout the axis and depend on a small parameter. As was pointed out in [62], these estimates are important for applications (they are of interest to us in connection with the material presented in the next section). In [62, 63] a final result was obtained, which is formulated below and which indicates both the upper and the lower bound of the norm of the bounded solutions.

In what follows, we denote by $C(R^n)$ the space of continuous functions, bounded throughout the axis, with values in R^n and by $C^k(R^n)$ the set of functions $x(t) \in C(R^n)$ such that $x'(t), \ldots, x^{(k)}(t) \in C(R^n)$ at the same time. We suppose that the norms in $C(R^n)$ and $C^k(R^n)$ are defined in an ordinary way.

We consider in $C(R^n)$ the operator $L(\varepsilon)x = \dot{x} + A(\varepsilon)x$ with coefficients analytic with respect to ε. We suppose that $A(0)$ has eigenvalues λ with

Re $\lambda = 0$, but when $\varepsilon \neq 0$, $A(\varepsilon)$ has no eigenvalues of this kind. Therefore, for $\varepsilon \neq 0$ the functions

$$\|L^{-1}(\varepsilon)\|_{C(R^n) \to C^1(R^n)} \quad \text{and} \quad \sup_{-\infty < \omega < \infty} \|[i\omega I + A(\varepsilon)]^{-1}\|_{E^n}$$

are defined, where E^n is a complex n-dimensional space, which, as will follow from the constructions carried out below, have polar singularities at zero with orders p and q of the poles respectively.

THEOREM 26.1. *The relation $p = q$ holds true.*

Technically the proof is very cumbersome. We shall first prove an auxiliary statement which is of interest in itself.

Let us consider the scalar differential operator

$$L(\varepsilon)x(t) = x^{(n)}(t) + a_1(\varepsilon)x^{(n-1)}(t) + \ldots + a_n(\varepsilon)x(t), \qquad (26.1)$$

acting from $C^n(R)$ into $C(R)$, whose coefficients analytically depend on the parameter $\varepsilon \in [0, \varepsilon_0]$. We denote by $P(\lambda, \varepsilon)$ its characteristic polynomial. Suppose that for $\varepsilon = 0$ it turns into a polynomial $P_0(\lambda)$, all the roots of which lie on the imaginary axis. Everywhere in what follows we assume that for $0 < \varepsilon \leq \varepsilon_0$ the real parts of the roots of the polynomial $P(\lambda, \varepsilon)$ are nonzero. Clearly, in this case, for the corresponding values of ε, the equation $L(\varepsilon)x(t) = f(t)$ has a unique solution $x(t) \in C^n(R)$ for every right-hand side $f(t) \in C(R)$. We have thus defined the operator $L^{-1}(\varepsilon)$ which acts continuously from $C(R)$ into $C^n(R)$.

We are interested in estimates of the form

$$K_2 q(\varepsilon) \leq \|L^{-1}(\varepsilon)\| \leq K_1 q(\varepsilon), \qquad (26.2)$$

where $q(\varepsilon)$ is a function of the parameter ε and the positive constants K_1 and K_2 do not depend on ε. In the sequel we shall use precisely these letters to denote constants of this type in similar situations, without any stipulations.

It follows from [13, 19] that the set $\{\lambda_1(\varepsilon), \ldots, \lambda_n(\varepsilon)\}$ of roots of the polynomials $P(\lambda, \varepsilon)$ can be divided into nonintersecting cosets $\Gamma_1, \ldots, \Gamma_l$ according to the following principle. We place two roots $\lambda_{j_1}(\varepsilon)$ and $\lambda_{j_2}(\varepsilon)$ into the same class if there exists a limit

$$\alpha = \lim_{\varepsilon \to 0} \left|\frac{\operatorname{Re} \lambda_{j_1}(\varepsilon)}{\operatorname{Re} \lambda_{j_2}(\varepsilon)}\right|, \qquad 0 < \alpha < \infty,$$

and, in addition,

$$|\operatorname{Im} \lambda_{j_1}(\varepsilon) - \operatorname{Im} \lambda_{j_2}(\varepsilon)| \leq K |\operatorname{Re} \lambda_{j_1}(\varepsilon)|.$$

We fix $\lambda_{j_k}(\varepsilon) \in \Gamma_k$. Then we set

$$q_k(\varepsilon) = |\operatorname{Re} \lambda_{j_k}(\varepsilon)|^{-\gamma_k} \prod_{\lambda_j(\varepsilon) \notin \Gamma_k} |\lambda_j(\varepsilon) - \lambda_{j_k}(\varepsilon)|^{-\sigma_j},$$

where $k = 1,\ldots,l$. Here σ_j are the multiplicities of $\lambda_j(\varepsilon)$ and γ_k is the number of roots in Γ_k with due account of their multiplicities.

THEOREM 26.2. *In inequalities (26.2) we can take*

$$q(\varepsilon) = \max_{1 \leq k \leq l} q_k(\varepsilon). \tag{26.3}$$

26.2. Justification of Theorem 26.2

We divide the proof into a number of stages.

Stage 1. We denote by $C(E^n)$ and $C^k(E^n)$ the spaces similar to $C(R^n)$ and $C^k(R^n)$ respectively. We also suppose that in all these spaces the norms have been introduced in an ordinary way.

Below it is convenient to consider the differential operator (26.1) as an operator from $C^n(E)$ into $C(E)$. Obviously, it is sufficient to justify the theorem for this case.

Let us consider the differential operator

$$M(\varepsilon)y(t) = \dot{y}(t) + A(\varepsilon)y(t), \tag{26.4}$$

acting from $C^1(E^n)$ into $C(E^n)$. Here

$$A(\varepsilon) = \begin{pmatrix} 0 & -1 & 0 & \cdots & 0 \\ \cdot & \cdot & \cdot & \cdots & \cdot \\ 0 & 0 & 0 & \cdots & -1 \\ a_n(\varepsilon) & a_{n-1}(\varepsilon) & a_{n-2}(\varepsilon) & \cdots & a_1(\varepsilon) \end{pmatrix}.$$

LEMMA 26.1. *The inequalities*

$$K_2 \|M^{-1}(\varepsilon)\| \leq \|L^{-1}(\varepsilon)\| \leq K_1 \|M^{-1}(\varepsilon)\| \tag{26.5}$$

hold true.

Obviously, only the left-hand side of (26.5) must be proved. We set

$$e_1(\varepsilon) = (0,\ldots,0,1), \quad e_2(\varepsilon) = A(\varepsilon)e_1(\varepsilon),$$
$$e_3(\varepsilon) = A^2(\varepsilon)e_1(\varepsilon),\ldots, \quad e_n(\varepsilon) = A^{n-1}(\varepsilon)e_1(\varepsilon)$$

and compose a matrix

$$V(\varepsilon) = (e_1(\varepsilon),\ldots,e_n(\varepsilon)),$$

whose columns are constituted by these vectors. Note that

$$K_2 \leq |\det V^{-1}(\varepsilon)| \leq K_1. \tag{26.6}$$

Let us consider now the differential equations

$$\dot{y} + A(\varepsilon)y = \varphi(t)e_k(\varepsilon), \qquad k = 1, \ldots, n, \tag{26.7}$$

where $\varphi(t) \in C(E)$. It is clear that by virtue of (26.2), in order to prove the left-hand side of (26.5), it is sufficient to establish the validity of the estimates

$$\|y_k(t)\| \leq K\|L^{-1}(\varepsilon)\| \cdot \|\varphi(t)\|, \qquad k = 1, \ldots, n, \tag{26.8}$$

where $y_k(t)$ are the solutions of the differential equations (26.7) bounded throughout the axis. To prove (26.8), we note that the solutions $y_k(t)$ can be represented as

$$y_k(t) = A^{k-1}(\varepsilon)y_1(t), \qquad k = 2, \ldots, n,$$

where $y_1(t)$ is the solution of the first equation of (26.7) which is equivalent to the equation

$$L(\varepsilon)x(t) = \varphi(t).$$

Stage 2. Suppose that different roots of the polynomial $P_0(\lambda)$ are exhausted by the scalars $i\omega_1, \ldots, i\omega_s$. Then the roots of the polynomial $P(\lambda, \varepsilon)$ have expansions of the form

$$\lambda_{kj}(\varepsilon) = i\omega_k + \varepsilon^{r_{kj}}\mu_{kj}(\varepsilon), \qquad k = 1, \ldots, s, \qquad j = 1, \ldots, n_k, \tag{26.9}$$

where n_k are natural numbers. In (26.9) we have $r_{kj} > 0$ and $\mu_{kj}(\varepsilon)$ are series in fractional powers of ε. We set

$$m_k = \sum_{j=1}^{n_k} \sigma_{kj}, \qquad k = 1, \ldots, s,$$

where σ_{kj} is the multiplicity of $\lambda_{kj}(\varepsilon)$.

We introduce into consideration the differential operators

$$L_k(\varepsilon)x(t) = x^{(m_k)}(t) + c_{k1}(\varepsilon)x^{(m_k-1)}(t) + \ldots + c_{km_k}(\varepsilon)x(t)$$

with complex coefficients which act from $C^{m_k}(E)$ into $C(E)$. We suppose that the numbers $\lambda_{kj}(\varepsilon)$, $j = 1, \ldots, n_k$, are the roots of their characteristic polynomials $P_k(\lambda, \varepsilon)$.

LEMMA 26.2. *The inequality*

$$K_2 l(\varepsilon) \leq \|L^{-1}(\varepsilon)\| \leq K_1 l(\varepsilon),$$

where $l(\varepsilon) = \max_{1 \leq k \leq s} \|L_k^{-1}(\varepsilon)\|$, *holds true.*

In order to prove this lemma, it is sufficient, referring the reader to Lemma 26.1, to pass to the equivalent differential operator (26.4) and then reduce this operator to a block-diagonal form by means of a uniform (with respect to ε) invertible linear substitution.

Thus, the proof of Theorem 26.2 has been reduced to the verification of inequalities (26.2) for the differential operator

$$S(\varepsilon)x(t) = x^{(n)}(t) + c_1(\varepsilon)x^{(n-1)}(t) + \ldots + c_n(\varepsilon)x(t) \qquad (26.10)$$

with complex coefficients in the case where the roots of its characteristic polynomial $R(\lambda, \varepsilon)$ turn into a single scalar $i\omega_0$ for $\varepsilon = 0$.

Stage 3. We denote by $i\varkappa(\varepsilon)$ the common part of the expansions of the imaginary parts of the roots of the polynomial $R(\lambda, \varepsilon)$. Furthermore, let ε^r be the least common multiplier of the quantities $\lambda_j(\varepsilon) - i\varkappa(\varepsilon)$, where $\lambda_j(\varepsilon)$ is an arbitrary root of $R(\lambda, \varepsilon)$. Obviously, the roots of $R(\lambda, \varepsilon)$ can be divided into groups by the principle of the equality of coefficients for ε^r. In this connection, it is convenient to number the roots of the polynomial $R(\lambda, \varepsilon)$ by two indices, where the first index characterizes the group to which the root belongs and the second index shows the order of the elements in the group.

We represent the roots of the polynomial $R(\lambda, \varepsilon)$ as

$$\lambda_{kj}(\varepsilon) = i\varkappa(\varepsilon) + \varepsilon^r \mu_k + \varepsilon^{r+\alpha_{kj}} \nu_{kj}(\varepsilon), \qquad (26.11)$$

where $k = 1, \ldots, p$, $j = 1, \ldots, n_k$. Here n_k are natural numbers, μ_k are distinct, $\nu_{kj}(\varepsilon)$ are series in the fractional positive powers of ε and, finally, $\alpha_{kj} > 0$. Then we set

$$m_k = \sum_{j=1}^{n_k} \sigma_{kj}, \qquad k = 1, \ldots, p,$$

where σ_{kj} is the multiplicity of $\lambda_{kj}(\varepsilon)$, and introduce the differential operators

$$S_k(\varepsilon)x(t) = x^{(m_k)}(t) + c_{k1}(\varepsilon)x^{(m_k-1)}(t) + \ldots + c_{km_k}(\varepsilon)x(t),$$

acting from $C^{m_k}(E)$ into $C(E)$. We suppose that the numbers $\lambda_{kj}(\varepsilon)$, $j = 1, \ldots, n_k$, are the roots of their characteristic polynomials $R_k(\lambda, \varepsilon)$. Let us now introduce the functions

$$\psi_k(\varepsilon) = \|S_k^{-1}(\varepsilon)\| \cdot \prod_{\substack{s \neq k \\ 1 \leq j \leq n_s}} |\lambda_{sj}(\varepsilon) - \lambda_{k1}(\varepsilon)|^{-\sigma_{sj}}, \qquad (26.12)$$

where $k = 1, \ldots, p$.

LEMMA 26.3. *The inequalities*

$$K_2 \psi(\varepsilon) \leq \|S^{-1}(\varepsilon)\| \leq K_1 \psi(\varepsilon),$$

where $\psi(\varepsilon) = \max\limits_{1 \leq k \leq p} \psi_k(\varepsilon)$, *hold true.*

In order to prove the lemma, we make the substitutions

$$x(t) = z(\tau) \exp(-i\varkappa(\varepsilon)t), \qquad \tau = \varepsilon^r t,$$

in the differential operator (26.10). As a result we get the differential operator

$$\tilde{S}(\varepsilon) z(\tau) = z^{(n)}(\tau) + \tilde{c}_1(\varepsilon) z^{(n-1)}(\tau) + \ldots + \tilde{c}_n(\varepsilon) z(\tau).$$

The numbers

$$\tilde{\lambda}_{kj}(\varepsilon) = \mu_k + \varepsilon^{\alpha_{kj}} \nu_{kj}(\varepsilon), k = 1, \ldots, p, \qquad j = 1, \ldots, n_k,$$

where μ_k and $\nu_{kj}(\varepsilon)$ appear in (26.11), are the roots of its characteristic polynomial $\tilde{R}(\lambda, \varepsilon)$. It is easy to see that

$$K_2 \varepsilon^{-rn} \|\tilde{S}^{-1}(\varepsilon)\| \leq \|S^{-1}(\varepsilon)\| \leq K_1 \varepsilon^{-rn} \|\tilde{S}^{-1}(\varepsilon)\|.$$

It remains to refer the reader to (26.11), (26.12), and Lemma 26.2 (here the numbers μ_k are not necessarily pure imaginary, but this is inessential for the validity of the lemma).

Stage 4. Let Γ_k be the cosets of the roots of the polynomial $R(\lambda, \varepsilon)$. As before, we denote the number of these classes by l and the number of roots, with due account of their multiplicities in every class Γ_k, by γ_k.

Let us consider the differential operators

$$S_k(\varepsilon) x(t) = x^{(\gamma_k)}(t) + c_{k1}(\varepsilon) x^{(\gamma_k - 1)}(t) + \ldots + c_{k\gamma_k}(\varepsilon) x(t),$$

acting from $C^{\gamma_k}(E)$ into $C(E)$. Here the elements of Γ_k serve as the roots of their characteristic polynomials $R_k(\lambda, \varepsilon)$. In every Γ_k we choose a root $\lambda_{j_k}(\varepsilon)$ and then set

$$\delta_k(\varepsilon) = \|S_k^{-1}(\varepsilon)\| \cdot \prod_{\lambda_j \notin \Gamma_k} |\lambda_j(\varepsilon) - \lambda_{j_k}(\varepsilon)|^{-\sigma_j},$$

where $k = 1, \ldots, l$ and, as before, σ_j are the multiplicities of the roots of the polynomial $R(\lambda, \varepsilon)$. Applying now recurrently Lemma 26.3, we infer that

$$K_2 \delta(\varepsilon) \leq \|S^{-1}(\varepsilon)\| \leq K_1 \delta(\varepsilon),$$

where $\delta(\varepsilon) = \max_{1 \leq k \leq l} \delta_k(\varepsilon)$.

In order to complete the proof of the theorem, it remains to show that

$$K_2 |\operatorname{Re} \lambda_{j_k}(\varepsilon)|^{-\gamma_k} \leq \|S_k^{-1}(\varepsilon)\| \leq K_1 |\operatorname{Re} \lambda_{j_k}(\varepsilon)|^{-\gamma_k}, \qquad k = 1, \ldots, l.$$

The last inequalities can be substantiated by means of a normalization of time.

26.3. Proof of Theorem 26.1

In the initial differential operator $L(\varepsilon)$, we carry out an invertible linear substitution, uniform with respect to ε, such that the matrix of this operator reduces to a block-diagonal form and every matrix block is associated with a single eigenvalue for $\varepsilon = 0$. Let us consider a differential operator generated by an arbitrary block. In the general case, this differential operator is equivalent to a system of several scalar higher-order differential operators. Carrying out a suitable normalization of time, we see that the dimension of the differential operator under consideration can be lowered. Thus, applying Theorem 26.1 when necessary, we calculate the order of $\|L^{-1}(\varepsilon)\|$ as a result of this procedure.

To complete the proof, it remains to point out that all these transformations can be carried out in the operator $i\omega I + A(\varepsilon)$ at the same time.

27. Stability of Solutions of Linear Parabolic Equations

27.1. Statement of the Problem

In the real Banach space E we consider the differential equation

$$\dot{x} + A(t, \varepsilon)x = 0, \qquad (27.1)$$

where $A(t, \varepsilon)$ is a closed linear operator dependent on two numerical parameters $-\infty < t < \infty$ and $0 \leq \varepsilon \leq \varepsilon_0$ with a constant domain of definition D which is dense in E. In what follows we assume that $A(t, \varepsilon)$ can be represented as a series in powers of ε:

$$A(t, \varepsilon) = A_0 + \varepsilon A_1(t) + \varepsilon^2 A_2(t) + \ldots, \qquad (27.2)$$

where A_0 does not depend on t and $A_1(t)$, $A_2(t),\ldots$ are operator trigonometric polynomials, i.e.,

$$A_k(t) = A_{k0} + \sum_{j=1}^{n_k}(A_{kj}^1 \cos\omega_j t + A_{kj}^2 \sin\omega_j t), \qquad k=1,2,\ldots.$$

Here n_k are natural numbers, A_{k0}, A_{kj}^1, A_{kj}^2 are linear operators whose domains of definition contain D. Let us formulate the constraints that are below supposed to be imposed on the coefficients of series (27.2). Using these constraints we can accurately describe the nature of its convergence.

We suppose that there exists a real number λ_0 such that for $\operatorname{Re}\lambda \geq \lambda_0$ the operator $A_0 + \lambda I$ is continuously invertible in H and

$$\|[A_0 + \lambda I]^{-1}\|_H \leq M_0[1 + |\lambda|]^{-1},$$

where the constant M_0 does not depend on λ, and denote by H, as we did in Sec. 25, the complexification of E. We also suppose that the operators $A_k(t)[A_0 + \lambda_0 I]^{-1}$ are bounded in the norm in E and

$$A_0[A_0 + \lambda_0 I]^{-1} + \varepsilon A_1(t)[A_0 + \lambda_0 I]^{-1} + \varepsilon^2 A_2(t)[A_0 + \lambda_0 I]^{-1} + \ldots$$

converges uniformly with respect to t and ε in the uniform operator topology. Finally, let $0 < \alpha < 1$. We set

$$c_k = \max_{j,\sigma}\{\|A_{kj}^\sigma[A_0 + \lambda_0 I]^{-1}\|_E \cdot |\omega_j|^\alpha\},$$

where $j = 1,\ldots,n_k$ and $\sigma = 1,2$. We suppose that

$$c_1 + \ldots + c_k + \ldots < \infty.$$

We assume additionally that the operator $[A_0 + \lambda_0 I]^{-1}$ is completely continuous. We can easily verify that in this case, for every ε, the differential equation (27.1) belongs to the class introduced in 25.1. The dependence on ε will be that required in Theorem 25.4.

Let us now state the problem. We suppose that the operator A_0 has m eigenvalues (with due account of their multiplicities) with zero real parts. As concerns its other points of the spectrum, we suppose that they lie in the right open half-plane of the complex plane. We perform the substitution

$$x = y\exp(-a_0 t) \qquad (27.3)$$

in the differential equation (27.1) and, as a result, obtain a new differential equation

$$\dot{y} + [A(t,\varepsilon) - a_0 I]y = 0, \qquad (27.4)$$

in which we choose the positive constant a_0 such that the operator $A_0 - a_0 I$ will have exactly m eigenvalues with negative real parts and the other points of the spectrum will have positive real parts as before.

It follows from Theorem 25.1 that the differential operator

$$L_0(\varepsilon)y(t) = \dot{y}(t) + [A(t,\varepsilon) - a_0 I]y(t) \qquad (27.5)$$

is regular for every $0 \le \varepsilon \le \varepsilon_0$ and a sufficiently small ε_0. From this and from Theorem 25.2 it follows that there is an exponential dichotomy of solutions for the differential equation (27.4). We denote by $P_+(t,\varepsilon)$ and $P_-(t,\varepsilon)$ the projectors by means of which it can be performed. It must be immediately pointed out that these projectors do not depend on the choice of a_0 and, by virtue of Theorem 25.4, are continuous in ε uniformly with respect to t in the uniform operator topology. Furthermore, let $E_+(t,\varepsilon)$ and $E_-(t,\varepsilon)$ be the ranges of the projectors $P_+(t,\varepsilon)$ and $P_-(t,\varepsilon)$ respectively.

From the definition of the concept of the exponential dichotomy and the form of substitution (27.3) we find out that the following facts are true for the initial differential equation (27.1). The solutions $x_+(t,\varepsilon)$ with the initial conditions $x_+(\tau,\varepsilon) \in E_+(\tau,\varepsilon)$, defined at the arbitrary moment of time τ, belong to $E_+(t,\varepsilon)$ as time increases and

$$\|x_+(t,\varepsilon)\|_E \le N_0 \|x_+(\tau,\varepsilon)\|_E \exp[-\mu_0(t-\tau)], \qquad (27.6)$$

where the positive constants N_0 and μ_0 do not depend on ε. The solutions $x_-(t,\varepsilon)$ with the initial conditions $x_-(\tau,\varepsilon) \in E_-(\tau,\varepsilon)$, defined at the arbitrary moment of time τ, are defined throughout the number line and belong to $E_-(t,\varepsilon)$ for every t. It is clear that the dimension of the set of this kind of solutions is m. We cannot a priori say anything definite about the nature of the growth of the norms of these solutions.

In this section, we want to elucidate the structure of the solutions of the differential equation (27.1) that belong to the subspace $E_-(t,\varepsilon)$ of the space E for all t. To be more precise, we shall construct the Lyapunov–Floquet asymptotic theory for this set of solutions.

27.2. The Algorithmic Part

On the assumptions that we have made, the differential equation

$$\dot{x} + A_0 x = 0 \qquad (27.7)$$

has exactly m linearly independent solutions $v_1(t), \ldots, v_m(t)$ that correspond to the eigenvalues of the operator A_0 with zero real parts which are defined throughout the number line and whose norms grow not faster than certain powers of t with an increase or decrease of time.

We introduce into consideration the row matrix

$$V_0(t) = (v_1(t), \ldots, v_m(t)). \tag{27.8}$$

By a suitable choice of the solutions $v_1(t), \ldots, v_m(t)$ we write it as

$$V_0(t) = (e_1(t), \ldots, e_m(t)) \exp D_0 t, \tag{27.9}$$

where $e_1(t), \ldots, e_m(t)$ are harmonic functions or constant elements and D_0 is the upper Jordan matrix of order m with a unique eigenvalue equal to zero. We suppose that the matrix D_0 has s zero rows with numbers $k_1 < \ldots < k_s$. In what follows we shall denote by $K(D_0)$ the class of matrices of order m whose nonzero elements can lie only in the rows with these numbers.

Next we introduce into consideration the formal series

$$x_k(t, \varepsilon) = e_k(t) + \varepsilon x_{k1}(t) + \varepsilon^2 x_{k2}(t) + \ldots, \tag{27.10}$$

$$D(\varepsilon) = D_0 + \varepsilon D_1 + \varepsilon^2 D_2 + \ldots, \tag{27.11}$$

where $x_{kj}(t)$, $j = 1, 2, \ldots$, $k = 1, \ldots, m$, are trigonometric polynomials with values in E and D_j, $j = 1, 2, \ldots$, are constant matrices belonging to $K(D_0)$. Using (27.10) and (27.11), we compose the formal expression

$$X(t, \tau, \varepsilon) = V(t, \varepsilon) \exp[D(\varepsilon)(t - \tau)], \tag{27.12}$$

in which $V(t, \varepsilon) = (x_1(t, \varepsilon), \ldots, x_m(t, \varepsilon))$.

We shall show later that from the linearly independent solutions of the differential equation (27.1) with the initial conditions from $E_-(\tau, \varepsilon)$, defined at the moment of time τ, we can compose a row matrix $X(t, \tau, \varepsilon)$ for which the formal relation (27.12) holds true. For the time being, just as in [60, 61], we shall describe the method of determining the unknown coefficients of series (27.10) and (27.11).

We seek them from the formal identity

$$\dot{V}(t, \varepsilon) + A(t, \varepsilon) V(t, \varepsilon) + V(t, \varepsilon) D(\varepsilon) = 0, \tag{27.13}$$

in which we understand the result of the action of the operator $A(t, \varepsilon)$ onto $V(t, \varepsilon)$ as a row matrix whose elements are the values of the operator on the elements of the original row matrix. The reason why we come to the formal identity (27.13) is clear from what we said above. We can also write this formal identity in another form. In order to do this, we first introduce some notations. We set

$$u(t,\varepsilon) = \operatorname{colon}(x_1(t,\varepsilon),\ldots,x_m(t,\varepsilon)),$$

$$D_I^*(\varepsilon) = \begin{pmatrix} d_{11}(\varepsilon)I & d_{21}(\varepsilon)I & \cdots & d_{m1}(\varepsilon)I \\ \cdots & \cdots & \cdots & \cdots \\ d_{1m}(\varepsilon)I & d_{2m}(\varepsilon)I & \cdots & d_{mm}(\varepsilon)I \end{pmatrix},$$

$$[A(t,\varepsilon)]_{\text{diag}}\, u(t,\varepsilon) = \operatorname{colon}(A(t,\varepsilon)x_1(t,\varepsilon),\ldots,A(t,\varepsilon)x_m(t,\varepsilon)).$$

Here $d_{kj}(\varepsilon)$, $k,j=1,\ldots,m$, denote the elements of $D(\varepsilon)$.

In these notations the formal identity (27.13) can be written as

$$\dot{u}(t,\varepsilon) + [A(t,\varepsilon)]_{\text{diag}}\, u(t,\varepsilon) + D_I^*(\varepsilon)u(t,\varepsilon) = 0. \qquad (27.14)$$

Isolating the coefficients in like powers of ε in identity (27.14) and equating them to zero, we get the sequence of differential equations

$$\dot{x}_{kj}(t) + A_0 x_{kj} + g_{kj}(t) = 0, \qquad j=1,2,\ldots, \quad k=1,\ldots,m, \qquad (27.15)$$

which must be satisfied by the coefficients of series (27.10).

Arguments similar to those given in [60, 61] allow us to infer that the trigonometric polynomials $g_{kj}(t)$ in (27.15) have the following structure. We denote by $f_{kj}(t)$ their parts which depend on $x_{\mu\nu}(t)$, $\mu = 1,\ldots,m$, $\nu = 1,\ldots,j-1$, on the elements of the matrices D_ν, $\nu = 1,\ldots,j-1$, and also on $x_{\mu j}(t)$, $\mu = 1,\ldots,k-1$, and on the elements $d_{\mu\nu}^j$, $\mu = 1,\ldots,m$, $\nu = 1,\ldots,k-1$, of the matrices D_j. As a result we find that

$$g_{kj}(t) = f_{kj}(t) + \sum_{r=1}^{s} d_{k_r k}^j e_{k_r}(t). \qquad (27.16)$$

Thus, we can regard (27.15) as a recurrent sequence of differential equations for the unknown trigonometric polynomials

$$x_{11}(t), x_{21}(t),\ldots, x_{m1}(t), x_{12}(t), x_{22}(t),\ldots, x_{m2}(t),\ldots.$$

In this case, by virtue of (27.16), the possible nonzero elements of the matrices $D_1, D_2,\ldots$ can be successively determined from the solvability conditions of the corresponding differential equations in the class of trigonometric polynomials. Recall that these conditions consist in the requirement of orthogonality in the mean of the trigonometric polynomials (27.16) to all trigonometric solutions of the differential equation

$$\dot{y} - A_0^* y = 0, \qquad (27.17)$$

adjoined to (27.7). Note that the differential equation (27.17) is regarded in the space E^*, which is a conjugate of E, and that the number of linearly independent trigonometric solutions of this equation is s.

In conclusion, we present a simple extension of the formalism suggested above, which we shall need in Sec. 28.

Let us consider the system

$$\dot{x} + A_0 x = \varepsilon^{1/2} A_1(\varphi) x + \varepsilon A_2(\varphi) x, \tag{27.18}$$

$$\dot{\varphi} = \omega + \varepsilon \psi + \varepsilon^2 \psi_*(\varphi, \varepsilon), \tag{27.19}$$

where the operators A_j, $j = 1, 2$, depend 2π-periodically on the vector argument $\varphi = (\varphi_1, \ldots, \varphi_m)$ and are operator trigonometric polynomials. A_1 consists of only the first harmonics and A_2 contains only zero and second harmonics. In (27.19) we have $\omega = (\omega_1, \ldots, \omega_m)$, where the positive numbers $\omega_1, \ldots, \omega_m$ are pairwise distinct, $\psi = (\psi_1, \ldots, \psi_m)$ and ψ_* is a vector function smooth with respect to the collection of variables which is 2π-periodic with respect to φ.

The main restrictions are that, first, the numbers $\pm i\omega_k$, $k = 1, \ldots, m$, are eigenvalues of the operator A_0 and the eigenvectors a_k, $\bar{a}_k$ corresponding to them are assumed to be chosen so that

$$l_j(a_k) = \delta_{kj}, \qquad l_j(\bar{a}_k) = 0,$$

where $A_0^* l_j = -i\omega_j l_j$, $j = 1, \ldots, m$ (the other eigenvalues A_0 have positive real parts); second,

$$\omega_k \neq n_1 \omega_1 + \ldots + n_m \omega_m, \qquad k = 1, \ldots, m, \tag{27.20}$$

where the integers $n_1, \ldots, n_m$ are related by the inequalities

$$2 \leq |n_1| + \ldots + |n_m| \leq 3.$$

Under the restrictions formulated, we set in (27.18)

$$x = [V_0(\varphi) + \varepsilon^{1/2} V_1(\varphi) + \varepsilon V_2(\varphi)] \exp \varepsilon D_1 t, \tag{27.21}$$

where

$$V_0(\varphi) = [a_1 \exp(i\varphi_1), \bar{a}_1 \exp(-i\varphi_1), \ldots, a_m \exp(i\varphi_m), \bar{a}_m \exp(-i\varphi_m)],$$

take into account (27.19), and try to determine the unknown row matrices V_1 and V_2, 2π-periodic with respect to φ, and the matrix D_1 of order $2m$. In doing this, we arrive at the equations

$$\sum_{k=1}^{m} \omega_k \frac{\partial v_{1j}}{\partial \varphi_k} + A_0 v_{1j} = A_1(\varphi) a_j \exp(i\varphi_j), \qquad j = 1, \ldots, m, \tag{27.22}$$

for the columns of the matrix

$$V_1 = (v_{11}, \bar{v}_{11}, \ldots, v_{1m}, \bar{v}_{1m}).$$

It is significant that the right-hand sides of (27.22) are trigonometric polynomials containing only zero and second harmonics. We shall seek the functions v_{1j} with the same dependence on φ. As a result, every one of the differential equations (27.22) disintegrates into a series of operator equations of the form

$$[A_0 + i(m_1\omega_{n_1} + m_2\omega_{n_2})I]u = f, \qquad (27.23)$$

where $|m_1| + |m_2| = 0$ or $|m_1| + |m_2| = 2$, whose solvability is guaranteed by inequality (27.20).

A new situation arises when we seek the matrices $V_2(\varphi)$. Here we attain the solvability of the equations similar to (27.23) at the expense of the elements of the matrix D_1 on which the right-hand side f linearly depends.

Two points must be emphasized. First, under condition (27.19) the right-hand side of (27.18) is not, in general, almost periodic. This circumstance is of no importance since the results of Sec. 25 remain valid for this class of parabolic equations. Second, if we disregard condition (27.20), then in (27.21) we must replace $\exp \varepsilon D_1 t$ by a fundamental matrix of a certain linear differential equation in R^{2m}

$$\dot{\eta} = (\varepsilon^{1/2} D_1 + \varepsilon D_2 + \ldots)\eta.$$

In this case, the matrices $D_1, D_2, \ldots$ are, in general, quasiperiodic.

It should be pointed out that the formalism that we have developed can be extended to wide classes of systems of the form (27.18), (27.19). We assign an exact meaning to this formalism just as we do below for equations of the form (27.1).

27.3. The Results

In $E_-(t,\varepsilon)$, for arbitrary fixed t and ε there exist m basic elements. We say that the basis is almost periodic if every one of the basic elements is almost periodic and continuous in ε uniformly with respect to t. The existence of an almost-periodic basis in $E_-(t,\varepsilon)$ is obvious, since for the zero value of ε, the projector $P_-(t,\varepsilon)$ turns into a constant operator acting onto the root subspace of the operator A_0 which corresponds to its eigenvalues with zero real parts.

The first proposition that we shall formulate refers to the problem of the existence of a smooth (with respect to ε) almost-periodic basis in $E_-(t,\varepsilon)$. We denote by

$$v_{kj}(t,\varepsilon) = e_k(t) + \varepsilon x_{k1}(t) + \ldots + \varepsilon^j x_{kj}(t) \qquad (27.24)$$

the jth partial sum of series (27.10) and by

$$s_{kj}(t,\varepsilon) = v_{kj}(t,\varepsilon) + \varepsilon^{j+1}\delta_{kj}(t,\varepsilon) \qquad (27.25)$$

the perturbations of (27.24) by terms of the $(j+1)$th order of smallness relative to ε, assuming that the functions $\delta_{kj}(t,\varepsilon)$ and $\dot{\delta}_{kj}(t,\varepsilon)$ are almost periodic and continuous in ε uniformly with respect to t.

THEOREM 27.1. *For any $j \geq 1$, there exists in $E_-(t,\varepsilon)$ an almost-periodic basis of the form (27.25).*

Let us now consider the problem of the structure of solutions $x_-(t,\varepsilon)$ of the differential equation (27.1). For arbitrarily fixed t and ε, we expand $x_-(t,\varepsilon)$ with respect to the basic elements of (27.25). It is convenient to write the result as

$$x_-(t,\varepsilon) = S_j(t,\varepsilon)\eta(t,\varepsilon), \qquad (27.26)$$

where

$$S_j(t,\varepsilon) = (s_{1j}(t,\varepsilon), \ldots, s_{mj}(t,\varepsilon)), \qquad (27.27)$$

and $\eta(t,\varepsilon)$ are elements of the m-dimensional Euclidean space R^m. Next we introduce into consideration a differential equation

$$\dot{\eta} = D_j(\varepsilon)\eta + \varepsilon^{j+1}\Delta_j(t,\varepsilon)\eta \qquad (27.28)$$

in R^m, where

$$D_j(\varepsilon) = D_0 + \varepsilon D_1 + \ldots + \varepsilon^j D_j \qquad (27.29)$$

is the jth partial sum of series (27.11) and the elements of the square matrices $\Delta_j(t,\varepsilon)$ of order m are almost periodic and continuous in ε uniformly with respect to t.

THEOREM 27.2. *For any $j \geq 1$, there exists a differential equation of the form (27.28) such that*

$$x_-(t,\varepsilon) = S_j(t,\varepsilon)U_j(t,\tau,\varepsilon)\eta(\tau,\varepsilon), \qquad (27.30)$$

where $U_j(t,\tau,\varepsilon)$ is a fundamental matrix of solutions of the differential equation (27.28), which turns into an identity matrix for $t = \tau$.

27.4. Proof of Theorem 27.1

The main idea of the proof is the construction of an auxiliary differential equation

$$\dot{x} + A(t,\varepsilon)x + \varepsilon^{j+1} A_j(t,\varepsilon)x = 0, \qquad (27.31)$$

where the finite-dimensional operator A_j is almost periodic with respect to t and continuous in ε uniformly with respect to t in the uniform operator topology. In this case, the addition A_j is chosen such that the elements (27.24) will be basic in the subspace $E_-^j(t,\varepsilon)$ which is similar to the subspace $E_-(t,\varepsilon)$ for Eq. (27.1). Then we show that

$$s_{kj}(t,\varepsilon) = v_{kj}(t,\varepsilon) + [P_-(t,\varepsilon) - P_-^j(t,\varepsilon)]v_{kj}(t,\varepsilon) \qquad (27.32)$$

can be taken as function (27.25).

In order to construct the operator A_j, we introduce into consideration a row matrix

$$(h_1(t),\ldots,h_m(t))\exp(-D_0' t), \qquad (27.33)$$

similar to (27.9), where $h_1,\ldots,h_m$ are harmonic functions or constant elements, whose columns are linearly independent solutions of (27.17) growing in both directions of time variation not faster than certain powers of t. In this case we can choose the functionals $h_1,\ldots,h_m$ such that

$$\langle e_k(t), h_n(t)\rangle \equiv \delta_{kn}, \qquad -\infty < t < \infty, \qquad (27.34)$$

where $\langle *,*\rangle$ is the value of the functional on the element. Next we consider the system of linear algebraic equations

$$\sum_{k=1}^{m} \zeta_k \langle v_{kj}(t,\varepsilon), h_\mu(t)\rangle = \langle x, h_\mu(t)\rangle, \qquad \mu = 1,\ldots,m, \qquad (27.35)$$

where x is an arbitrary element from E. By virtue of relations (27.34), the solutions $f_{kj}(t,*,\varepsilon)$ of this system are linear almost-periodic functionals which are continuous in ε uniformly with respect to t in the uniform topology, and

$$f_{kj}(t, v_{\mu j}(t,\varepsilon), \varepsilon) = \delta_{k\mu}, \qquad k, \mu = 1,\ldots,m. \qquad (27.36)$$

Next we denote by $\varkappa_{1j}(t,\varepsilon),\ldots,\varkappa_{mj}(t,\varepsilon)$ the columns of the matrix

$$-\varepsilon^{-(j+1)}[\dot{V}_j(t,\varepsilon) + A(t,\varepsilon)V_j(t,\varepsilon) + V_j(t,\varepsilon)D_j(\varepsilon)],$$

where $V_j(t,\varepsilon) = (v_{1j}(t,\varepsilon),\ldots,v_{mj}(t,\varepsilon))$. It is clear that these functions are almost periodic with respect to t and continuous in ε uniformly with respect to t. We now define the operator A_j by the formula

$$A_j(t,\varepsilon)x = \sum_{k=1}^{m} f_{kj}(t,x,\varepsilon)\varkappa_{kj}(t,\varepsilon). \tag{27.37}$$

It follows from (27.36), (27.37) that in the subspace $E_-^j(t,\varepsilon)$ the functions $v_{kj}(t,\varepsilon)$, $k = 1,\ldots,m$, form an almost-periodic basis.

In order to complete the proof of Theorem 27.1, it remains to point out that the required properties of the functions

$$\delta_{kj}(t,\varepsilon) = \varepsilon^{-(j+1)}[P_-(t,\varepsilon) - P_-^j(t,\varepsilon)]v_{kj}(t,\varepsilon)$$

follow from (25.11), (25.12) and the representation

$$P_-^j(t,\varepsilon)x - P_-(t,\varepsilon)x = \varepsilon^{j+1}\int_{-\infty}^{\infty} Y_1(t,\sigma,\varepsilon)A_j(\sigma,\varepsilon)Y_2(\sigma,t,\varepsilon)x\,d\sigma,$$

where $Y_1(t,\tau,\varepsilon)$, $Y_2(t,\tau,\varepsilon)$ are the Green operators of Eq. (27.4) and

$$\dot{y} + [A(t,\varepsilon) - a_0 I]y + \varepsilon^{j+1}A_j(t,\varepsilon)y = 0$$

respectively.

27.5. Proof of Theorem 27.2

It follows from (27.26) that the coordinates $\eta_1(t,\varepsilon),\ldots,\eta_m(t,\varepsilon)$ of the vector $\eta(t,\varepsilon)$ can be found from the system

$$\sum_{k=1}^{m} \eta_k \langle s_{kj}(t,\varepsilon), h_\mu(t)\rangle = \langle x_-(t,\varepsilon), h_\mu(t)\rangle, \tag{27.38}$$

similar to system (27.35). Therefore, the function $\eta(t,\varepsilon)$ is continuously differentiable with respect to t. Next we write (27.26) as

$$x_-(t,\varepsilon) = V_j(t,\varepsilon)\eta(t,\varepsilon) + \varepsilon^{j+1}B_j(t,\varepsilon)\eta(t,\varepsilon), \tag{27.39}$$

where the elements of the matrix $B_j(t,\varepsilon)$ and their derivatives with respect to t are almost periodic and continuous in ε uniformly with respect to t. Differentiating now (27.39), after simple transformations we arrive at the formula

$$S_j(t,\varepsilon)[\dot{\eta}(t,\varepsilon) - D_j(\varepsilon)\eta(t,\varepsilon)] =$$
$$= -\varepsilon^{j+1}[\dot{B}_j(t,\varepsilon) + B_j(t,\varepsilon)D_j(\varepsilon) + A(t,\varepsilon)B_j(t,\varepsilon)]\eta(t,\varepsilon). \tag{27.40}$$

It follows from (27.40) that in order to find the coordinates of the vector $\xi(t,\varepsilon) = \dot{\eta}(t,\varepsilon) - D_j(\varepsilon)\eta(t,\varepsilon)$, we can write a system of equations similar to (27.38), and then the statement of Theorem 27.2 is obvious.

27.6. Strong Stability and Instability

Let us consider the differential equation

$$\dot{\eta} = D_j(\varepsilon)\eta \qquad (27.41)$$

in R^m. We say that the differential equation (27.41) is strongly stable (unstable) if for every $R > 0$ we can indicate $\varepsilon(R) > 0$ such that the solutions of the differential equation

$$\dot{\eta} = D_j(\varepsilon)\eta + \varepsilon^{j+1} D\eta$$

are stable (unstable) for $0 < \varepsilon \leq \varepsilon(R)$, whatever the matrix D, whose norm does not exceed R.

Below, when we speak of the possibility of completing the algorithm for investigating the stability of the solutions of the differential equation (27.1), we shall mean that the differential equation

$$\dot{\eta} = D_{j_0}(\varepsilon)\eta \qquad (27.42)$$

is strongly stable or unstable for certain subscripts j_0. It should be pointed out that every one of these properties is invariant with respect to the choice of j_0. This follows immediately from the definitions presented above.

Theorem 27.2 implies the following proposition.

THEOREM 27.3. *Suppose that the algorithm can be completed. Then there exists $\varepsilon_0 > 0$ such that for $0 < \varepsilon \leq \varepsilon_0$ the solutions of the differential equation (27.1) are exponentially stable or exponentially unstable, depending on the similar properties of the differential equation (27.42).*

Note that we can judge whether the differential equation (27.41) is stable or unstable if we use any one of the known criteria of stability of solutions of differential equations with constant coefficients. In this case, care must be taken that the additions of the $(j+1)$th order of smallness relative to ε do not affect the property of stability or unstability. The following simple example demonstrates the dangers that may be encountered here.

Let us consider in R^2 the differential equation

$$\dot{x} + \varepsilon A(\varepsilon)x = 0, \qquad (27.43)$$

where

$$A(\varepsilon) = \begin{pmatrix} 0 & 1 \\ 0 & 0 \end{pmatrix} + \varepsilon \begin{pmatrix} 1 & 0 \\ 0 & 1 \end{pmatrix} + \sum_{j=2}^{\infty} \varepsilon^j \begin{pmatrix} -(-1)^j & 0 \\ (-1)^j & 0 \end{pmatrix}.$$

Systems of Reaction–Diffusion Type

Let
$$A_1(\varepsilon) = \begin{pmatrix} 0 & 1 \\ 0 & 0 \end{pmatrix} + \varepsilon \begin{pmatrix} 1 & 0 \\ 0 & 1 \end{pmatrix} \tag{27.44}$$

and, for $k \geq 2$,
$$A_k(\varepsilon) = A_1(\varepsilon) + \sum_{j=2}^{k} \varepsilon^j \begin{pmatrix} -(-1)^j & 0 \\ (-1)^j & 0 \end{pmatrix}. \tag{27.45}$$

We denote by $\lambda_{1k}(\varepsilon)$ and $\lambda_{2k}(\varepsilon)$, $k = 1, 2, \ldots$, the eigenvalues of matrices (27.44) and (27.45). It is easy to verify that
$$\lambda_{11}(\varepsilon) = \lambda_{21}(\varepsilon) = \varepsilon,$$

and, for $k \geq 2$,
$$\lambda_{1k}(\varepsilon) = 2\varepsilon + o(\varepsilon), \qquad \lambda_{2k}(\varepsilon) = -\frac{(-1)^k}{2}\varepsilon^k + o(\varepsilon^k). \tag{27.46}$$

Thus, the solutions of the differential equations
$$\dot{x} + \varepsilon A_k(\varepsilon) x = 0 \tag{27.47}$$

are exponentially stable for odd k and exponentially unstable for even k if, of course, $0 < \varepsilon \leq \varepsilon_k$ and ε_k are sufficiently small.

In conclusion, we shall consider the possibility of using Theorem 26.1. If we speak of a strong stability, the method of its application is clear. In the case of instability it is sometimes expedient to carry out the substitution
$$\eta \to \eta \exp \varepsilon^\gamma t$$

in (27.41), where $\gamma > 0$ is a suitable constant, and then apply Theorem 26.1.

By way of example, we shall consider the differential operators
$$L_k(\varepsilon) = \frac{d}{dt} + \varepsilon A_k(\varepsilon), \qquad k = 2, 3, \ldots,$$

where $A_k(\varepsilon)$ are matrices (27.45). By virtue of (27.46), each of them is regular, and after the normalization of time we can apply Theorem 26.2 to them. It follows from this theorem that the norms $L_k^{-1}(\varepsilon)$ have orders $\varepsilon^{-(k+2)}$. Thus, we see the reason why a change of stability occurs in (27.47) with the growth of k.

28. Bifurcation of Invariant Tori of Parabolic Systems with Small Diffusion

28.1. Singularly Perturbed Parabolic System and Its Quasinormal Form

The problem of bifurcation of periodic auto-oscillations of the van der Pol system with small diffusion under Dirichlet's boundary conditions on an interval was considered in [8]. This work is academic in many respects. In this section we prove two theorems on the existence, asymptotics, and stability of smooth invariant tori (that bifurcate from the zero equilibrium position) of parabolic systems with small diffusion under Neumann's boundary conditions. Results of this kind are important for applications. Their justification belongs to Yu. S. Kolesov and is based essentially on the properties of linear parabolic systems with coefficients close to constants (with respect to time) presented in Sec. 27 and on the method of proving theorems on the preservation of invariant tori under small perturbations developed in [84].

We consider for $u \in R^n$, $n \geq 2$, the parabolic boundary-value problem

$$\frac{\partial u}{\partial t} = \varepsilon D \Delta u + A_0 u + \varepsilon A_1 u + F(u), \tag{28.1}$$

$$\left. \frac{\partial u}{\partial \nu} \right|_{\partial \Omega} = 0 \tag{28.2}$$

with a small positive parameter ε. The eigenvalues of the matrix D lie on the right of the imaginary axis; the smooth nonlinear vector function F has at zero an order of smallness higher than the first; the matrix A_0 has m pairs of different eigenvalues $\pm i\omega_s$, $\omega_s > 0$, $s = 1, \ldots, m$, on the imaginary axis and for the other points of the spectrum λ_p, $p = 1, \ldots, n - 2m$, the real parts of this matrix are negative. In (28.2), we denote by ν the direction of the outer normal to the boundary $\partial \Omega$ of the bounded domain $\Omega \subset R^N$, $N \geq 1$, of the class $C^{2+\alpha}$, $0 < \alpha < 1$.

CONDITION 28.1. The eigenvalues of the matrix A_0 that lie on the imaginary axis have no resonances up to the fourth order, i.e.,

$$\omega_s \neq n_1 \omega_1 + \ldots + n_m \omega_m, \quad s = 1, \ldots, m,$$

where the integers $n_1, \ldots, n_m$ are related by the inequalities

$$2 \leq |n_1| + \ldots + |n_m| \leq 4.$$

To clarify the meaning of further constructions, first, under Condition 28.1, we consider the ordinary differential equation

$$\dot{u} = A_0 u + \varepsilon A_1 u + F_2(u, u) + F_3(u, u, u) + \ldots \quad (28.3)$$

in R^n in whose right-hand side the nonlinearity of F is presented as the sum of terms of the second order of smallness, of the third order of smallness etc. We set

$$u = \varepsilon^{1/2} \sum_{s=1}^{m} [\xi_s a_s \exp(i\omega_s t) + \bar{\xi}_s \bar{a}_s \exp(-i\omega_s t)] +$$
$$+ \varepsilon u_2(\tau, t) + \varepsilon^{3/2} u_3(\tau, t) + \ldots, \quad (28.4)$$

where $A_0 a_s = i\omega_s a_s$, $\xi_s = \xi_s(\tau)$, $\tau = \varepsilon t$, and the dots stand for terms of a higher order of smallness with respect to ε.

Substituting (28.4) into (28.3) and equating the coefficients of the first powers of ε on the left-hand and right-hand sides, we obtain

$$\frac{\partial u_2}{\partial t} = A_0 u_2 + \sum_{s,k=1}^{m} \{F_2(a_s, a_k)\xi_s \xi_k \exp i(\omega_s + \omega_k)t +$$

$$+ F_2(a_s, \bar{a}_k)\xi_s \bar{\xi}_k \exp i(\omega_s - \omega_k)t + F_2(\bar{a}_s, a_k)\bar{\xi}_s \xi_k \exp i(\omega_k - \omega_s)t +$$
$$+ F_2(\bar{a}_s, \bar{a}_k)\bar{\xi}_s \bar{\xi}_k \exp[-i(\omega_s + \omega_k)t]\}, \quad (28.5)$$

where τ is regarded as a parameter. Relation (28.5) uniquely defines the function $u_2(\tau, t)$ as the sum of the zero and second harmonics.

Then we equate the coefficients of the exponents of order $\varepsilon^{3/2}$. We arrive at an equation similar to (28.5) but with a nonhomogeneity which is the sum of the first and third harmonics. In the same form we seek $u_3(\tau, t)$. However, we now face a new problem, namely, for the equations for the amplitudes of the first harmonics to be solvable, the ordinary orthogonality conditions must be satisfied. Together, they are equivalent to the equation

$$\frac{d\zeta}{d\tau} = (A_1 a, b)\zeta + \zeta d|\zeta|^2, \quad (28.6)$$

where $\bar{\zeta}$ satisfies the complex conjugate equation;

$$A_0^* b_s = -i\omega_s b_s, \quad (a_s, b_s) = 1, \quad (A_1 a, b) = \text{diag }\{(A_1 a_s, b_s)\},$$

ζ is a vector with coordinates ζ_s, $|\zeta|^2$ is a vector with coordinates $|\zeta_s|^2$, d is a matrix of order m which can be naturally called a matrix Lyapunov quantity, and the vector multiplication in the last term is coordinatewise.

Going back to the accepted terminology, we call Eq. (28.6) a normal form of Eq. (28.3). We know that it can be used to investigate the problem of invariant tori (that bifurcate from the zero solution) of the original equation (28.3) with reasonable completeness. In connection with the way in which we have derived the normal form (28.6) of Eq. (28.3), we must point out that in the general case the auto-oscillations, which are in a sufficiently small neighborhood of its zero equilibrium position, have amplitudes of order $\varepsilon^{1/2}$. Also note that the fourth-order nonresonance that was not used in the process of deriving the normal form of (28.6) appreciably affects the asymptotic behavior of bifurcating tori.

By analogy with (28.4), we seek the stationary solutions of the boundary-value problem (28.1), (28.2) in the form

$$u = \varepsilon^{1/2} \sum_{s=1}^{m} [\zeta_s a_s \exp(i\omega_s t) + \bar{\zeta}_s \bar{a}_s \exp(-i\omega_s t)] +$$
$$+ \varepsilon u_2(\tau, t, x) + \varepsilon^{3/2} u_3(\tau, t, x) + \ldots, \qquad (28.7)$$

where we now have $\zeta_s = \zeta_s(\tau, x)$, $x \in R^N$. Substituting (28.7) into (28.1), (28.2) and acting just as we did in the case of the ordinary differential equation (28.3), we arrive at the boundary-value problem

$$\frac{\partial \zeta}{\partial \tau} = (Da, b)\Delta\zeta + (A_1 a, b)\zeta + \zeta d|\zeta|^2, \qquad (28.8)$$

$$\left.\frac{\partial \zeta}{\partial \nu}\right|_{\partial\Omega} = 0, \qquad (28.9)$$

where $(Da, b) = \text{diag}\{(Da_s, b_s)\}$, and obtain the equation and the boundary condition for $\bar{\zeta}$ by means of a complex conjugation. We call the boundary-value problem (28.8), (28.9) a quasinormal form of the original boundary-value problem (28.1), (28.2). It is natural to expect that, in accordance with (28.7), the tori of a certain kind of structure of the quasinormal form (28.8), (28.9) are associated with the tori of the boundary-value problem (28.1), (28.2). This is also the fact for another two restrictions and after the refinement of a number of concepts.

CONDITION 28.2. For all $z > 0$, the eigenvalues of the matrix $A_0 - zD$ are in the left complex half-plane and generically diverge from the imaginary axis, i.e.,

$$\text{Re}\,(Da_s, b_s) > 0, \quad s = 1, \ldots, m. \qquad (28.10)$$

CONDITION 28.3. For every s, we have

$$\lambda_p - \lambda_k \neq i\omega_s, \quad p, k = 1, \ldots, n - 2m. \qquad (28.11)$$

Condition 28.2 appeared for the first time in [50], where it was called the condition of absence of an explosive diffusion instability, since at a gross violation of at least one of inequalities (28.10) the boundary-value problem (28.1), (28.2) with a rejected nonlinearity has solutions that grow with the coefficient of the exponent of order unity. Inequalities (28.11) characterize the generality of the position, i.e., are not restrictive.

In what follows, we take C^α as the phase space of the boundary-value problems (28.1), (28.2) and (28.8), (28.9). C^α is a Banach space of the functions of the variable x, continuous in the closure of Ω, which satisfy Hölder's condition with exponent α. Recall that, owing to this concept, the main notions from the qualitative theory of systems of ordinary differential equations are extended, practically without changes, to the boundary-value problems under consideration.

DEFINITION 28.1. We say that the boundary-value problem (28.8), (28.9) has an m-dimensional self-similar torus if it has an m-parametric family of solutions

$$\xi(\varphi, x) = \xi_0(x)\exp i\varphi, \qquad \frac{d\varphi}{d\tau} = \psi, \qquad \psi \in R^m, \qquad (28.12)$$

where none of the coordinates ξ_{0s} of the vector ξ_0 is identically equal to zero, $\exp i\varphi$ is a vector with coordinates $\exp i\varphi_s$, $s = 1,\ldots,m$, and the vector multiplication is coordinatewise.

In the boundary-value problem (28.8), (28.9), we make the substitution

$$\xi = \eta \exp i\psi\tau. \qquad (28.13)$$

As a result, all its terms on the right-hand side are preserved and a term $-i\psi\eta$ is added. Linearizing this boundary-value problem and its complex conjugate on the equilibrium position $(\xi_0(x), \bar{\xi}_0(x))$, we obtain

$$\frac{\partial h}{\partial \tau} = L_0 h, \qquad \left.\frac{\partial h}{\partial \nu}\right|_{\partial\Omega} = 0, \qquad (28.14)$$

where the dimension of $h = (h_1, h_2)$ is $2m$ and the structure of the elliptic operator L_0 is clear. Note that the linear boundary-value problem (28.14) has linearly independent solutions

$$(i\xi_{0s}(x)e_s, -i\bar{\xi}_{0s}(x)e_s), \qquad s = 1,\ldots,m, \qquad (28.15)$$

where e_s are unit vectors in R^m.

DEFINITION 28.2. We say that the self-similar torus (28.12) of the quasinormal form (28.8), (28.9) is dichotomous if a trichotomy takes place for the solutions of the linear boundary-value problem (28.14), i.e., if the solutions can be divided into three groups, namely, exponentially growing with an increase in τ and exponentially decreasing with a decrease in τ; bounded with respect to $-\infty < \tau < \infty$; and exponentially decreasing with an increase in τ. The set of solutions bounded for all τ include only linear combinations of its solutions (28.15). We speak of the orbital exponential stability in the case where the first of the indicated groups is empty.

Naturally, we can judge the properties of the solutions of the boundary-value problem (28.14) from the spectrum of the operator L_0.

THEOREM 28.1. *Suppose that under Conditions 28.1–28.3 the quasinormal form (28.8), (28.9) has a dichotomous (orbitally exponentially stable) self-similar torus (28.12). Then there exists $\varepsilon_0 > 0$ such that for $0 < \varepsilon \leq \varepsilon_0$ the boundary-value problem (28.1), (28.2) has a dichotomous (orbitally exponentially stable) torus*

$$u(\varphi, x, \varepsilon) =$$
$$= \varepsilon^{1/2} \sum_{s=1}^{m} [\xi_{0s}(x) a_s \exp(i\varphi_s) + \bar{\xi}_{0s}(x) \bar{a}_s \exp(-i\varphi_s)] +$$
$$+ \varepsilon u_2(\varphi, x) + \varepsilon^{3/2} u_*(\varphi, x, \varepsilon), \qquad (28.16)$$
$$\frac{d\varphi}{dt} = \omega + \varepsilon \psi + \varepsilon^2 \psi_*(\varphi, \varepsilon). \qquad (28.17)$$

Here ω is a vector with coordinates ω_s; u_2 is formed, with due account of (28.12), from the term of order ε appearing on the right-hand side of (28.7); the 2π-periodic functions u_* and ψ_* are uniformly bounded with respect to ε with a sufficient number of derivatives with respect to φ; and with respect to x the function $u_* \in C^{2+\alpha}$.

DEFINITION 28.3. We speak of the self-similar torus with a T_0-periodically varying amplitude if in (28.12) we have $\xi_0 = \xi_0(\tau, x)$ and

$$\frac{\partial}{\partial \tau}[\xi_0(\tau, x) \exp(i\gamma\tau)] \not\equiv 0 \qquad (28.18)$$

for any $\gamma \in R^m$.

The concept of a dichotomous (orbitally exponentially stable) torus with a periodically varying amplitude can be introduced in the same way as in the case of torus (28.12). Note that here, by virtue of condition (28.18), a boundary-value problem similar to (28.14) will have T_0-periodic coefficients.

THEOREM 28.2. *Suppose that under Conditions 28.1–28.3 the quasinormal form (28.8), (28.9) has a dichotomous (orbitally exponentially stable)* self-similar torus with a T_0-periodically varying amplitude. Then there exists $\varepsilon_0 \geq 0$ such that for $0 < \varepsilon \leq \varepsilon_0$ the boundary-value problem (28.1), (28.2) has a dichotomous (orbitally exponentially stable) torus

$$u(\varphi_0, \varphi, x, \varepsilon) =$$

$$= \varepsilon^{1/2} \sum_{s=1}^{m} [\xi_{0s}(\varphi_0, x) a_s \exp(i\varphi_s) + \bar{\xi}_{0s}(\varphi_0, x) \bar{a}_s \exp(-i\varphi_s)] +$$

$$+ \varepsilon u_2(\varphi_0, \varphi, x) + \varepsilon^{3/2} u_*(\varphi_0, \varphi, x, \varepsilon),$$

$$\frac{d\varphi_0}{dt} = \varepsilon + \varepsilon^2 \psi_0(\varphi_0, \varphi, \varepsilon), \qquad \frac{d\varphi}{dt} = \omega + \varepsilon \psi + \varepsilon^2 \psi_*(\varphi_0, \varphi, \varepsilon).$$

Here the functions u_* and ψ_* are T_0-periodic with respect to the scalar argument φ_0 and 2π-periodic with respect to the vector argument φ, and their other properties are similar to those indicated in Theorem 28.1.

If we assume that the quasinormal form (28.8), (28.9) has an invariant torus with a complicated dynamics on it, then the problem of the existence of an invariant torus of the boundary-value problem (28.1), (28.2) cannot practically be solved since the difficulties encountered here are approximately equivalent to those we encounter in the general problem of the preservation of an invariant torus when an arbitrary nonlinear system is perturbed.

28.2. Trichotomy of the Solutions of a Linearized Boundary-Value Problem

Theorems 28.1 and 28.2 can be proved in the same way, and, therefore, we shall only prove Theorem 28.1. Its demonstration is based on a complex of different ideas, among which two ideas are principal; we shall describe them.

We denote by $1/\text{mes}\,\Omega$, $g_1(x), g_2(x), \ldots$ the eigenfunctions of the operator $-\Delta$ subject to Neumann's boundary conditions which are normalized by unity in the mean square and which correspond to its eigenvalues numbered in the order of increase. Suppose that $v \in R^{2m(M+1)}$ is a vector whose coordinates are the projections of the solution of the boundary-value problem (28.14) onto the system of functions

$$1/\text{mes}\,\Omega, g_1(x), \ldots, g_M(x), \qquad (28.19)$$

and w is the projection of the solution onto the orthogonal complement. In these variables the boundary-value problem (28.14) can be rewritten in the operator form

$$\frac{dv}{d\tau} = A_{11}v + A_{12}w, \qquad \frac{dw}{d\tau} = A_{21}v + A_{22}w. \qquad (28.20)$$

Therefore, it is expedient to call the equation

$$\frac{dv}{d\tau} = A_{11}v \qquad (28.21)$$

in the space $R^{2m(M+1)}$ (supplied with Euclidean topology, a fact that should be emphasized since in the sequel we choose M sufficiently large) a Galerkin approximation of the boundary-value problem (28.14). To establish the relationship between them, we assume from the very beginning that the substitution

$$h \to h\exp(-\gamma_0 \tau) \qquad (28.22)$$

has been carried out in (28.14). Here the constant $\gamma_0 > 0$ is such that the eigenvalues of the operator L_0 with negative real parts remain the same for the operator $L_0 + \gamma_0 I$.

We denote by Π_0 a parabolic operator generated by the boundary-value problem (28.14) with due account of the substitution (28.22). We shall consider it in the space $C^{\alpha,\alpha/2}$ that consists of functions of the variables x and τ that satisfy Hölder's condition with exponent α with respect to x and with exponent $\alpha/2$ with respect to τ varying throughout the axis. It is clear that the operator Π_0^{-1} is determined and

$$\|\Pi_0^{-1}\|_{C^{\alpha,\alpha/2} \to C^{2+\alpha,1+\alpha/2}} \leq C_0. \qquad (28.23)$$

We denote by Π_Γ the differential operator $d/d\tau - A_{11}$ acting into the space $C(-\infty, \infty)$ of continuous functions bounded for $-\infty < \tau < \infty$. We are interested in the problem of the existence of the operator Π_Γ^{-1} acting from $C(-\infty, \infty)$ into $C^1(-\infty, \infty)$.

LEMMA 28.1. *For all sufficiently large M, the operator Π_Γ is regular and*

$$\|\Pi_\Gamma^{-1}\|_{C(-\infty,\infty) \to C^1(-\infty,\infty)} \leq C_\Gamma, \qquad (28.24)$$

where the constant C_Γ does not depend on M.

To prove this lemma, we shall consider the nonhomogeneous system

$$\begin{aligned}\frac{dv}{d\tau} &= A_{11}v + A_{12}w + f(\tau),\\ \frac{dw}{d\tau} &= A_{21}v + A_{22}w.\end{aligned} \qquad (28.25)$$

We set $\Pi_M = d/d\tau - A_{22}$. For large M the principal part of the operator A_{22} is the Laplace operator considered in the orthogonal complement of the system of functions (28.19). Therefore, the operator Π_M is invertible, and this makes it possible to pass to the equation

$$\frac{dv}{d\tau} = A_{11}v + A_{12}\Pi_M^{-1}A_{21}v + f(\tau) \qquad (28.26)$$

in the problem on the solutions of system (28.25) bounded for $-\infty < \tau < \infty$.

Since we have an estimate similar to (28.23) in Sobolev's spaces of functions, we can uniquely determine the function $v(\tau)$, bounded throughout the axis, from Eq. (28.26) (since it is equivalent to system (28.25)). Here

$$\|v(\tau)\|_{C^1(-\infty,\infty)} \le C_\Gamma \|f(\tau)\|_{C(-\infty,\infty)}, \qquad (28.27)$$

where the constant C_Γ does not depend on M.

In order to complete the proof, we introduce into consideration the equation

$$\frac{dv}{d\tau} - A_{11}v - A_{12}\Pi_M^{-1}A_{21}v = f(\tau) - \mu A_{12}\Pi_M^{-1}A_{21}v, \qquad (28.28)$$

where $0 \le \mu \le 1$. According to (28.27), the operator appearing on the left-hand side is invertible. But for large M the norm of the operator Π_M^{-1} is small, and, therefore, for all μ under consideration, we determine from (28.28) the function $v(\tau)$ which satisfies an estimate of type (28.27). In order to obtain inequality (28.24), it remains to set $\mu = 1$. We have proved the lemma.

LEMMA 28.2. *Suppose that the substitution* (28.22) *has been carried out in the boundary-value problem* (28.14). *Then, for all sufficiently large M, the dimensions of the subspaces of the exponentially growing solutions of the boundary-value problem* (28.14) *and of its Galerkin approximation* (28.21) *are the same.*

In order to prove the lemma, we shall consider the parabolic operator generated by the system

$$\frac{dv}{d\tau} = A_{11}v + \mu A_{12}w,$$

$$\frac{dv}{d\tau} = A_{21}v + A_{22}w.$$

It follows from the arguments that we presented when proving Lemma 28.1 that for sufficiently large M it is regular for all $0 \le \mu \le 1$, and, therefore, according to Theorem 25.4, the dimensions of the subspaces of the exponentially growing solutions are the same. It remains to set $\mu = 0$ and point out that for large M the spectrum of the operator A_{22} lies in the left complex half-plane. We have proved the lemma.

Thus, the first principal idea is the relationship between the boundary-value problem (28.14) and its Galerkin approximation (28.21). The second principal idea is connected with certain heuristic considerations.

On the right-hand side of (28.7) we take terms up to order ε inclusive and take into account relations (28.12) in them. We denote the result by $u_*(\varphi, x, \varepsilon)$, where

$$\frac{d\varphi}{dt} = \omega + \varepsilon\psi, \qquad (28.29)$$

and carry out linearization of the boundary-value problem (28.1), (28.2) on the functions $u = u_*(\varphi, x, \varepsilon)$. Rejecting the inessential orders of smallness with respect to ε, we arrive at the linear boundary-value problem

$$\frac{\partial h}{\partial t} = \varepsilon D \Delta h + A_0 h + \varepsilon^{1/2} B_1(\varphi, x) h + \varepsilon B_2(\varphi, x) h, \qquad (28.30)$$

$$\left.\frac{\partial h}{\partial \nu}\right|_{\partial \Omega} = 0, \qquad (28.31)$$

where the matrices B_1 and B_2, smooth with respect to the set of variables, are periodic with respect to φ with period 2π. Since Eq. (28.30) is singular, the method of revealing the stability properties presented at the end of 27.2 cannot be directly applied to the boundary-value problem (28.30), (28.31). However, let us see what we shall have if we apply it formally.

We introduce into consideration the row matrix

$$V_0(\varphi) =$$
$$= (a_1 \exp(i\varphi_1), \bar{a}_1 \exp(-i\varphi_1), \ldots, a_m \exp(i\varphi_m), \bar{a}_m \exp(-i\varphi_m))$$

and compose the expression

$$V(\varphi, x, \varepsilon) \exp(tL(\varepsilon)), \qquad (28.32)$$

where

$$V(\varphi, x, \varepsilon) = V_0(\varphi) + \varepsilon^{1/2} V_1(\varphi, x) + \varepsilon V_2(\varphi, x) + \ldots, \qquad (28.33)$$

$$L(\varepsilon) = \varepsilon L_0 + \varepsilon^2 L_1 + \ldots. \qquad (28.34)$$

Substituting (28.32) into (28.30), (28.31), we successively equate the coefficients of like powers of ε. From the relationships of the terms on the right-hand side of (28.7) on the harmonics pointed out in 28.1 it follows that the equation for $V_1(\varphi, x)$ is solvable and from the conditions of solvability of the equation for $V_2(\varphi, x)$ we deduce that the operator L_0 appearing on the right-hand side of (28.34) and subject to determination is the same as that in the boundary-value problem (28.14). (The equation for $V_3(\varphi, x)$ is also solvable and the operator L_1, which is a fourth-order operator, can be found from the solvability condition for the equation for $V_4(\varphi, x)$.) Therefore, we can make an assumption that the behavior of the solutions of the boundary-value problem (28.30), (28.31) and of the half-group $\exp(\varepsilon t L_0)$ are similar as $t \to \infty$.

LEMMA 28.3. *Suppose that substitution (28.22) has been carried out in the boundary-value problem (28.30), (28.31). Then there exists $\varepsilon_0 > 0$ such that for $0 < \varepsilon \leq \varepsilon_0$ the parabolic operator $\Pi(\varepsilon)$ generated by it possesses the regularity properties, and*

$$\|\Pi^{-1}(\varepsilon)\|_{C^{\alpha,\alpha/2} \to C^{2+\alpha,1+\alpha/2}} \leq \frac{C_0}{\varepsilon}, \tag{28.35}$$

where the constant $C_0 > 0$ does not depend on ε.

In order to prove this lemma, we first make substitution

$$h \to [I + \varepsilon^{1/2} B_3(\varphi, x)] h \tag{28.36}$$

in the boundary-value problem (28.30), (28.31). Here the smooth matrix B_3, 2π-periodic with respect to φ, possesses the property

$$\sum_{s=1}^{m} \frac{\partial B_3}{\partial \varphi_s} \omega_s = A_0 B_3 - B_3 A_0 + B_1.$$

It is well known that, by virtue of Condition 28.3, such a matrix B_3 can always be found. As a result of the substitution (28.36), the matrix B_1 appearing on the right-hand side of (28.30) is cancelled but terms of order $\varepsilon^{3/2}$ appear. Rejecting them for a while, we arrive at the boundary-value problem

$$\frac{\partial h}{\partial t} = \varepsilon D \Delta h + A_0 h + \varepsilon B_4(\varphi, x) h, \tag{28.37}$$

$$\left.\frac{\partial h}{\partial \nu}\right|_{\partial \Omega} = 0. \tag{28.38}$$

We write the boundary-value problem (28.37), (28.38) as the system (28.20) in which, of course, we must replace τ by t. We choose the constant M large enough for the solutions of the equation

$$\frac{dw}{dt} = A_{22} w \tag{28.39}$$

to be exponentially damped (naturally, with the coefficient of the exponent of order ε, which, however, by virtue of Condition 28.2, increases with the growth of M). We apply the algorithm for the investigation of stability presented in Sec. 27 to the "ordinary part" of this system, i.e., to an equation of the form (28.21) in $R^{n(M+1)}$. As a result of its application, the critical and noncritical variables connected, respectively, with the root subspace of the matrix A_0 corresponding to the eigenvalues lying on the imaginary axis and its root subspace corresponding to the eigenvalues with negative real parts are separated.

The formal algorithm described above allows us to realize that for the critical variables we get (in the time t) the Galerkin approximation of the boundary-value problem (28.14). This fact and Lemma 28.1 imply the regularity of the differential operator generated by the "ordinary part." The properties of the "infinite-dimensional" part of (28.39), which we have already pointed out (see the proof of Lemma 28.1), imply the regularity of the differential operator generated by the boundary-value problem (28.37), (28.38) and the validity of estimate (28.35) for it (here Condition 28.2 in conjunction with the general properties of parabolic systems [90] is valid). Thus, the terms of order $\varepsilon^{3/2}$ that we have rejected do not affect the regularity property. We have proved the lemma.

It follows from Lemma 28.3 and Theorem 25.2 that after the substitution (28.22) the solutions of the boundary-value problem (28.30), (28.31) possess the property of exponential dichotomy. In addition, it follows from the proofs of Lemmas 28.2 and 28.3 that the dimension of the subspace of its exponentially growing solutions coincides with that of a similar subspace of the solutions of the boundary-value problem (28.14). These important facts allow us to make additional inferences concerning the properties of the solutions of the boundary-value problem (28.30), (28.31).

Suppose that the operator $L_0 + \gamma_0 I$ has k eigenvalues with positive real parts. The solutions of the boundary-value problem (28.14) corresponding to them can be rewritten as

$$H_0(x)\exp[\tau(D_0 + \gamma_0 I)],$$

where D_0 is a $k \times k$ matrix and $H_0(x)$ is an $m \times k$ matrix. Let us now compose the expression

$$V_*(\varphi, x, \varepsilon)H_0(x)\exp[\varepsilon t(D_0 + \gamma_0 I)], \qquad (28.40)$$

where

$$V_* = V_0(\varphi) + \varepsilon^{1/2}V_1(\varphi, x) + \varepsilon V_2(\varphi, x) + \varepsilon^{3/2}V_3(\varphi, x).$$

The main fact, whose justification is carried out in the same way as that presented in Sec. 27, is the following: after the substitution (28.22), we can improve the right-hand side of the boundary-value problem (28.30), (28.31) by means of a finite-dimensional operator of order ε^2 such that, as a result, formula (28.40) gives all its exponentially growing solutions. Carrying out the inverse of substitution (28.22), we infer that the formula

$$V_*(\varphi, x, \varepsilon) H_0(x) \exp(\varepsilon t D_0) \qquad (28.41)$$

gives the solutions of the original, but improved, boundary-value problem which increase exponentially or are quasiperiodic.

Thus, a trichotomy of solutions holds for the improved boundary-value problem. By virtue of Lemma 28.3 and the results given in Sec. 25, the solutions of the original boundary-value problem also possess the trichotomy property. It is carried out successively with respect to ε by means of the projectors $P_\pm(\varphi, \varepsilon)$ and $P_0(\varphi, \varepsilon)$, bounded and 2π-periodic with respect to φ, and is interpreted as follows: the solutions belonging to $P_-(\varphi, \varepsilon) C^\alpha$ and $P_+(\varphi, \varepsilon) C^\alpha$ grow correspondingly and are damped with the coefficients of the exponents of order ε, and the solutions belonging to $P_0(\varphi, \varepsilon) C^\alpha$ have characteristic exponents whose order does not exceed ε^2.

For the sequel it is important to have an analytic description of the zero subspace of the operator $P_0(\varphi, \varepsilon)$. To clarify this problem, we return to the improved boundary-value problem but consider a problem in which the inverse of substitution (28.22) is carried out. Recall that formula (28.41) gives a good idea of the structure of its quasiperiodic solutions. Let us consider the adjoined boundary-value problem. Its quasiperiodic solutions have the same structure. With the aid of the quasiperiodic solutions of these two boundary-value problems we can construct, in an ordinary manner, m functionals whose zeros constitute a zero subspace of the projector $P_0(\varphi, \varepsilon)$ of the improved boundary-value problem. If we return to the original boundary-value problem (28.30), (28.31), we see that, by virtue of the results of Sec. 25, only insignificant deformations of these functionals occur.

Thus, the zero subspace of the operator $P_0(\varphi, \varepsilon)$ can be described by the relations

$$l_s(\varphi, h, \varepsilon) = 0, \qquad s = 1, \ldots, m. \qquad (28.42)$$

Note that it follows from the constructions that we have carried out that the principal parts of the functionals l_s (their values for $\varepsilon = 0$) can be taken independent of φ.

We sum up with the conclusion that the gist of the second idea is the turning of formalism (28.32)–(28.34) to advantage in a sufficiently nontrivial way. Incidentally, all the constructions carried out in this subsection are connected with linear equations and, hence, it is not difficult to understand the whole complex of ideas.

28.3. Justification of the Theorems

Using Condition 28.1, we continue the series appearing on the right-hand side of (28.7) up to order ε^2 inclusive and in each of its terms take relations (28.12) into account. As above, we denote the result by $u_*(\varphi, x, \varepsilon)$, where φ is the solution of Eq. (28.29). In the neighborhood of this approximate torus we introduce coordinates φ and h, setting

$$u = u_*(\varphi, x, \varepsilon) + \varepsilon h, \qquad h = Ph, \qquad (28.43)$$

where $P = P_+(\varphi, \varepsilon) + P_-(\varphi, \varepsilon)$. Here and in what follows we preserve the notations for the projectors, introduced in 28.2, which realize the trichotomy of the solutions of the boundary-value problem (28.1), (28.2) linearized on the more precise approximate torus. We substitute (28.43) into (28.1), (28.2) and write the result as

$$\frac{\partial u_*}{\partial \varphi} \frac{d\varphi}{dt} + \varepsilon \frac{\partial P}{\partial t} h + \varepsilon P \frac{\partial h}{\partial t} =$$
$$= \varepsilon B(\varphi, x, \varepsilon) h + \varepsilon^2 F_1(\varphi, x, h, \varepsilon) + \varepsilon^{5/2} F_2(\varphi, x, \varepsilon). \qquad (28.44)$$

Here B is an elliptic operator appearing upon the linearization on the approximate torus (recall that its principal part is proportional to ε) and the vector-function F_1 has the second order of smallness on h.

By virtue of Theorem 25.3,

$$\frac{\partial P}{\partial t} = BP - PB \qquad (28.45)$$

under condition (28.29). But we expect that upon the substitution of (28.43), instead of (28.29), we will get an equation of the form

$$\frac{d\varphi}{dt} = \omega + \varepsilon\psi + \varepsilon^{3/2}\psi_1(\varphi, h, \varepsilon) + \varepsilon^2 \psi_2(\varphi, \varepsilon), \qquad (28.46)$$

where $\psi_1(\varphi, 0, \varepsilon) \equiv 0$. Let us show that this is really the case.

We apply the projector P_0 to relation (28.44). In order to prove the validity of relation (28.46), we must use relation (28.45) and make note of the fact that the determinant of the matrix $P_0 \partial u_*/\partial \varphi$ is of order $\varepsilon^{1/2}$.

Relation (28.46) makes it possible to rewrite (28.44) as an equation in the range of the projector P:

$$\frac{\partial h}{\partial t} = B(\varphi, x, \varepsilon)h + \varepsilon \Phi_1(\varphi, x, h, \varepsilon) + \varepsilon^{3/2} \Phi_2(\varphi, x, \varepsilon), \qquad (28.47)$$

where the properties of Φ_1 and Φ_2 are similar to those of F_1 and F_2 respectively. Recall that now we have (28.46) instead of (28.29). Therefore, the linear part of (28.47) can differ, in principle, from the linear part of (28.44) by a finite-dimensional operator of order ε^2 acting in the range of the operator P (this follows from the structure of the projector P_0 described in 28.2). This circumstance is inessential for the sequel.

It is technically inconvenient to consider Eq. (28.47) under Neumann's boundary conditions in the subspace of the space C^α dependent on φ and ε, and, therefore, by projecting onto a certain number of the first eigenfunctions of the Laplace operator, we write (28.47) as a system that relates finite-dimensional and infinite-dimensional variables. Then we use relations (28.42) in order to exclude m finite-dimensional variables. As a result, we get a certain equation of the form (28.47) in a suitable contraction of C^α.

We set $h = 0$ in ψ_1 from (28.46) and in Φ_1 from (28.47). As a result we get the equations

$$\frac{\partial h}{\partial t} = B(\varphi, x, \varepsilon)h + \varepsilon^{3/2} \Phi_2(\varphi, x, \varepsilon), \qquad (28.48)$$

$$\frac{d\varphi}{dt} = \omega + \varepsilon \psi + \varepsilon^2 \psi_2(\varphi, \varepsilon). \qquad (28.49)$$

We first consider separately the homogeneous equation

$$\frac{\partial h}{\partial t} = B(\varphi, x, \varepsilon)h. \qquad (28.50)$$

As in [84], we denote by $\varphi_t(\varphi, \varepsilon)$ the solution of (28.49) with the initial condition φ defined for $t = 0$. By $U_\tau^t(\varphi, \varepsilon)$ we denote the Cauchy resolving operator of Eq. (28.50) and set

$$G_0(\tau, \varphi, \varepsilon) = \begin{cases} U_\tau^0(\varphi, \varepsilon) P_+(\varphi_\tau(\varphi, \varepsilon), \varepsilon), & \tau < 0, \\ -U_\tau^0(\varphi, \varepsilon) P_-(\varphi_\tau(\varphi, \varepsilon), \varepsilon), & \tau > 0, \end{cases} \qquad (28.51)$$

where $P_\pm$ are projectors that realize the exponential dichotomy of the solutions of Eq. (28.50) under condition (28.49). Note that the replacement of (28.29) by (28.49) is inessential i.e., these projectors possess the properties described in 28.2. As is shown in [84], the invariant torus of system (28.48), (28.49) can be explicitly represented with the aid of (28.51):

$$h_1(\varphi, x, \varepsilon) = \varepsilon^{3/2} \int_{-\infty}^{\infty} G_0(\tau, \varphi, \varepsilon) \Phi_2(\varphi_\tau(\varphi, \varepsilon), x, \varepsilon) \, d\tau.$$

Next we substitute $h = h_1(\varphi, x, \varepsilon)$ into ψ_1 from (28.46) and into Φ_1 from (28.47) and then repeat the procedure, i.e., introduce the Green function $G_1(\tau, \varphi, \varepsilon)$, similar to (28.51), of the problem on invariant tori, construct $h_2(\varphi, x, \varepsilon)$, and so on. It is shown in [84] (for ordinary differential equations, but this is inessential since the solutions of parabolic systems are compact) that the iteration process described converges, making it possible to determine the invariant torus $h = h(\varphi, x, \varepsilon)$ of system (28.46), (28.47), and, as can be easily seen, its norm in $C^{2+\alpha}$ is of order $\varepsilon^{1/2}$. This and (28.43), (28.46) imply (28.16), (28.17). Note that with respect to φ the smoothness of the torus is arbitrary, finite. This follows from the constructions carried out in 28.2, by virtue of which all the encountered Green functions depend on φ sufficiently smoothly.

Let us discuss the stability properties of the constructed invariant torus $h = h(\varphi, x, \varepsilon)$ of system (28.46), (28.47). Let us suppose that the self-similar torus (28.12) is orbitally exponentially stable. Under this condition $P_- = 0$ (here and in what follows $P_\pm(\varphi, \varepsilon)$ are projectors that realize the exponential dichotomy of the solutions of Eq. (28.47) linearized for $h = h(\varphi, x, \varepsilon)$, in which $\varphi = \varphi_t(\varphi, \varepsilon)$ is determined from (28.46) for $h = h(\varphi, x, \varepsilon)$) and, as a consequence, the solution of (28.47) with the initial conditions from a sufficiently small neighborhood of the torus $h = h(\varphi, x, \varepsilon)$ are wound around it according to an exponential law with the coefficient of the exponent of order ε.

We can prove this fact by employing, for instance, the technique developed by Perron, connected with the use of integral inequalities (it can be found in many books where Lyapunov's theorem on the stability with respect to the first approximation is substantiated). In our case its application does not cause any difficulties since, by virtue of what was said in 28.2, the third and fourth terms in (28.46) do not affect the character of damping of the solutions of the linearized (on the invariant torus) equation (28.47).

Now if torus (28.12) is dichotomous, then $P_- \neq 0$. Under this condition, in order to prove the instability of the torus $h = h(\varphi, x, \varepsilon)$, we can again apply Perron's scheme for the substantiation of Lyapunov's theorem on the instability with respect to the first approximation (it is contained in [66], for instance).

In conclusion, we want to point out the following. If, in (28.12), some of the coordinates $\xi_{0s_1}, \ldots, \xi_{0s_k}$ of the amplitude of the self-similar torus are zero, then the statements of Theorems 28.1 and 28.2 remain valid, but we shall speak of tori of smaller dimensions. In this case, the proof given above remains valid practically without any changes.

28.4. Principle of Averaging

Thus, with the use of Theorems 28.1 and 28.2, we have established a correspondence between the special solutions of the quasinormal form (28.8), (28.9) and of the original boundary-value problem (28.1), (28.2). It turns out that when one additional condition is satisfied, formula (28.7) defines a relationship between any solutions of the boundary-value problem (28.1), (28.2) and those of its quasinormal form.

For the sake of simplicity, we shall only consider an example of a system of the reaction–diffusion type:

$$\frac{\partial u}{\partial t} = \varepsilon D \frac{\partial^2 u}{\partial x^2} + A_0 u + \varepsilon A_1 u + F(u), \tag{28.52}$$

$$\left.\frac{\partial u}{\partial x}\right|_{x=0} = \left.\frac{\partial u}{\partial x}\right|_{x=\pi} = 0. \tag{28.53}$$

Here, as before, $u \in R^n$, $n \geq 2$, the eigenvalues of the matrix D lie on the right of the imaginary axis, the smooth vector-function F has at zero an order of smallness higher than the first. The simplicity of the boundary-value problem (28.52), (28.53) consists in that

$$A_0 a = i\omega_0 a, \qquad A_0^* b = -i\omega_0 b, \qquad \omega_0 > 0, \qquad (a,b) = 1,$$

and the other eigenvalues of the matrix A_0 are on the left of the imaginary axis. We also suppose that

$$\mathrm{Re}\,(A_1 a, b) > 0, \qquad \mathrm{Re}\, d < 0,$$

where d is the first Lyapunov quantity of the ordinary differential equation resulting from (28.52) for $\varepsilon = 0$. As a phase space of the boundary-value problem (28.52), (28.53) it is convenient to take the Hilbert space $E = W_2^1(0, \pi)$ in which its properties can be easily revealed.

Substituting, as before, into (28.52), (28.53), the relation

$$u = \varepsilon^{1/2}[\zeta(\tau, x) a \exp(i\omega_0 t) + \bar{\zeta}(\tau, x) \bar{a} \exp(-i\omega_0 t)] +$$

$$+ \varepsilon u_2(\tau,t,x) + \varepsilon^{3/2} u_3(\tau,t,x) + \ldots, \qquad (28.54)$$

where $\tau = \varepsilon t$, and equating the coefficients in like powers of ε, we arrive, at the third step (see the beginning of the section), at the quasinormal form

$$\frac{\partial \zeta}{\partial \tau} = (Da,b)\frac{\partial^2 \zeta}{\partial x^2} + (A_1 a,b)\zeta + \zeta d|\zeta|^2, \qquad (28.55)$$

$$\left.\frac{\partial \zeta}{\partial x}\right|_{x=0} = \left.\frac{\partial \zeta}{\partial x}\right|_{x=\pi} = 0. \qquad (28.56)$$

When studying some problems of statistical physics, Ginsburg and Landau obtained, in 1950, an equation similar to (28.55). Therefore, it is sometimes called the Ginsburg–Landau equation in the literature. It is also called the Kuramoto–Tsudsuki equation after two Japanese physicists who, proceeding from the physical representations developed by Ginsburg and Landau, erroneously assumed that it defined the dynamics of a parabolic system with the diffusion coefficients of order unity when its homogeneous equilibrium position loses stability in an oscillatory way. By now, many works have been carried out in which, with the aid of computers, different properties of stationary solutions of the boundary-value problem (28.55), (28.56) have been revealed.

A description of the restrictions under which the boundary-value problem (28.55), (28.56) will be studied now follows. The first condition is similar to Condition 28.2.

CONDITION 28.4. *For all $z > 0$ the eigenvalues of the matrix $A_0 - zD$ lie on the left of the imaginary axis, and the eigenvalues $\pm i\omega_0$ of the matrix A_0 generically recede from the imaginary axis for small z, i.e., Re $(Da,b) > 0$.*

CONDITION 28.5. *Suppose that P is a projector onto the root subspace of the matrix A_0, corresponding to the eigenvalues with negative real parts, in the direction of a two-dimensional root subspace corresponding to its eigenvalues $\pm i\omega_0$. We suppose that the eigenvalues PDP have positive real parts and that for all $z > 0$ the eigenvalues $P(A_0 - zD)P$ lie on the left of the imaginary axis.*

Note that Condition 28.4 is a certain definite correspondence between the "diffusional" and "ordinary" parts of (28.52). Condition 28.5, which was first formulated in [50], strengthens this correspondence. Therefore, when it is satisfied, we have a statement, more general than Theorems 28.1 and 28.2, similar to the principle of averaging on a finite time interval for systems of ordinary differential equations.

THEOREM 28.3. *Suppose that Conditions 28.4, 28.5 are fulfilled. Then there are positive constants ε_0, r_0 and $\gamma_1 < \gamma_2$ such that for $0 < \varepsilon \leq \varepsilon_0$ every* solution of the boundary-value problem (28.52), (28.53), belonging to the ball $S(r_0)$ of a phase space with center at zero for all $t \geq 0$, for $\gamma_1/\varepsilon \leq t \leq \gamma_2/\varepsilon$ differs by a quantity of order $\varepsilon^{3/2}$ from the function calculated from formula (28.54), in which the first and second terms are preserved, and $\zeta(\tau, x)$ is a suitable solution of the quasinormal form of (28.55), (28.56).

Before proving the theorem, we shall give a remark of a general nature. Employing, for instance, the fractional-powers apparatus of an elliptic operator and using Condition 28.4, we can easily find the estimates of the solutions of the boundary-value problem (28.52), (28.53) under consideration in smooth metrics.

Here are the main features of the proof. We say that the variables $u_1 = (I-P)u$ are critical and the variables $u_2 = Pu$ are noncritical. First, with the use of an appropriate substitution, we normalize the "ordinary" part of Eq. (28.52), upon which the noncritical variables are completely separated. It is convenient to retain the notations of the critical and noncritical variables.

Then, using Lagrange's formula, we find u_2 from the equation for the noncritical variables and substitute it into the equation for the critical variables. Note that the characteristics order of time is ε^{-1}. It is precisely this time in which, by virtue of Condition 28.5, the noncritical variables become of order $\varepsilon^{3/2}$ and the critical ones, by virtue of the inequality Re $d < 0$ (we do not use the inequality Re $(A_1 a, b) > 0$) become of order $\varepsilon^{1/2}$. This circumstance makes it possible to carry out their normalization. It remains to introduce a rotating coordinate system

$$u_1 = a\eta \exp(i\omega_0 t) + \bar{a}\bar{\eta}\exp(-i\omega_0 t),$$

to replace t by t/ε and carry out the operation of averaging, whose legitimacy can be justified by applying a fully standard scheme with the use of Condition 28.4.

Allowing ourselves a certain freedom of speech, we can say that under the conditions of Theorem 28.3, with an accuracy to within $\varepsilon^{3/2}$, the quasinormal form (28.55), (28.56) defines the flow on the integral manifold defined by (28.54).

28.5. A Boundary-Value Problem with a Terminal Turn Point

Theorems 28.1–28.3, in their totality, constitute the method of quasinormal forms suggested by Yu. S. Kolesov, which is the main technique of constructing stationary solutions for parabolic systems with a small diffusion. In accordance with [55], here and in 28.6 we discuss the range of its application. We shall not touch upon the hypothesis of Theorem 28.3, which is a uniform averaging principle, but shall concentrate attention on the hypotheses of Theorems 28.1 and 28.2, which constitute strong averaging principles.

We shall consider the cases where Condition 28.4 is violated with a first-order contingence at a certain point $z_0 \geq 0$. We agree to say that there is a *terminal turn point* if $z_0 = 0$ and an *interior turn point* if $z_0 > 0$.

Suppose that
$$D = D_0 + \varepsilon^{1/2} D_1 \tag{28.57}$$
in (28.52), where the matrix D_1 is arbitrary and the eigenvalues of D_0 lie on the right of the imaginary axis. The main restriction is that for all $z > 0$ the eigenvalues of the matrix $A_0 - zD_0$ are in the left complex half-plane, but, in contrast to Condition 28.4,
$$\text{Re}\,(D_0 a, b) = 0, \qquad \text{Im}\,(D_0 a, b) \neq 0, \qquad \text{Re}\,(D_0 a_1, b) > 0,$$
where a_1 is the solution of the algebraic equation
$$A_0 a_1 - i\omega_0 a_1 = D_0 a - (D_0 a, b)a.$$

It is interest to note that we encounter this in the classical predator–prey problem with due account of the migration factor and with natural constraints imposed on the parameters. The numerical investigation of this problem was carried out in [15] in which, for biological reasons, the Malthusian coefficient of a linear growth of the prey was taken relatively large. If we diminish it in an appropriate way, then the resulting boundary-value problem is similar, in its properties, to (28.52), (28.53).

It turned out, however, that in this case a peculiar situation is possible, namely, alongside a stable homogeneous equilibrium position there are stable periodic oscillations with a complex spatial structure. It follows that the quasinormal form, in which we must replace D by D_0, does not always sufficiently fully describe the dynamics of the boundary-value problem (28.52), (28.53). Indeed, for $\text{Re}\,(A_1 a, b) < 0$ the zero solution of the boundary-value problem (28.55), (28.56) is globally stable, i.e., the original problem seems not to have cycles of order $\varepsilon^{1/2}$.

We shall explain the mathematical reason for this phenomenon.

We consider the linear parabolic equation
$$\frac{\partial u}{\partial t} = \varepsilon D \frac{\partial^2 u}{\partial x^2} + A_0 u + \varepsilon A_1 u \tag{28.58}$$

under the boundary conditions (28.53). We denote by $u(t,x,\varepsilon)$ the solution of the boundary-value problem (28.58), (28.53).

LEMMA 28.4. *Under condition (28.57) we have the asymptotically exact estimate*

$$\|u(t,x,\varepsilon)\|_{E_1} \leq \frac{M_0}{\varepsilon^{3/2}t}\|u|_{t=0}\|_E \exp(\varepsilon\gamma_0 t), \qquad (28.59)$$

where $t > 0$, $E_1 = W_2^3$, and γ_0, M_0 are constants.

To prove the lemma, it is sufficient to use the Fourier method.

Inequality (28.59) implies that under condition (28.57) the matrix diffusion coefficient of the boundary-value problem (28.52), (28.53) is of order $\varepsilon^{3/2}$. It is natural that such a strong singularity essentially affects the dynamic properties.

Thus, in the boundary-value problem (28.52), (28.53) with a terminal turn point the diffusion is too small relative to linearity. In this connection, instead of (28.52), (28.53) we consider the model boundary-value problem (which is of interest by itself)

$$\frac{\partial u}{\partial t} = \varepsilon^{1/2} D \frac{\partial^2 u}{\partial x^2} + A_0 u + \varepsilon A_1 u + F(u), \qquad (28.60)$$

$$\left.\frac{\partial u}{\partial x}\right|_{x=0} = \left.\frac{\partial u}{\partial x}\right|_{x=\pi} = 0, \qquad (28.61)$$

where the matrix D has the form (28.57). For the linear system (28.58) corresponding to it in an inequality similar to (28.59), we must replace $\varepsilon^{3/2}$ by ε, i.e., the boundary-value problem (28.60), (28.61) is "normally" coercive. Thus $\varepsilon^{1/2}$ and not ε is the natural order of smallness of the diffusion under condition (28.57).

We shall investigate the boundary-value problem (28.60), (28.61) in Sec. 31. We shall also show there that under condition (28.57) the analogs of Theorems 28.1 and 28.2 are valid under an additional condition of stability that does not follow from the linearization of the quasinormal form. It follows, in particular, from the results that in one of the variants, because of the terminal turn point, there appear a large number of stable periodic oscillations of the same kind.

28.6. A Boundary-Value Problem with an Interior Turn Point

In this case, we suppose that

$$D = D_0 + \varepsilon^{1/2} D_1 + \varepsilon D_2 \qquad (28.62)$$

in (28.52). Here the eigenvalues of the matrix D_0 lie on the right of the imaginary axis; for all $z > 0$, $z \neq z_0$, $z_0 > 0$, the eigenvalues of the matrix $A_0 - zD_0$ are in the left complex half-plane and for $z = z_0$ it has a simple zero eigenvalue; the matrices D_1 and D_2 are arbitrary, but $(D_1\alpha, \beta) = 0$, where

$$(A_0 - z_0 D_0)\alpha = 0, \qquad (A_0^* - z_0 D_0^*)\beta = 0, \qquad (\alpha, \beta) = 1,$$

and, finally,

$$(D_0\alpha, \beta) = 0, \qquad (D_0\alpha_1, \beta) > 0, \qquad \text{Re}\,(D_0 a, b) > 0,$$

where α_1 is the solution of the algebraic equation

$$(A_0 - z_0 D_0)\alpha_1 = D_0\alpha.$$

LEMMA 28.5. *Suppose that in* (28.58) *the matrix D has form* (28.62). *Then we have the asymptotically exact estimate*

$$\|u(t, x, \varepsilon)\|_{E_1} \leq \frac{M_0}{\varepsilon^2 t} \|u|_{t=0}\|_E \exp(\varepsilon\gamma_0 t), \qquad (28.63)$$

where $t > 0$ and γ_0, M_0 are constants.

Before discussing the consequences of inequality (28.63), we shall consider a generalization of the method of quasinormal forms which seems to be reasonable at first glance. Since, under condition (28.62), high and low modes (with subscripts of order $\varepsilon^{-1/2}$) strongly interact in the boundary-value problem (28.52), (28.53), it is expedient to represent its solutions as

$$u = u(t, x, y), \qquad y = \sqrt{\frac{z_0}{\varepsilon}}\left(x - \frac{\pi}{2}\right). \qquad (28.64)$$

It is very likely that at the qualitative level the original boundary-value problem is equivalent to the problem

$$\frac{\partial u}{\partial t} = DL(\varepsilon)u + A_0 u + \varepsilon A_1 u + F(u), \qquad (28.65)$$

where, in addition to the boundary conditions (28.53), we set the conditions of boundedness with respect to y at infinity. In (28.65),

$$L(\varepsilon) = \varepsilon\frac{\partial^2}{\partial x^2} + (\varepsilon z_0)^{1/2}\frac{\partial^2}{\partial x \partial y} + z_0\frac{\partial^2}{\partial y^2}. \qquad (28.66)$$

We reject the nonlinear term in (28.65) and set $\varepsilon = 0$. The resulting linear problem

$$\frac{\partial u}{\partial t} = z_0 D_0 \frac{\partial^2 u}{\partial y^2} + A_0 u$$

Systems of Reaction–Diffusion Type

has the solution

$$\zeta(x)a\exp(i\omega_0 t) + \bar{\zeta}(x)\bar{a}\exp(-i\omega_0 t) + \eta(x)\alpha\exp(iy) + \bar{\eta}(x)\alpha\exp(-iy),$$

where the complex-valued functions ζ and η are arbitrary. Therefore, we can try to seek the stationary solutions of the boundary-value problem (28.65) in the form

$$u = \varepsilon^{1/2}[\zeta(\tau, x)a\exp(i\omega_0 t) + \bar{\zeta}(\tau, x)\bar{a}\exp(-i\omega_0 t) +$$
$$+ \eta(\tau, x)\alpha\exp(iy) + \bar{\eta}(\tau, x)\alpha\exp(-iy)] + \varepsilon u_2(t, \tau, x, y) + \ldots, \quad (28.67)$$

where $\tau = \varepsilon t$. Substituting (28.67) into (28.65) and taking into account (28.62) and (28.66), we arrive, at the third step, at the system

$$\frac{\partial \zeta}{\partial \tau} = (D_0 a, b) \frac{\partial^2 \zeta}{\partial x^2} + (A_1 a, b)\zeta + (d_{11}|\zeta|^2 + d_{12}|\eta|^2)\zeta, \quad (28.68)$$

$$\frac{\partial \eta}{\partial \tau} = z_0(D_0\alpha_1, \beta)\frac{\partial^2 \eta}{\partial x^2} + iz_0^{3/2}(D_0\alpha_2 + D_1\alpha_1, \beta)\frac{\partial \eta}{\partial x} +$$
$$+ [(A_1\alpha, \beta) - z_0(D_2\alpha, \beta) - z_0^2(D_1\alpha_2, \beta)]\eta + (d_{21}|\zeta|^2 + d_{22}|\eta|^2)\eta, \quad (28.69)$$

where, by virtue of the structure of the right-hand side of (28.67), it is reasonable to set the boundary conditions

$$\left.\frac{\partial \zeta}{\partial x}\right|_{x=0} = \left.\frac{\partial \zeta}{\partial x}\right|_{x=\pi} = 0, \quad (28.70)$$

$$\left.\frac{\partial \eta_1}{\partial x}\right|_{x=0} = \left.\frac{\partial \eta_1}{\partial x}\right|_{x=\pi} = 0, \quad \left.\eta_2\right|_{x=0} = \left.\eta_2\right|_{x=\pi} = 0. \quad (28.71)$$

Here $\eta = \eta_1 + i\eta_2$, $d_{11} = d$, and d_{12}, d_{21}, d_{22} are the other Lyapunov quantities, the last two of which are real, and, finally, α_2 is the solution of the algebraic equation

$$(A_0 - z_0 D_0)\alpha_2 = D_1\alpha.$$

It seems natural that qualitatively the boundary-value problem (28.68)–(28.71) must satisfactorily reflect the main dynamic properties of the original boundary-value problem (28.52), (28.53). However, the statement of Lemma 28.5 casts some doubts upon the truth of this fact.

To be more precise, our hypothesis is as follows: the auto-oscillations of the boundary-value problem (28.52), (28.53) with an interior turn point, which have a sufficiently large domain of attraction, are essentially nonlocal, i.e., a so-called diffusion explosion occurs. From the viewpoint of the general mathematics, this is quite natural, since a turn point usually fundamentally changes the character of a problem.

29. Applications of Bifurcation Theorems

29.1. Spatially Nonhomogeneous Periodic Solutions

Recall that, under the conditions formulated in 28.4, the quasinormal form (28.55), (28.56) corresponding to the boundary-value problem (28.52), (28.53) has a spatially homogeneous self-similar cycle $\zeta = \zeta_0 \exp(i\alpha_0 \tau)$, where

$$\zeta_0 = \sqrt{-\operatorname{Re}(A_1 a, b)/\operatorname{Re} d}, \qquad \alpha_0 = \operatorname{Im}(A_1 a, b) + \zeta_0^2 \operatorname{Im} d. \qquad (29.1)$$

Let us find out whether this cycle is stable, and if it is not, whether there are spatially nonhomogeneous cycles in a small neighborhood. Just as in [22], we shall only consider a simple interesting case of a boundary-value problem of the form (28.52), (28.53),

$$\frac{\partial u}{\partial t} = \varepsilon D \frac{\partial^2 u}{\partial x^2} + A_0 u + \varepsilon A_1 u + F(u), \qquad (29.2)$$

$$\left.\frac{\partial u}{\partial x}\right|_{x=0} = \left.\frac{\partial u}{\partial x}\right|_{x=\pi} = 0, \qquad (29.3)$$

when $u \in R^2$ and $D = \operatorname{diag}\{d_1, d_2\}$, $d_1, d_2 > 0$.

In order to find the amplitude $\zeta(x)$ and the frequency α of an arbitrary self-similar cycle, we use the quasinormal form (28.55), (28.56) and arrive at the boundary-value problem

$$(Da, b)\frac{d^2\zeta}{dx^2} + [(A_1 a, b) - i\alpha]\zeta + \zeta d|\zeta|^2 = 0, \qquad (29.4)$$

$$\left.\frac{d\zeta}{dx}\right|_{x=0} = \left.\frac{d\zeta}{dx}\right|_{x=\pi} = 0, \qquad (29.5)$$

which we shall investigate when Condition 28.4 is fulfilled. This means that $\operatorname{Re}(Da, b) > 0$ and, for all $z > 0$, the eigenvalues of the matrix $A_0 - zD$ lie on the left of the imaginary axis. In this case the first restriction always holds since $\operatorname{Re}(Da, b) = (d_1 + d_2)/2$ (we can easily make sure of this by a simple verification), and the second condition can be written out in the explicit form

$$\varkappa(d_1 - d_2) > -2\sqrt{d_1 d_2}, \qquad \varkappa = \frac{\operatorname{Re} a_1 \bar{a}_2}{\operatorname{Im} a_1 \bar{a}_2}, \qquad a = (a_1, a_2). \qquad (29.6)$$

We shall first consider the problem of the stability of the spatially homogeneous periodic solution of the boundary-value problem (29.2), (29.3).

THEOREM 29.1. *Suppose that inequality (29.6) is satisfied. Then there exists $\varepsilon_0 > 0$ such that for $0 < \varepsilon \le \varepsilon_0$ the spatially homogeneous periodic solution of the boundary-value problem (29.2), (29.3) is exponentially orbitally stable (unstable) if*

$$d_1 + d_2 - 4\operatorname{Re}(A_1 a, b)(x_0 y_0 - 1)/(1 + x_0^2) > 0 \quad (<0),$$

where $x_0 = \varkappa(d_1 - d_2)/(d_1 + d_2)$, $y_0 = \operatorname{Im} d/\operatorname{Re} d$.

The proof of the theorem reduces to the analysis of the spectrum of the operator

$$Lh = (Da, b)\frac{d^2 h}{dx^2} + [(A_1 a, b) - i\alpha_0]h + d\zeta_0^2(2h + \bar{h}), \qquad (29.7)$$

$$\left.\frac{dh}{dx}\right|_{x=0} = \left.\frac{dh}{dx}\right|_{x=\pi} = 0, \qquad (29.8)$$

where the constants ζ_0 and α_0 are defined by (29.1).

If $x_0 y_0 > 1$, then, for $D = D_0$, where the diagonal elements d_{10}, d_{20} of the matrix D_0 are related as

$$d_{10} + d_{20} = 4\operatorname{Re}(A_1 a, b)(x_0 y_0 - 1)/(1 + x_0^2),$$

to the zero eigenvalue of operator (29.7), (29.8) there corresponds, in addition to the eigenfunction $(i\xi_0, -i\xi_0)$, one more eigenfunction

$$(\alpha_1 + i\alpha_2, \alpha_1 - i\alpha_2)\cos x, \qquad \alpha_1 = 1 - x_0 y_0, \qquad \alpha_2 = x_0 + y_0;$$

$\varkappa$ can, of course, be determined from the matrix D_0.

Let us suppose that

$$D = D_0 - \nu D_1, \qquad D_1 = \operatorname{diag}\{d_{11}, d_{21}\}, \qquad 0 < \nu \ll 1, \qquad (29.9)$$

in (29.2), (29.3), and

$$\mu = 2\operatorname{Re}\left[\frac{1 + i\frac{\alpha_2}{\alpha_1}}{1 - ix_0}(D_1 a, b)\right]\nu > 0. \qquad (29.10)$$

THEOREM 29.2 *There exist positive constants $\mu_1, \mu_2, \varepsilon_0$ such that for $\mu_1 \le \mu \le \mu_2$ and $0 < \varepsilon \le \varepsilon_0$, under conditions (29.6) and (29.9), (29.10), the boundary-value problem (29.2), (29.3) has two spatially nonhomogeneous stable periodic solutions whose asymptotics is given below.*

We shall first present the algorithmic part. It follows from the ramification theory that under the conditions of this theorem the solutions $\zeta(x)$ and α of the boundary-value problem (29.4), (29.5) can be sought in the form

$$\zeta(x, \mu) = \zeta_0(1 + \sqrt{\mu}h_1 + \mu h_2 + \mu^{3/2}h_3 + \ldots), \qquad (29.11)$$

$$\alpha(\mu) = \alpha_0 + \delta_1\mu + \delta_2\mu^2 + \ldots, \tag{29.12}$$

where $h_1 = c(\alpha_1 + i\alpha_2)\cos x$. By virtue of (29.11), (29.12) we have the boundary-value problem

$$(d_{10} + d_{20})\frac{d^2h_2}{dx^2} - \frac{2\,\mathrm{Re}\,(A_1a,b)(\alpha_1 + i\alpha_2)}{1+x_0^2}(h_2 + \bar{h}_2 + h_1^2 + 2|h_1|^2) -$$

$$-\frac{2(-x_0+i)}{1+x_0^2}\delta_1 = 0,$$

$$\left.\frac{dh_2}{dx}\right|_{x=0} = \left.\frac{dh_2}{dx}\right|_{x=\pi} = 0$$

for determining h_2 and δ_1 from which it follows that

$$\delta_1 = -c^2\,\mathrm{Re}\,(A_1a,b)(1+y_0^2)\alpha_1\alpha_2, \qquad h_2 = v + (w_1 + iw_2)\cos 2x,$$

where

$$v = -\frac{c^2}{2}\left(\frac{3}{2}\alpha_1^2 + \frac{1}{2}\alpha_2^2 - y_0\alpha_1\alpha_2\right),$$

$$w_1 = \frac{c^2}{12}(3\alpha_1^2 - \alpha_2^2), \qquad w_2 = \frac{c^2}{24}\cdot\frac{\alpha_2}{\alpha_1}(9\alpha_1^2 + \alpha_2^2).$$

From the condition of solvability of the boundary-value problem for h_3, we find out that

$$c^2 = \frac{3(x_0y_0 - 1)(1+x_0^2)}{Q(x_0,y_0)\,\mathrm{Re}\,(A_1a,b)},$$

where

$$Q(x_0,y_0) = (30y_0^4 - 9y_0^2 + 1)x_0^4 + (12y_0^5 - 126y_0^3 + 22y_0)x_0^3 +$$

$$+(-45y_0^4 + 186y_0^2 - 9)x_0^2 + (58y_0^3 - 102y_0)x_0 + (y_0^4 - 21y_0^2 + 6).$$

This polynomial proves to be positive for $x_0y_0 > 1$, and, therefore,

$$c = \pm\sqrt{\frac{3(x_0y_0 - 1)(1+x_0^2)}{Q(x_0,y_0)\,\mathrm{Re}\,(A_1a,b)}}.$$

We have thus constructed the solution $\zeta(x,\mu)$ of the boundary-value problem (29.4), (29.5) for $\alpha = \alpha(\mu)$, where the asymptotic expansions (29.11), (29.12) are valid for ζ and α. In order to now use Theorem 28.1, it remains to study the spectral boundary-value problem

$$(Da,b)\frac{d^2\eta}{dx^2} + [(A_1a,b) - i\alpha(\mu)]\eta + d(\zeta^2(x,\mu)\bar{\eta} + 2|\zeta(x,\mu)|^2\eta) = \lambda\eta, \tag{29.13}$$

Systems of Reaction–Diffusion Type

$$\frac{d\eta}{dx}\bigg|_{x=0} = \frac{d\eta}{dx}\bigg|_{x=\pi} = 0. \tag{29.14}$$

Obviously, we must consider the eigenvalues of λ which vanish for $\mu = 0$. We first use the substitution $\eta = \zeta(x,\mu)u$ to reduce it to a more convenient form:

$$(Da, b)\left[\frac{d^2 u}{dx^2} + \frac{2}{\zeta}\frac{d\zeta}{dx} \cdot \frac{du}{dx}\right] + d|\zeta|^2(u + \bar{u}) = \lambda u \tag{29.15}$$

$$\frac{du}{dx}\bigg|_{x=0} = \frac{du}{dx}\bigg|_{x=\pi} = 0. \tag{29.16}$$

We introduce a series of row matrices

$$V = V_0 + \sqrt{\mu}V_1 + \mu V_2 + \ldots, \tag{29.17}$$

supposing that

$$V_0 = (i, (\alpha_1 + i\alpha_2)\cos x), \tag{29.18}$$

and a real matrix series

$$B = \sqrt{\mu}B_1 + \mu B_2 + \ldots. \tag{29.19}$$

In accordance with the methods presented in Sec. 27, we seek the coefficients of series (29.17) and (29.19) from the identities

$$(Da, b)\left(\frac{d^2 V}{dx^2} + \frac{2}{\zeta}\frac{d\zeta}{dx}\frac{dV}{dx}\right) + d|\zeta|^2(V + \bar{V}) = VB, \tag{29.20}$$

$$\frac{dV}{dx}\bigg|_{x=0} = \frac{dV}{dx}\bigg|_{x=\pi} = (0,0). \tag{29.21}$$

Since we have arrived at them proceeding from the problem on the cycle stability, it follows that

$$\det B = 0. \tag{29.22}$$

Identities (29.20), (29.21) imply that

$$V_1 = (0, v_1), \quad B_1 = \begin{pmatrix} 0 & b_{12}^1 \\ 0 & 0 \end{pmatrix},$$

$$V_2 = (0, v_2), \quad B_2 = \begin{pmatrix} 0 & b_{12}^2 \\ 0 & b_{22}^2 \end{pmatrix},$$

where

$$b_{22}^2 = -(1 + x_0^2)\alpha_1/(\alpha_1 - \alpha_2 x_0) < 0. \tag{29.23}$$

Recalling (29.22), we get the required result from inequality (29.23).

Thus, the methods of the ramification theory can be used to construct self-similar cycles of the quasinormal form (28.55), (28.56), which essentially depend on x, when the self-similar cycle of this form, which depends on x, loses stability upon a suitable variation of the diffusion coefficients. Thus, Theorem 29.2 allows us to describe formally the process of a spatial desynchronization of spatially homogeneous auto-oscillations in homogeneous media with diffusion (in Khokhlov's terminology, this is the process of transformation of auto-oscillations into autowaves), and this is important for applications [1].

Approximately in the same way, we solve the problem of a spatial desynchronization of double-frequency homogeneous auto-oscillations in media with small diffusion. However, there is a new feature here, namely, they can be transformed both into spatially nonhomogeneous double-frequency auto-oscillations [20] when Theorem 28.1 is valid and into triple-frequency auto-oscillations [21] when we are under the conditions of Theorem 28.2.

29.2. Desynchronization of Auto-Oscillations of Hutchinson's Equation

Naturally, Theorem 29.2, which describes the process of desynchronization of spatially homogeneous auto-oscillations, is general and applicable to many classes of systems of the reaction–diffusion type. As a nontrivial example which served as a reason for the creation of the method of quasinormal forms, we shall consider the well-known Hutchinson equation [94] with diffusion

$$\frac{\partial N}{\partial t} = \varepsilon D \frac{\partial^2 N}{\partial x^2} + \left(\frac{\pi}{2} + \varepsilon\right)[1 - N(t-1, x)]N, \qquad (29.24)$$

$$\left.\frac{\partial N}{\partial x}\right|_{x=0} = \left.\frac{\partial N}{\partial x}\right|_{x=1} = 0, \qquad (29.25)$$

which is a base equation in mathematical ecology.

It is shown in [50] that its quasinormal form is the boundary-value problem

$$\left(1 + i\frac{\pi}{2}\right)\frac{\partial \zeta}{\partial \tau} = D\frac{\partial^2 \zeta}{\partial x^2} + i\zeta + \frac{\pi(1-3i)}{10}\zeta|\zeta|^2, \qquad (29.26)$$

$$\left.\frac{\partial \zeta}{\partial x}\right|_{x=0} = \left.\frac{\partial \zeta}{\partial x}\right|_{x=1} = 0. \qquad (29.27)$$

We omit the corresponding calculations, since they are similar to those presented earlier for a system of the reaction–diffusion type without time lag.

Suppose that the restrictions (similar to Conditions 28.4, 28.5)

$$\operatorname{Re} P'\left(i\frac{\pi}{2}\right) > 0, \tag{29.28}$$

where $P(\lambda) = \lambda + \frac{\pi}{2}\exp(-\lambda)$, are satisfied and the roots of the equations

$$P(\lambda) + z = 0, \tag{29.29}$$

$$\frac{P(\lambda)}{\lambda^2 + \frac{\pi^2}{4}}\left[1 - \frac{z}{\lambda^2 + \frac{\pi^2}{4}}\left(\frac{\lambda + i\frac{\pi}{2}}{P'(i\frac{\pi}{2})} + \frac{\lambda - i\frac{\pi}{2}}{P'(-i\frac{\pi}{2})}\right)\right] + \frac{z}{\lambda^2 + \frac{\pi^2}{4}} = 0 \tag{29.30}$$

lie in the left complex half-plane for all $z > 0$. Conditions (29.28), (29.29) can be verified trivially. We shall discuss condition (29.30) in greater detail.

We suppose that it does not hold true. Then there exists $z > 0$ and $\lambda = i\omega$, where $\omega > 0$ and $\omega \neq \frac{\pi}{2}$, such that

$$\frac{1}{z} = -\frac{\pi}{2}\left[\frac{\cos\omega}{\omega^2 + \frac{\pi^2}{4} - \pi\omega\sin\omega} + \frac{\pi}{(\omega^2 - \frac{\pi^2}{4})(1 + \frac{\pi^2}{4})}\right], \tag{29.31}$$

$$\frac{\left(\omega^2 - \frac{\pi^2}{4}\right)\left(1 - \frac{\pi}{2}\frac{\sin\omega}{\omega}\right)}{\omega^2 + \frac{\pi^2}{4} - \pi\omega\sin\omega} = \frac{2}{1 + \frac{\pi^2}{4}}. \tag{29.32}$$

For $\omega > \pi/2$, (28.31) implies the inequality

$$\left(\omega^2 - \frac{\pi^2}{4}\right)\left(1 + \frac{\pi^2}{4}\right) \geq \pi\left(\omega^2 + \frac{\pi^2}{4} - \pi\omega\sin\omega\right). \tag{29.33}$$

Comparing (29.32) and (29.33), we get $\pi\sin\omega \geq 2(\pi - 2)\omega$, and this cannot hold true for the ω under consideration. Suppose now that $\omega < \pi/2$. It follows from (29.31) that

$$\lim_{\omega \to \frac{\pi}{2} - 0}\frac{1}{z} = -\frac{1}{2}\frac{\pi^2 - 4}{\pi^2 + 4},$$

i.e., the right-hand side of (29.31) is negative for $\frac{\pi}{2} - \delta < \omega < \frac{\pi}{2}$ and a sufficiently small δ. Numerical verification shows [50] that the right-hand side of (29.31) is negative for all $0 < \omega < \pi/2$.

We shall try to show that the statements of Theorems 28.1 and 28.2 are valid under conditions (29.28) and (29.29) for the boundary-value problem (29.24), (29.25).

We take the functions $N(s, x)$, which are continuous in $-1 \leq s \leq 0$ and belong to W_2^1 with respect to $0 \leq x \leq 1$, as a phase space of the boundary-value problem (29.24), (29.25). Let us suppose, for instance, that we have constructed its approximate cycle. We perform a linearization on it, then reject the diffusion term and apply the algorithm for the stability investigation [61], similar to that presented in Sec. 27, to the resulting time lag relation. We consider x to be a parameter. Next we linearize the boundary-value problem with due account of the diffusion. We write it as [50, 51]

$$\frac{\partial u}{\partial t} = \frac{\partial u}{\partial s},$$

$$\left.\frac{\partial u}{\partial s}\right|_{s=0} = \varepsilon D \left.\frac{\partial^2 u}{\partial x^2}\right|_{s=0} + l(t, x, u(t, s, x), \varepsilon),$$

$$\left.\frac{\partial u}{\partial x}\right|_{x=0, s=0} = \left.\frac{\partial u}{\partial x}\right|_{x=1, s=0} = 0.$$

First, using the formulas obtained upon the realization of the algorithm for the investigation of stability, we pass to the "critical" variables $v(t, x)$ and the "noncritical" variables $w(t, s, x)$ in this system, i.e., operate with the variable s. Then we pass to the variable x. With this aim in view, we carry out the projection onto the functions

$$1, \cos \pi x, \ldots, \cos M \pi x$$

in the resulting system, passing in this way to variables independent of and dependent on x. Suppose that M is of order $M_0 \varepsilon^{-1/4}$, where the constant M_0 is sufficiently large. Then, clearly, the variables dependent on x, taken separately, are exponentially damped. After a suitable exponential substitution, the operator generated by the variables independent of x possesses the regularity property. It is significant that in a complete system the variables that we have considered are connected by operators of order ε. This circumstance allows us to establish the regularity of the whole operator. The subsequent constructions are similar to those presented in Sec. 28.

It should also be pointed out that, by virtue of condition (29.30), the principle of averaging on a finite time interval is valid for the boundary-value problem (29.24), (29.25).

Recall [65] that the boundary-value problem (29.24), (29.25) has the spatially homogeneous periodic solution

$$N = 1 + \sqrt{\frac{40\varepsilon}{3\pi - 2}} \cos \frac{\pi}{2}\tau + \frac{4\varepsilon}{3\pi - 2}(\sin \pi\tau + 2\cos \pi\tau) + O(\varepsilon^{3/2}), \quad (29.34)$$

$$\tau = \left(1 + \frac{4\varepsilon}{\pi(3\pi - 2)} + O(\varepsilon^2)\right)^{-1} t, \quad (29.35)$$

which is associated, in a quasinormal form (a fact that can be immediately verified) with a self-similar cycle $\zeta_0 \exp(i\alpha_0)$, where

$$\zeta_0^2 = 10/(3\pi - 2), \qquad \alpha_0 = 2/(3\pi - 2). \tag{29.36}$$

Linearizing the boundary-value problem (29.26), (29.22) on this cycle, we infer that the periodic solution (29.34), (29.35) of the boundary-value problem (29.24), (29.25) is stable (unstable) for

$$D - 2/\pi(3\pi - 2) > 0 \qquad (< 0). \tag{29.37}$$

Let us now suppose that

$$D = \frac{2}{\pi(3\pi - 2)} - \mu, \qquad 0 < \mu_1 \le \mu \le \mu_2, \tag{29.38}$$

in (29.24) and μ_1, μ_2 are sufficiently small. It is shown in [6] (see the similar place in 29.1) that under condition (29.37) and for $0 < \varepsilon \le \varepsilon_0(\mu_1, \mu_2)$ the boundary-value problem (29.24), (29.25) has two stable periodic solutions:

$$N_\pm = 1 + \sqrt{\varepsilon}\left(\zeta \exp\left(i\frac{\pi}{2}\tau\right) + \bar{\zeta} \exp\left(-i\frac{\pi}{2}\tau\right)\right) +$$

$$+ \varepsilon\left(\frac{\zeta^2}{2+i} \exp\left(i\pi\tau\right) + \frac{\bar{\zeta}^2}{2-i} \exp(-i\pi\tau)\right) + O(\varepsilon^{3/2} + \varepsilon^{1/2}\mu^{3/2}), \tag{29.39}$$

$$\tau = \left[1 + \varepsilon\left(\frac{4}{\pi(3\pi - 2)} - \frac{8\pi}{\pi + 2}\mu\right) + O(\varepsilon^2 + \varepsilon\mu^2)\right]^{-1} t, \tag{29.40}$$

where

$$\zeta = \sqrt{\frac{10}{3\pi - 2}}\left[1 \pm \sqrt{\mu}c(1 - 3i)\cos\pi x - \mu c^2\left(\frac{3(7\pi + 2)}{2(3\pi - 2)} + \right.\right.$$

$$\left.\left. + \frac{2 + 9i}{4}\cos 2\pi x\right)\right], \qquad c = (3\pi - 2)\sqrt{\frac{\pi}{15(\pi + 2)}}. \tag{29.41}$$

From the formulas given for spatially nonhomogeneous cycles, it becomes clear why the homogeneous auto-oscillations in the Hutchinson equation are desynchronized: on the average, the spatial nonhomogeneity provides for the damping of density oscillations (this follows from (29.41)).

Note that the numerical analysis of the boundary-value problem (29.24), (29.25) was performed in [14]. It turned out (we consider this to be significant) that the range of applicability of the asymptotic formulas (29.39), (29.40), (29.41) is sufficiently wide.

In conclusion, we want to point out [12] where the spatially nonhomogeneous cycles of the boundary-value problem (29.24), (29.25) considered on the square $\{0 \le x_1 \le 1, 0 \le x_2 \le 1\}$ were constructed. It turns out that under condition (29.37) it has four stable spatially nonhomogeneous cycles with a dependence on x_1 or x_2, respectively, and with asymptotics (29.39), (29.41), where x is replaced by x_1 or x_2, and four unstable cycles with an essential dependence on both variables x_1 and x_2. Here the biological explanation is as follows: the cycles on which the damping of auto-oscillations is more effective are stable.

29.3. Running Tori of Parabolic Systems with a Small Diffusion

In the preceding subsections, we spoke of stationary solutions of a quasinormal form which result from the desynchronization of spatially homogeneous auto-oscillations. Here we shall speak of another class of stationary solutions, which bifurcate from an unstable zero equilibrium position and become stable, growing in amplitude. We shall consider the boundary-value problem (28.8), (28.9) on the circle

$$\frac{\partial \zeta_1}{\partial \tau} = (Da_1, b_1)\frac{\partial^2 \zeta_1}{\partial x^2} + (A_1 a_1, b_1)\zeta_1 + (d_{11}|\zeta_1|^2 + d_{12}|\zeta_2|^2)\zeta_1, \quad (29.42)$$

$$\frac{\partial \zeta_2}{\partial \tau} = (Da_2, b_2)\frac{\partial^2 \zeta_2}{\partial x^2} + (A_1 a_2, b_2)\zeta_2 + (d_{21}|\zeta_1|^2 + d_{22}|\zeta_2|^2)\zeta_2, \quad (29.43)$$

$$\zeta_s|_{x=0} = \zeta_s|_{x=2\pi}, \quad \frac{\partial \zeta_s}{\partial x}\bigg|_{x=0} = \frac{\partial \zeta_s}{\partial x}\bigg|_{x=2\pi}, \quad s = 1, 2, \quad (29.44)$$

for $m = 2$ as a model example. We seek its self-similar tori in the form

$$\zeta_{10} \exp i(k_1 x + \alpha_1 \tau), \quad \zeta_{20} \exp i(k_2 x + \alpha_2 \tau). \quad (29.45)$$

Substituting (29.45) into (29.42), (29.43), we get the system

$$\begin{array}{l}\operatorname{Re} d_{11}\zeta_{10}^2 + \operatorname{Re} d_{12}\zeta_{20}^2 = -\operatorname{Re}(A_1 a_1, b_1) + k_1^2 \operatorname{Re}(Da_1, b_1),\\ \operatorname{Re} d_{21}\zeta_{10}^2 + \operatorname{Re} d_{22}\zeta_{20}^2 = -\operatorname{Re}(A_1 a_2, b_2) + k_2^2 \operatorname{Re}(Da_2, b_2)\end{array} \quad (29.46)$$

for the amplitudes ζ_{10}, ζ_{20} and the relations

$$\alpha_1 = \operatorname{Im}[(A_1 a_1, b_1) - k_1^2(Da_1, b_1) + d_{11}\zeta_{10}^2 + d_{12}\zeta_{20}^2],$$

$$\alpha_2 = \operatorname{Im}[(A_1 a_2, b_2) - k_2^2(Da_2, b_2) + d_{21}\zeta_{10}^2 + d_{22}\zeta_{20}^2]$$

for the frequencies α_1, α_2. Suppose that the determinant of system (29.46) is nonzero and its solution has positive coordinates. Then the self-similar tori

(29.45) are constructed. Note that the right-hand sides of system (29.46) include characteristic exponents of the zero solution of the boundary-value problem (29.42)–(29.44) (taken with the opposite sign), which correspond to the modes $\exp ik_1 x$ and $\exp ik_2 x$. This explains the origination of the solutions (29.45) — they branch off from the zero equilibrium position.

When analyzing the stability of the self-similar tori (29.45), it is convenient to make the substitutions

$$\zeta_1 \exp(-i(k_1 x + \alpha_1 \tau)) \to \zeta_1, \qquad \zeta_2 \exp(-i(k_2 x + \alpha_2 \tau)) \to \zeta_2$$

in (29.42)–(29.44), which bring them to the equilibrium positions (ζ_{10}, ζ_{20}). Therefore, the eigenvalues of the fourth-order matrices

$$\begin{pmatrix} \zeta_{10}^2 d_{11} & \zeta_{10}^2 \bar{d}_{11} & \zeta_{10}\zeta_{20} d_{12} & \zeta_{10}\zeta_{20} d_{12} \\ \zeta_{10}^2 d_{11} & \zeta_{10}^2 \bar{d}_{11} & \zeta_{10}\zeta_{20} \bar{d}_{12} & \zeta_{10}\zeta_{20} \bar{d}_{12} \\ \zeta_{10}\zeta_{20} d_{21} & \zeta_{10}\zeta_{20} d_{21} & \zeta_{20}^2 d_{22} & \zeta_{20}^2 d_{22} \\ \zeta_{10}\zeta_{20} \bar{d}_{21} & \zeta_{10}\zeta_{20} \bar{d}_{21} & \zeta_{20}^2 \bar{d}_{22} & \zeta_{20}^2 \bar{d}_{22} \end{pmatrix} +$$

$$+\mathrm{diag}\ \{-(2k_1 m_1 + m_1^2)(Da_1, b_1), (2k_1 m_1 - m_1^2)(D\bar{a}_1, \bar{b}_1),$$
$$-(2k_2 m_2 + m_2^2)(Da_2, b_2), (2k_2 m_2 - m_2^2)(D\bar{a}_2, \bar{b}_2)\},$$

where $m_1, m_2 = 0, \pm 1, \ldots$, are responsible for their stability, i.e., the problem reduces to an algebraic one.

It is clear that there exist boundary-value problems of type (29.42)–(29.44) with a large number of stable self-similar tori of the form (29.45).

30. Diffusion Chaos

30.1. Statement of the Problem

The boundary-value problem (29.24), (29.25) is significant for the understanding of the material presented below. Recall that its spatially homogeneous cycle (29.34), (29.35) loses stability with a decrease in diffusion and two stable spatially nonhomogeneous cycles (29.39)–(29.41) appear in its neighborhood, which are generated by self-similar cycles of the quasinormal form (29.26), (29.27) of the boundary-value problem (29.24), (29.25). The numerical analysis shows that with a decrease in D, these cycles are transformed into self-similar tori with T_0-periodically varying amplitudes. With a further decrease of D, the latter pass into more complicated tori and so on. Thus, only self-similar cycles of the quasinormal form (29.26), (29.27) are stable. Of course, this does not exclude the existence of a large number of unstable cycles (see [53], where unstable running waves on a circle are constructed). It is significant, however, that the attractor of the boundary-value problem (29.24), (29.25) is closely connected with a deepening process of desynchronization of auto-oscillations.

Below we shall try to explain that a gradual spatial complication of autowaves, which occurs with a decrease in diffusion, with an accuracy to within inessential details connected with the specific features of various boundary-value problems, is a sufficiently general situation. To be more precise, we shall show that at a sufficiently small diffusion we observe a so-called *diffusion chaos*.

To substantiate this circumstance, we consider the model system

$$\dot{u}_j = \varepsilon\mu D_0(u_{j+1} - 2u_j + u_{j-1}) + A_0 u_j + \varepsilon A_1 u_j + F(u_j), \tag{30.1}$$

where $j = 1, \ldots, n$, $u_0 = u_1$, $u_{n+1} = u_n$, $0 < \mu \ll 1$,

$$D_0 = \mathrm{diag}\,\{d_{10}, d_{20}\}, \qquad d_{10}, d_{20} > 0, \qquad d_{10} + d_{20} = 1.$$

Just as in the preceding section, we assume that

$$A_0 a = i\omega_0 a, \qquad A_0^* b = -i\omega_0 b, \qquad (a, b) = 1, \qquad \mathrm{Re}\,(A_1 a, b) > 0$$

and that the real part of the first Lyapunov quantity d of the ordinary differential equation, which results from (30.1) for $\varepsilon = 0$, is negative. Under the constraints that we have imposed, system (30.1) has a homogeneous harmonic cycle, when $u_1 = \ldots = u_n$, which is unstable if

$$x_0 y_0 - 1 > 0, \tag{30.2}$$

where

$$x_0 = -\,\mathrm{Im}\,(D_0 a, b)/\,\mathrm{Re}\,(D_0 a, b), \qquad y_0 = \mathrm{Im}\,d/\,\mathrm{Re}\,d. \tag{30.3}$$

Under condition (30.2), we pose a question concerning the attractor of system (30.1).

30.2. The Method of Investigation

In system (30.1) we carry out normalization, pass to polar coordinates, and reject higher orders of smallness. As a result we get the system

$$\dot{\rho}_j = \varepsilon\tau_0'\rho_j + \mathrm{Re}\,d \cdot \rho_j^3 + \varepsilon\mu g[\rho_{j+1}\cos(\omega_0(\varphi_{j+1} - \varphi_j) + \gamma) - \\ - 2\rho_j \cos\gamma + \rho_{j-1}\cos(\omega_0(\varphi_{j-1} - \varphi_j) + \gamma)], \tag{30.4}$$

$$\dot{\varphi}_j = 1 + \varepsilon \frac{\omega_0'}{\omega_0} + \frac{1}{\omega_0} \operatorname{Im} d \cdot \rho_j^2 + \varepsilon \mu \frac{g}{\omega_0} \Big[\frac{\rho_{j+1}}{\rho_j} \sin(\omega_0(\varphi_{j+1} - \varphi_j) + \gamma) -$$
$$- 2\sin\gamma + \frac{\rho_{j-1}}{\rho_j} \sin(\omega_0(\varphi_{j-1} - \varphi_j) + \gamma) \Big], \qquad (30.5)$$

where $\rho_0 = \rho_1$, $\rho_{n+1} = \rho_n$, $\varphi_0 = \varphi_1$, $\varphi_{n+1} = \varphi_n$, $2g = (1 + x_0^2)^{1/2}$, $\gamma = -\operatorname{sign} x_0 \cdot \arccos(1 + x_0^2)^{-1/2}$, $\tau_0' + i\omega_0' = (A_1 a, b)$.

Then, setting $\rho_j = \varepsilon^{1/2} \eta_j$, $\varepsilon t \to t$, we pass from (30.4), (30.5) to the slow variables:

$$\dot{\eta}_j = \tau_0' \eta_j + \operatorname{Re} d \cdot \eta_j^3 + \mu g [\eta_{j+1} \cos(\alpha_j + \gamma) -$$
$$- 2\eta_j \cos\gamma + \eta_{j-1} \cos(\gamma - \alpha_{j-1})], \qquad (30.6)$$

$$\dot{\alpha}_j = \operatorname{Im} d \cdot (\eta_{j+1}^2 - \eta_j^2) + \mu g \Big[\frac{\eta_{j+2}}{\eta_{j+1}} \sin(\alpha_{j+1} + \gamma) - \frac{\eta_{j+1}}{\eta_j} \sin(\alpha_j + \gamma) +$$
$$+ \frac{\eta_j}{\eta_{j+1}} \sin(\gamma - \alpha_j) - \frac{\eta_{j-1}}{\eta_j} \sin(\gamma - \alpha_{j-1}) \Big], \qquad (30.7)$$

where $\alpha_j = \omega_0(\varphi_{j+1} - \varphi_j)$.

Note that we can arrive at system (30.6), (30.7) if we pass to polar coordinates in the difference approximation of the quasinormal form and introduce the variables α_j in the same way.

For $\mu = 0$, system (30.6), (30.7) has a globally exponentially stable invariant torus $\eta_j = (-\tau_0' / \operatorname{Re} d)^{1/2}$, the behavior of the trajectories on which can be described by the equations $\dot{\alpha}_j = 0$. The general statements of [71] imply that for sufficiently small μ it has a similar integral manifold defined by the relations

$$\eta_j = (-\tau_0' / \operatorname{Re} d)^{1/2} [1 + \mu \psi_j(\alpha_1, \ldots, \alpha_{n-1}, \mu)], \qquad (30.8)$$

where the functions ψ_j, smooth with respect to the set of variables and 2π-periodic with respect to α_k, satisfy the asymptotic relations

$$\psi_j = \frac{g}{2\tau_0'} [\cos(\alpha_j + \gamma) - 2\cos\gamma + \cos(\gamma - \alpha_{j-1})] + O(\mu). \qquad (30.9)$$

We substitute (30.9) into (30.8) and the latter into (39.7). As a result, after the normalization of time $\mu(x_0 y_0 - 1)t \to 2t$ with an accuracy to within the terms of order μ, we get the equations

$$\dot{\alpha}_j = -(\sin\alpha_{j+1} - 2\sin\alpha_j + \sin\alpha_{j-1}) + \varkappa(\cos\alpha_{j-1} - \cos\alpha_{j+1}),$$
$$j = 1, \ldots, n-1, \qquad (30.10)$$

on the globally stable invariant torus of system (30.6), (30.7). In (30.10) we have $\alpha_0 = \alpha_n = 0$, $\varkappa = (x_0 + y_0)/(x_0 y_0 - 1)$.

We sum up what we have said as the following statement [27].

THEOREM 30.1. *There exist sufficiently small positive numbers ε_0, μ_0 and a positive number r_0 independent of them such that for $0 < \varepsilon \leq \varepsilon_0$, $0 < \mu \leq \mu_0$, system (30.1) has an n-dimensional torus to which all solutions with nonzero initial conditions tend from a ball with radius r_0 and the center at zero as $t \to \infty$. The behavior of the solutions on the invariant torus can be described with an accuracy to within terms of order μ by the system of phase equations (30.10).*

30.3. Dynamics of a System of Phase Equations

We have thus reduced the problem to the analysis of the attractor of system (30.10) whose advantage is that it does not include small parameters. In order to formulate the results of its investigation, we introduce into consideration the familiar formula [92] for Lyapunov's dimensionality of the attractor

$$d_L = s + \sum_{k=1}^{s} \frac{\lambda_k}{|\lambda_{s+1}|},$$

where λ_k are Lyapunov's exponents written out in order of decrease and s is a number such that

$$\sum_{k=1}^{s} \lambda_k > 0, \qquad \sum_{k=1}^{s+1} \lambda_k < 0.$$

The numerical analysis performed in [11] has shown that for $\varkappa = 3$ the attracting set of system (30.10) is a strange attractor and the approximate formula

$$d_L = 0.866(n-1) + 0.6$$

is valid with an accuracy to within 0.5 for $5 \leq n \leq 40$. We observe the same for other values of $\varkappa$.

In conclusion, we want to point out the work [10], in which the quasinormal form

$$\frac{\partial \zeta}{\partial \tau} = (Da, b)\frac{\partial^2 \zeta}{\partial x^2} + (A_1 a, b)\zeta + \zeta d|\zeta|^2 \qquad (30.11)$$

is considered on a circle. Since the domain is symmetric, in this case the finite-difference analog of the quasinormal form does not reflect its properties in full measure, and, therefore, it is inferred in [10] that the maximum dimension of the attractor of the quasinormal form (30.11) is realized for "average" values of diffusion.

However, the purposeful numerical analysis has shown that in this case, too, under condition (30.2), with a decrease of diffusion the dimension of the attractor in the difference approximation of the quasinormal form (30.11) grows with an increase in the number of approximating equations and becomes the largest for small values of the diffusion coefficients. The authors of [10] are right only as concerns the fact that in the difference approximation of the quasinormal form a large number of stable cycles exist alongside the strange attractor.

31. Dynamics of Systems with a Terminal Turn Point

31.1. The Main Result

Here the object of our investigation is the boundary-value problem (28.60), (28.61):

$$\frac{\partial u}{\partial t} = \varepsilon^{1/2}(D_0 + \varepsilon^{1/2}D_1)\frac{\partial^2 u}{\partial x^2} + A_0 u + \varepsilon A_1 u + F(u), \qquad (31.1)$$

$$\left.\frac{\partial u}{\partial x}\right|_{x=0} = \left.\frac{\partial u}{\partial x}\right|_{x=\pi} = 0. \qquad (31.2)$$

Recall that here $u \in R^N$, $N \geq 3$; the matrix D_0 is Hurwitzian, and D_1 is arbitrary; for all $z > 0$ of the matrix $A_0 - zD_0$ is Hurwitzian and for small z it has a simple eigenvalue $\lambda(z)$, $\lambda(0) = i\omega_0$, $\omega_0 > 0$, which deviates from the imaginary axis with a first-order contingence, i.e.,

$$\text{Re}\,(D_0 a, b) = 0, \qquad \text{Im}\,(D_0 a, b) \neq 0, \qquad \text{Re}\,(D_0 a_1, b) > 0,$$

where

$$A_0 a = i\omega_0 a, \qquad A_0^* b = -i\omega_0 b, \qquad (a, b) = 1, \qquad (\bar{a}, b) = 0,$$

and the vector a_1, which satisfies the condition $(a_1, b) = 0$, is the solution of the algebraic equation

$$A_0 a_1 - i\omega_0 a_1 = D_0 a - (D_0 a, b)a;$$

except for the pair $\pm i\omega_0$, all the other eigenvalues λ_k, $k = 1, \ldots, N-2$, of the matrix A_0 have negative real parts and satisfy the nonresonance condition

$$\lambda_k - \lambda_s \neq i\omega_0, \qquad k, s = 1, \ldots, N-2;$$

F is a vector-function of the class C^∞, which has an order of smallness higher than the first at zero; $\operatorname{Re} d < 0$, where d is the first Lyapunov quantity of the ordinary differential equation (31.1) for $\varepsilon = 0$.

From the constraints imposed on Eq. (31.1), it follows that we must speak of periodic solutions with frequency close to ω_0. We seek the correction of the frequency and the periodic solution itself as asymptotic series in integer powers of $\varepsilon^{1/2}$:

$$t = (1 + \alpha_1 \varepsilon^{1/2} + \alpha_2 \varepsilon + \ldots)\tau, \qquad (31.3)$$

$$u = \varepsilon^{1/2}[a \exp(i\omega_0 \tau) + \bar{a} \exp(-i\omega_0 \tau)]\zeta_n \cos nx + \varepsilon u_2 + \ldots, \qquad (31.4)$$

where ζ_n is an unknown "amplitude" of oscillations.

Substituting series (31.3), (31.4) into (31.1), (31.2) and equating the coefficients of ε, we make sure that

$$u_2 = n^2[a_1 \exp(i\omega_0 \tau) + \bar{a}_1 \exp(-i\omega_0 \tau)]\zeta_n \cos nx + f_2(\tau)\zeta_n^2 \cos^2 nx +$$

$$+ a\eta_1(x)\exp(i\omega_0\tau) + \bar{a}\bar{\eta}_1(x)\exp(-i\omega_0\tau),$$

$$\alpha_1 = n^2 \operatorname{Im}(D_0 a, b)/\omega_0,$$

where $f_2(\tau)$ is a $2\pi/\omega_0$-periodic vector-function, which contains only zero and second harmonics, and η_1 is an arbitrary function. At the third step, when we equate the coefficients of $\varepsilon^{3/2}$ in (31.1), we get an ordinary linear differential equation with a $2\pi/\omega_0$-periodic nonhomogeneity. Proceeding from its solvability in the class of periodic functions, we arrive at the so-called determining equation

$$(D_0 a, b)\left[\frac{d^2\eta_1}{dx^2} + n^2\eta_1\right] - [n^4(D_0 a_1, b) + n^2(D_1 a, b) +$$

$$+ \alpha_1 n^2(D_0 a, b) - (A_1 a, b) - i\omega_0 \alpha_2 - d\zeta_n^2 \cos^2 nx]\zeta_n \cos nx = 0,$$

which has a solution satisfying Neumann's boundary conditions if

$$\frac{3}{4}\zeta_n^2 = -\tau_n'/\operatorname{Re} d, \qquad \tau_n' = \operatorname{Re}[(A_1 a, b) - n^4(D_0 a_1, b) - n^2(D_1 a, b)],$$

$$\omega_0 \alpha_2 = n^4 \operatorname{Im}(D_0 a_1, b) + n^2[\alpha_1 \operatorname{Im}(D_0 a, b) + \operatorname{Im}(D_1 a, b)] -$$

$$- \operatorname{Im}(A_1 a, b) - \frac{3}{4}\operatorname{Im} d\zeta_n^2.$$

Thus, from the determining equation we derive an equation for the square of the amplitude of oscillations.

For the further realization of the algorithm, we solve the equations that we encounter on account of the corrections of the frequency and terms of the form

$$a\eta(x)\exp(i\omega_0\tau) + \bar{a}\bar{\eta}(x)\exp(-i\omega_0\tau)$$

appearing at every step.

In the statement formulated below, we suppose that, just as in 28.5, we take the Hilbert space $E = W_2^1(0,\pi)$ as the phase space of the boundary-value problem (31.1), (31.2).

THEOREM 31.1. *Suppose that, for a certain $n \geq 1$, we have the following: the quantity $\tau_n' > 0$, all numbers $R_{nm} = \tau_m' - \frac{4}{3}\tau_n'$, $m = 0, 1, \ldots$, are nonzero, and the number of positive numbers among them is m_0. Then there exists $\varepsilon_0 > 0$ such that for $0 < \varepsilon \leq \varepsilon_0$ the boundary-value problem (31.1), (31.2) has a solution, periodic with respect to t, with asymptotics (31.3), (31.4), which is orbitally exponentially stable for $m_0 = 0$ and dichotomous for $m_0 > 0$ with the dimension of the unstable manifold $2m_0 + 1$.*

The central feature of the proof of the theorem is the investigation of the properties of stability of the variational equation

$$\frac{\partial h}{\partial \tau} = (1 + \alpha_1 \varepsilon^{1/2} + \ldots + \alpha_{k-1}\varepsilon^{(k-1)/2})[\varepsilon^{1/2}(D_0 + \varepsilon^{1/2}D_1)\frac{\partial^2 h}{\partial x^2} +$$
$$+ A_0 h + \varepsilon^{1/2} A_1(\tau, x)h + \varepsilon A_2(\tau, x, \varepsilon)h], \quad (31.5)$$

which results upon the linearization of (31.1) on the intervals of series (31.3), (31.4) of lengths $k-1$ and k respectively, where $k \geq 4$. In what follows, for technical reasons, it is more convenient to use an equation in which the term $\varepsilon^{1/2}A_1(\tau, x)h$ is removed by means of a suitable substitution. Since the matrix $A_1(\tau, x)$ contains only the first harmonics with respect to τ, it follows, by virtue of the nonresonance of the eigenvalues of the matrix A_0, that such a substitution has the form

$$[I + \sqrt{\varepsilon}\Omega(\tau, x)]^{-1}h \to h, \quad (31.6)$$

where the $2\pi/\omega_0$-periodic matrix Ω is the solution of the equation

$$\frac{\partial \Omega}{\partial \tau} = A_0 \Omega - \Omega A_0 + A_1(\tau, x)$$

and also contains only first harmonics, and I is an identity matrix. Carrying out substitution (31.6), we arrive at the equation

$$\frac{\partial h}{\partial \tau} = (1 + \alpha_1 \varepsilon^{1/2} + \ldots + \alpha_{k-1}\varepsilon^{(k-1)/2})\Big[\varepsilon^{1/2}(D_0 + \varepsilon^{1/2}D_1)\frac{\partial^2 h}{\partial x^2} + A_0 h +$$

$$+ \varepsilon B_1(\tau,x,\varepsilon)h + \varepsilon B_2(\tau,x,\varepsilon)\frac{\partial^2 h}{\partial x^2} + \varepsilon B_3(\tau,x,\varepsilon)\frac{\partial h}{\partial x}\Big]. \quad (31.7)$$

We introduce an orthoprojector in the phase space E:

$$P_v = \sum_{r=0}^{r_0} P_r v, \quad P_r v = v_r c_r \cos rx, \quad v_r = (1+r^2)c_r \int_0^\pi v \cos rx\,dx, \quad (31.8)$$

where $c_0 = 1/\sqrt{\pi}$, $c_r = \sqrt{2/(r^2+1)\pi}$, $r \geq 1$, and choose r_0 later. Setting $h = v + w$, $v = Ph$, $w = (I - P)h$, we pass from Eq. (31.7) to the system

$$\dot{v}_r = (1 + \alpha_1 \varepsilon^{1/2} + \ldots + \alpha_{k-1}\varepsilon^{(k-1)/2})[-\varepsilon^{1/2}r^2(D_0 + \varepsilon^{1/2}D_1)v_r + A_0 v_r +$$

$$+ \varepsilon P_r\Big(B_1(\tau,x,\varepsilon) + B_2(\tau,x,\varepsilon)\frac{\partial^2}{\partial x^2} + B_3(\tau,x,\varepsilon)\frac{\partial}{\partial x}\Big)(v+w)], \quad (31.9)$$

$$\dot{w} = (1 + \alpha_1 \varepsilon^{1/2} + \ldots + \alpha_{k-1}\varepsilon^{(k-1)/2})[\varepsilon^{1/2}(D_0 + \varepsilon^{1/2}D_1)\frac{\partial^2 w}{\partial x^2} + A_0 w +$$

$$+ \varepsilon(I - P)\Big(B_1(\tau,x,\varepsilon) + B_2(\tau,x,\varepsilon)\frac{\partial^2}{\partial x^2} + B_3(\tau,x,\varepsilon)\frac{\partial}{\partial x}\Big)(v+w)]. \quad (31.10)$$

We apply the algorithm of the separation of critical and noncritical variables from Sec. 27 to the system of ordinary differential equations that resulted from (31.9) for $w = 0$. For $w = 0$ we set in (31.9)

$$v = V(\tau,\varepsilon)\exp C(\varepsilon)\tau,$$

where the rectangular $(r_0 + 1)N \times (2r_0 + 2)$ matrix V, $2\pi/\omega_0$-periodic with respect to τ, and the square matrix C of order $2r_0 + 2$ can be expanded in asymptotic series in powers of $\sqrt{\varepsilon}$. In this case, $C(0) = 0$, and the matrix $V(\tau,0)$ has the form

$$(\varkappa_0, \bar{\varkappa}_0, \varkappa_1, \bar{\varkappa}_1, \ldots, \varkappa_{r_0}, \bar{\varkappa}_{r_0}),$$

where $\varkappa_m$ is a vector with components $P_m \varkappa_m = a\exp(i\omega_0\tau)$, $P_k\varkappa_m = 0$, $k \neq m$.

Equating the coefficients of $\varepsilon^{1/2}$ and ε, we get the equation

$$\dot{y} = \varepsilon^{1/2}C_1 y + \varepsilon C_2 y + \varepsilon C_3 y \quad (31.11)$$

for the critical variables with an accuracy to within quantities of order $\varepsilon^{3/2}$. Here $y = $ colon $(y_0, \bar{y}_0, \ldots, y_{r_0}, \bar{y}_{r_0})$, and the matrices C_1 and C_2 consist of blocks diag $\{g, \bar{g}\}$, where g is equal, respectively, to

$$(n^2 - r^2)(D_0 a, b) \quad \text{and} \quad -r^4(D_0 a_1, b) - r^2[(D_1 a, b) + \alpha_1(D_0 a, b)].$$

To determine the matrix C_3, we consider the auxiliary equation

$$\dot{v} = (1 + \alpha_2\varepsilon)A_0 v + \varepsilon B_1(\tau, x, 0)v, \tag{31.12}$$

dependent on x as on a parameter. The result of the application of the algorithm given in Sec. 27 will be the matrix

$$\begin{pmatrix} q_1 & q_2 \\ \bar{q}_2 & \bar{q}_1 \end{pmatrix},$$

where $q_1(x) = (A_1 a, b) + i\omega_0\alpha_2 + 2d\zeta_n^2 \cos^2 nx$, $q_2(x) = d\zeta_n^2 \cos^2 nx$. It remains to point out that since the matrices $B_2(\tau, x, 0)$ and $B_3(\tau, x, 0)$ contain only the first harmonics with respect to τ, the matrix C_3 results upon the application of the algorithm from Sec. 27 to the system

$$\dot{v}_r = (1 + \alpha_2\varepsilon)A_0 v_r + \varepsilon P_r B_1(\tau, x, 0)v, \qquad r = 0, 1, \ldots, r_0.$$

The relationship between this system and Eq. (31.12) is obvious. Therefore, in accordance with (31.8),

$$C_3 y = \text{colon}\,(p_0, \bar{p}_0, p_1, \bar{p}_1, \ldots, p_{r_0}, \bar{p}_{r_0}),$$

where

$$p_m = \int_0^\pi \sum_{r=0}^{r_0} (1 + m^2) c_m c_r (q_1 y_r + q_2 \bar{y}_r) \cos rx \cos mx\, dx.$$

By virtue of the structure of the matrix $C_1 + \sqrt{\varepsilon}(C_2 + C_3)$, we can use the substitution $(I + \sqrt{\varepsilon}Y)^{-1}y \to y$ to transform it to a block diagonal form with an accuracy to within terms of order ε. As a result, system (31.11) disintegrates into the scalar equations

$$\dot{y}_r = [\varepsilon^{1/2}(n^2 - r^2)(D_0 a, b) + \varepsilon\beta_r]y_r, \qquad r \neq n, \tag{31.13}$$

(the equation for $\bar{y}_r$ can be obtained from (31.13) by means of complex conjugation) and the system

$$\dot{y}_n = \varkappa(y_n + \bar{y}_n), \qquad \dot{\bar{y}}_n = \bar{\varkappa}(y_n + \bar{y}_n),$$

where $\varkappa = d\zeta_n^2$ and

$$\beta_r = (1 + r^2)c_r^2 \int_0^\pi q_1(x) \cos^2 rx\, dx -$$

$$- r^4(D_0 a_1, b) - r^2[(D_1 a, b) + \alpha_1(D_0 a, b)]. \tag{31.14}$$

Taking into account the formula for $q_1(x)$ in (31.14), we infer that one characteristic exponent of Eq. (31.11) is of order $\varepsilon^{3/2}$ and the asymptotic expansion of the real parts of the other characteristic exponents begins with εR_{nr}.

The further analysis is based on the theory of the exponential dichotomy of the solutions of linear parabolic systems presented in Sec. 25, and, therefore, we assume that the substitution $h \to h\exp(-\varepsilon\delta\tau)$ has been carried out in (31.7), where the constant $\delta > 0$ is so small that the signs of the real parts of all characteristic exponents of (31.11), which are of order ε, are preserved.

$C(E)$ used below is a Banach space of functions, continuous and bounded for $-\infty < \tau < \infty$, with values in E, the norm in which has been introduced as follows. Suppose that

$$u(\tau,x) = \sum_{m=0}^{\infty} u_m(\tau) c_m \cos mx, \qquad \gamma_m = \sup_{\tau} \|u_m(\tau)\|_{R^N},$$

where $\|*\|_{R^N}$ is the Euclidean norm. Then, by definition,

$$\|u\|_{C(E)} = \left(\sum_{m=0}^{\infty} \gamma_m^2\right)^{1/2}. \tag{31.15}$$

We introduce a multiplier $0 \le \mu \le 1$ before w in (31.9) and before v in (31.10). The resulting system generates a differential operator L in $C(E)$. We shall show that for μ under consideration and sufficiently small ε it is invertible and

$$\|\Pi_0 L^{-1}\|_{C(E) \to C(E)} + \|L^{-1}\|_{C(E) \to C(E)} \le M/\varepsilon, \tag{31.16}$$

where $\Pi_0 = d^2/dx^2$ with the Neumann boundary conditions. Here and in what follows we denote by M, M_1 etc. constants, independent of ε, μ, and r_0, whose exact values are inessential.

It follows from (31.9), (31.10) that $L = \text{diag}\{L_1, L_2\}$ for $\mu = 0$. Furthermore, the analysis of Eqs. (31.13) implies the existence of $\varepsilon_0(r_0)$ such that

$$\|\Pi_0 L_1^{-1} P\|_{C(E) \to C(E)} + \|L_1^{-1} P\|_{C(E) \to C(E)} \le M_1/\varepsilon \tag{31.17}$$

for $0 < \varepsilon \le \varepsilon_0(r_0)$. Employing, then, Fourier's method, the properties of the matrix $A_0 - zD_0$, and considering the number r_0 to be sufficiently large, we infer that the spectrum of the operator L_2 lies in the left complex half-plane and that

$$\|\Pi_0 L_2^{-1}(I-P)\|_{C(E)\to C(E)} \le M_2/\varepsilon r_0^2, \tag{31.18}$$

$$\|L_2^{-1}(I-P)\|_{C(E)\to C(E)} \le M_3/\varepsilon r_0^4. \tag{31.19}$$

Let us consider now a nonhomogeneous system with nonhomogeneities from $C(E)$ which corresponds to (31.9), (31.10). Using the operators L_1 and L_2, we pass from it to a system of integral equations. Substituting the first equation into the second, we see that by virtue of (31.17), (31.19), for a sufficiently large value of r_0, we can apply to it the principle of contraction mappings. Hence we obtain the required fact, i.e., inequality (31.16).

It follows from what has been said and from the results of Sec. 25 that an exponential dichotomy takes place for the solutions of the boundary-value problem (31.5), and the properties of its stability can be found from (31.11).

The concluding stage is the proof of the existence of a periodic solution of the boundary-value problem (31.1), (31.2). It can be carried out as follows. In (31.1) we make the change of time

$$t = (1 + \alpha_1 \varepsilon^{1/2} + \ldots + \alpha_{k-1}\varepsilon^{(k-1)/2} + \delta_k \varepsilon^{k/2})\tau,$$

where δ_k is the bounded function ε that must be defined, and set $u = u_k(\tau, x, \varepsilon) + \varepsilon^{(k-1)/2} h$, where $u_k(\tau, x, \varepsilon)$ is an interval of series (31.4) of length k. As a result we arrive at the equation

$$\Pi h = F(\tau, x, \varepsilon, \delta_k, h), \tag{31.20}$$

where Π is a linear parabolic operator, and

$$\|F(\tau, x, \varepsilon, \delta_k, h_1) - F(\tau, x, \varepsilon, \delta_k, h_2)\|_{C(E)} \leq$$

$$\leq M_4 \varepsilon^{(k-1)/2} \|h_1 - h_2\|_{C(E)}, \quad \|F(\tau, x, \varepsilon, \delta_k, 0)\|_{C(E)} \leq M_5 \varepsilon, \tag{31.21}$$

where $M_4 = M_4(R)$, $R = \max\{\|h_1\|_{C(E)}, \|h_2\|_{C(E)}\}$.

It follows from our constructions that the equation $\Pi h = 0$ has an approximate periodic solution $\frac{\partial}{\partial \tau} u_k$. We perturb this equation by terms of order $\varepsilon^{(k+1)/2}$ according to the rule presented in Sec. 27. It can be made exact, a fact that we assume below. Naturally, we can also take into consideration the corresponding addition on the right-hand side of (31.20), whose general properties will not change.

We regard Π as an operator in the space of vector-functions, continuous and $2\pi/\omega_0$-periodic with respect to τ, with values in E and norm (31.15). We know that the necessary and sufficient condition of solvability of the equation $\Pi h = J(\tau, x)$ in the indicated class of functions is the condition of orthogonality in the mean of the right-hand side J to the periodic solution g of the conjugate boundary-value problem. In this case, the solution h_J is unique if it itself is orthogonal in the mean to the function g. Moreover, the coerciveness inequality

$$\|h_{\mathcal{J}}\|_{C(E)} + \|\frac{\partial^2}{\partial x^2}h_{\mathcal{J}}\|_{C(E)} \leq \frac{M_6}{\varepsilon}\|\mathcal{J}\|_{C(E)} \qquad (31.22)$$

holds true, which follows from the trichotomy of the solutions of the equation $\Pi h = 0$ and estimate (31.16).

Regarding the right-hand side of (31.20) as the nonhomogeneity $\mathcal{J}$ and correcting it as needed (subtracting from $\mathcal{J}$ the projection onto $\frac{\partial}{\partial \tau}u_k$ in the direction of g), we pass to an integral equation inverting the operator Π. By virtue of inequalities (31.21), (31.22), the right-hand side of this integral equation transforms into inself a ball of radius $1 + M_5 M_6$, where M_5 is from (31.21) and M_6 is from (31.22), and satisfies the Lipschitz condition with a constant $q < 1$, which makes it possible to determine the vector-function h, $2\pi/\omega_0$-periodic with respect to τ. Its smoothness with respect to τ, x follows from estimate (31.22). It also follows from this estimate that with respect to the variable δ_k the right-hand side satisfies Lipschitz' condition with a certain universal constant. Equating to zero the correction on the right-hand side of (31.20), we get the equation $\delta_k = p(\varepsilon) + \varepsilon^{1/2}\Lambda(\varepsilon, \delta_k)$ for δ_k, where the function $p(\varepsilon)$ is bounded and Λ satisfies Lipschitz' condition with respect to δ_k. We have completed the proof of the theorem.

31.2. Explanation of the Phenomenon of Diffusion Bufferness

We say that we observe a diffusion bufferness if the system of reaction–diffusion type under consideration has a sufficiently large number of stable cycles, which depend on a space variable in a complicated way, and the dependence on the time variable is close to harmonic.

This interesting phenomenon was revealed for the first time in [15], by means of numerical methods, in the process of investigating a problem of the fox–mouse type, in which the migration of the predator is taken into account. In this work, the authors pose a question concerning the theoretical explanation of the phenomenon of diffusion bufferness on the model example of the boundary-value problem

$$\frac{\partial u}{\partial t} = \varepsilon(D_0 + \mu D_1)\frac{\partial^2 u}{\partial x^2} + A_0 u + \varepsilon A_1 u + F(u), \qquad (31.23)$$

$$\left.\frac{\partial u}{\partial x}\right|_{x=0} = \left.\frac{\partial u}{\partial x}\right|_{x=\pi} = 0. \qquad (31.24)$$

Here $0 < \varepsilon \ll 1$, $|\mu| \ll 1$, the matrix D_1 is arbitrary, and the same constraints are imposed on the matrices D_0, A_0 and the nonlinearity of F as in 31.1.

The analysis of the boundary-value problem (31.23), (31.24) is based on the concept of the heuristic quasinormal form

$$\frac{\partial \zeta}{\partial \tau} = -\varepsilon(D_0 a_1, b)\frac{\partial^4 \zeta}{\partial x^4} + [\mu(D_1 a, b) + i\varkappa]\frac{\partial^2 \zeta}{\partial x^2} +$$
$$+ (A_1 a, b)\zeta + \zeta d|\zeta|^2, \qquad (31.25)$$

$$\left.\frac{\partial \zeta}{\partial x}\right|_{x=0} = \left.\frac{\partial \zeta}{\partial x}\right|_{x=\pi} = \left.\frac{\partial^3 \zeta}{\partial x^3}\right|_{x=0} = \left.\frac{\partial^3 \zeta}{\partial x^3}\right|_{x=\pi} = 0, \qquad (31.26)$$

where $\tau = \varepsilon t$, and $\zeta = \zeta(\tau, x)$ is a complex function. The boundary-value problem is derived from the original problem as follows. In (31.23), (31.24) we set

$$u = \varepsilon^{1/2}[\zeta(\tau, x) a \exp(i\omega_0 t) + \bar\zeta(\tau, x)\bar a \exp(-i\omega_0 t)] + \varepsilon u_2(t, \tau, x) + \ldots, \qquad (31.27)$$

where all terms are $2\pi/\omega_0$-periodic with respect to t. In this way we obtain an ordinary quasinormal form. To pass from this form to the boundary-value problem (31.25), (31.26), we must use the method described in Sec. 27 in order to approximate more exactly the linear part of the boundary-value problem (31.23), (31.24).

LEMMA 31.1. *Suppose that the inequality*

$$1 + \frac{2}{3}\tau_0' \operatorname{Im} d/\varkappa \operatorname{Re} d > 0 \qquad (31.28)$$

is satisfied for $\tau_0' > 0$. Then for any sufficiently large natural numbers $n_1 < n_2$ and any $\delta \in (0, \frac{2}{\sqrt{3}}\sqrt{\tau_0' \operatorname{Re}(D_0 a_1, b)})$ we can indicate sufficiently small positive numbers $\varepsilon_0 = \varepsilon_0(n_1, n_2, \delta)$, $\mu_0 = \mu_0(n_1, n_2, \delta)$ such that for $0 < \varepsilon \leq \varepsilon_0$, $-\delta \varepsilon^{1/2} \leq \mu \leq \mu_0$ the boundary-value problem (31.25), (31.26) has an exponentially orbitally stable self-similar cycle

$$\zeta_n(\tau, x) = \zeta_n(x) \exp(i\alpha_n \tau), \qquad n = n_1, n_1 + 1, \ldots, n_2,$$

where

$$\zeta_n(x) = \zeta_0 \cos nx + O(|\mu| + \varepsilon + \sigma), \qquad \zeta_0^2 = -\frac{4}{3}\tau_0'/\operatorname{Re} d, \qquad (31.29)$$

$$\sigma = 1/n^2, \qquad \alpha_n = n^2[-\varkappa + O(|\mu| + \varepsilon + \sigma)]. \qquad (31.30)$$

Mainly, the proof of the lemma reduces to the analysis of the boundary-value problem (31.25), (31.26) for $\varepsilon = \mu = 0$.

THEOREM 31.2. *Under the condition* (31.28), *every cycle* (31.29), (31.30) *of the boundary-value problem* (31.25), (31.26) *is associated with an orbitally exponentially stable (periodic with respect to t) solution of the original boundary-value problem* (31.23), (31.24), *whose principal asymptotics can be constructed by means of* (31.27).

This theorem can be justified with the use of the same scheme as was used for proving Theorem 31.1. The only difference is that we must now take $r_0 \sim \Delta_0 \varepsilon^{-1/4}$, where $\Delta_0 > 0$ is sufficiently large.

Let us suppose that on the left-hand side of inequality (31.28) the coefficient standing after τ_0' is negative. Then, with an increase in τ_0', inequality (31.28) changes to the opposite. It turns out that in this case cycles (31.29), (31.30) of the boundary-value problem (31.25), (31.26) lose stability in an oscillatory way, giving it to the tori bifurcating from them. In this case (see Sec. 28), the boundary-value problem (31.23), (31.24) also has stable two-dimensional tori.

We have thus given a sufficiently complete theoretical explanation of the phenomenon of diffusion bufferness. Only the problem on the rigid appearance of auto-oscillations of the boundary-value problem (31.23), (31.24) for $\tau_0' < 0$ does not admit of investigation. This problem arises in the process of numerical experiments connected with the predator–prey problem [15]. The heuristic quasinormal form (31.25), (31.26) that we have introduced allows us to make some inferences about the nature of this phenomenon.

Indeed, for $\mu \leq -c_0 \varepsilon^{1/2}$, where c_0 is a sufficiently large positive constant, a stochastization of the stationary solutions of the boundary-value problem (31.25), (31.26) obviously occurs. Obviously, in the framework of the boundary-value problem (31.23), (31.24) this factor sometimes manifests itself more vividly, and this can lead, even for relatively small values of c_0, to the origination of auto-oscillations for $\tau_0' < 0$, whose principal asymptotics cannot be determined from (31.27). This proposition directly follows from the general theoretic propositions given in 28.5, according to which the singularly perturbed parabolic boundary-value problem (31.23), (31.24) has a turn point, called an extreme point, and this necessarily leads to qualitatively new dynamic effects. In this connection, we must emphasize that already under the conditions of Theorem 31.2 such an effect is observed, namely, the boundary-value problem (31.23), (31.24) has very many stable periodic solutions.

The results of this section were obtained in [30, 33].

References

1. "Autowave processes in systems with diffusion," *Collection of Scientific Articles*, ed. M. T. Grekhova. Gor'kii (1981).

2. A. A. Andronov, A. A. Witt, and S. E. Khaikin, *Theory of Oscillations*, Pergamon Press, Oxford, and Addison-Wesley, Reading, Mass. (1966).

3. D. V. Anosov, "On limit cycles of systems of differential equations with a small parameter multiplying the highest derivatives," *Amer. Math. Soc. Transl.*, Ser. 2, 33, 233–275 (1963).

4. V.I. Arnol'd, *Geometric Methods in the Theory of Ordinary Differential Equations*, Springer-Verlag, Berlin–New York (1982).

5. N. N. Bautin and E. A. Leontovich, *Methods and Rules for the Qualitative Study of Dynamical Systems on the Plane* [in Russian], Nauka, Moscow (1976).

6. N. N. Bogolyubov and Yu. A. Mitropol'skii, *Asymptotic Methods in the Theory of Nonlinear Oscillations*. Gordon and Breach, New York (1962).

7. A. B. Vasil'eva and V. F. Butuzov, *Asymptotic Expansions of Solutions of Singularly Perturbed Equations* [in Russian], Nauka, Moscow (1973).

8. A. B. Vasil'eva, S. A. Kashchenko, Yu. S. Kolesov, and N. Kh. Rozov, "Bifurcation of self-oscillations of nonlinear parabolic equations with small diffusion," *Math. USSR Sb.*, **58**, No. 2, 491–503 (1987).

9. N. K. Gavrilov, "On bifurcations of a periodic motion in the vicinity of intrinsic resonance 1:3," in: *Investigation of Stability and Oscillation Theory* [in Russian], Yaroslavl (1977), pp. 192–199.

10. A. V. Gaponov-Grekhov and M. I. Rabinovich, "Autostructures. Chaotic dynamics of ensembles," in: *Nonlinear Waves. Structures and Bifurcations* [in Russian], Nauka, Moscow (1987), pp. 7–44.

11. S. D. Glyzin, "Numerical justification of Landau–Kolesov hypothesis on the nature of turbulence," in: *Mathematical Models in Biology and Medicine* [in Russian], No. 3, Vilnius (1989), pp. 31–36.

12. S. D. Glyzin and A. Yu. Kolesov, "Steady-state solutions of Hutchinson's equation with small diffusion in the case of a square," in: *Qualitative Methods of Investigation of Operator Equations* [in Russian], Yaroslavl (1988), pp. 21–26.

13. N. Dunford and J. T. Schwartz, *Linear Operators. General Theory.* Interscience, New York–London (1958).

14. A. A. Zakharov, "Numerical verification of two hypotheses of Yu. S. Kolesov," in: *Nonlinear Oscillations in Ecology Problems* [in Russian], Yaroslavl (1985), pp. 44–54.

15. A. A. Zakharov and Yu. S. Kolesov, "Spatially nonhomogeneous solutions in the predator–prey problem," in: *Nonlinear Oscillations and Ecology* [in Russian], Yaroslavl (1984), pp. 3–15.

16. A. K. Zvonkin and M. A. Shubin, "Nonstandard analysis and singular perturbations," *Russian Math. Surveys,* **39** (1984).

17. G. P. Ivanitskii, V. I. Krinskii, and E. E. Sel'kov, *Mathematical Biophysics of a Cell* [in Russian], Nauka, Moscow (1978).

18. A. M. Il'in, A. S. Kalashnikov, and O. A. Oleinik, "Second-order linear equations of parabolic type," *Russian Math. Surveys,* **17**, No. 3, 1–143 (1962).

19. T. Kato, *Perturbation Theory for Linear Operators.* Springer-Verlag, Berlin–New York (1966).

20. S. A. Kashchenko and Yu. S. Kolesov, "Diffusion instability of a torus," *Soviet Math. Dokl.,* **31** (1985).

21. S. A. Kashchenko and Yu. S. Kolesov, " On the diffusion instability of a torus," in: *Mathematical and Computational Methods in Biology* [in Russian], Pushchino (1985), pp. 82–84.

References

22. A. Yu. Kolesov, "Migration effects in single-species biocenose," in: *Nonlinear Oscillations and Ecology* [in Russian], Yaroslavl (1984), pp. 34–61.

23. A. Yu. Kolesov, "On the domain principle in the problem of oscillations in the number of mammals," in: *Nonlinear Oscillations in Ecology Problems* [in Russian], Yaroslavl (1985), pp. 11–22.

24. A. Yu. Kolesov, "Stability of relaxation auto-oscillations in systems with diffusion," *Soviet Phys. Dokl.*, **32** (1987).

25. A. Yu. Kolesov, "Properties of a certain linear system connected with the stability of the relaxation cycle in media with diffusion," *Russian Math. Surveys*, **43**, No. 3, 209–210 (1988).

26. A. Yu. Kolesov, *Stability of a Homogeneous Cycle in Systems with Diffusion* [in Russian], Dissertation of a Candidate of Phys. Math. Sci., Steklov Inst. Math. Akad. Nauk SSSR, Moscow (1987).

27. A. Yu. Kolesov, "Description of the phase instability of a system of harmonic oscillations weakly connected by diffusion," *Dokl. Akad. Nauk SSSR*, **300**, No. 4, 831–835 (1988).

28. A. Yu. Kolesov, "Relaxation parabolic systems," *Soviet Math. Dokl.*, **39**, No. 3, 610–613 (1989).

29. A. Yu. Kolesov, "The structure of a neighborhood of a homogeneous cycle in a medium with diffusion," *Math. USSR Izv.*, **34**, No. 2. 355–372 (1990).

30. A. Yu. Kolesov, " Theoretical explanation of the phenomenon of diffusion bufferness," in: *Dynamics of Biological Populations* [in Russian], Gor'kii (1990), pp. 26–29.

31. A. Yu. Kolesov, "Duck-trajectories of relaxation systems connected with the violation of the conditions of normal switching," *Mat. Sb.*, **180**, No. 10, 1428–1438 (1989).

32. A. Yu. Kolesov, "On the instability of duck-cycles arising during the passage of an equilibrium of a multidimensional relaxation system through the disruption manifold," *Russian Math. Surveys*, **44**, No. 5, 203–205 (1989).

33. A. Yu. Kolesov, "Bifurcation of periodic solutions of singularly perturbed parabolic systems of first-degree structural instability," *Ukr. Mat. Zh.*, **42**, No. 8, 1037–1042 (1990).

34. A. Yu. Kolesov, "Specific relaxation cycles of systems of Lotka–Volterra type," *Math. USSR Izv.*, **38**, No. 3, 503–523 (1992).

35. A. Yu. Kolesov and Yu. S. Kolesov, "Constructing a normal form in the neighborhood of a cycle by means of the Krylov–Bogolyubov–Mitropol'skii method," *Differ. Uravn.*, **24**, No. 5, 891–894 (1988).

36. A. Yu. Kolesov, Yu. S. Kolesov, E. F. Mishchenko, and N. Kh. Rozov, "Asymptotic integration of a system in variations of a multidimensional relaxation cycle. I, II," *Differ. Uravn.*, **23**, No. 11, 1270–1276; No. 12, 1366–1374 (1987).

37. A. Yu. Kolesov, Yu. S. Kolesov, and N. Kh. Rozov, "Relaxation system in the neighborhood of a breakaway point: reduction to a regular case," *Russian Math. Surveys*, **43** (1988).

38. A. Yu. Kolesov, Yu. S. Kolesov, E. F. Mishchenko, and N. Kh. Rozov, "On the chaos phenomena in three-dimensional relaxation systems," *Mat. Zametki*, **46**, No. 2, 153–155 (1989).

39. A. Yu. Kolesov, Yu. S. Kolesov, E. F. Mishchenko, and N. Kh. Rozov, "Multidimensional relaxation oscillations in media with diffusion," *Soviet Math. Dokl.*, **38**, No. 2, 431–434 (1989).

40. A. Yu. Kolesov and E. F. Mishchenko, "Asymptotics of relaxation oscillations," *Math. USSR Sb.*, **65**, No. 1, 1–17 (1990).

41. A. Yu. Kolesov and E. F. Mishchenko, "A Pontryagin problem," *Differ. Equ.*, **25** (1989).

42. A. Yu. Kolesov and E. F. Mishchenko, "Existence and stability of a relaxation torus," *Russian Math. Surveys*, **44** (1989).

43. A. Yu. Kolesov and E. F. Mishchenko, "The Pontryagin phenomenon of delay and stable duck-cycles of multidimensional relaxation systems with one slow variable," *Mat. Sb.*, **181**, No. 5, 579–588 (1990).

44. Yu. S. Kolesov, "The Schauder principle and the stability of periodic solutions," *Dokl. Akad. Nauk SSSR*, **188**, No. 6, 1234–1236 (1969).

45. Yu. S. Kolesov, "Investigation of the stability of solutions of second-order parabolic equations in a critical case," *Izv. Akad. Nauk SSSR, Ser. Mat.*, **33**, No. 6, 1356–1372 (1969).

46. Yu. S. Kolesov, "Periodic solutions of relay systems with distributed parameters," *Math. USSR Sb.*, **12** (1970).

47. Yu. S. Kolesov, "On the stability of linear differential equations of parabolic type with almost periodic coefficients," *Tr. Mat. Obshch.*, **36**, 3–27 (1978).

48. Yu. S. Kolesov, "Mathematical models of ecology," in: *Investigations of Stability and Oscillation Theory* [in Russian], Yaroslavl (1979), pp. 3–40.

49. Yu. S. Kolesov, "Method of normal forms for systems with time lag," *Litovskii Mat. Sb.*, **20**, No. 4, 73–78 (1980).

50. Yu. S. Kolesov, "The parasite–host problem," in: *Dynamics of Biological Populations* [in Russian], Gor'kii (1984), pp. 16–29.

51. Yu. S. Kolesov, "On a certain bifurcation theorem in the theory of self-oscillations of distributed systems," *Differ. Uravn.*, **21**, No. 10, 1709–1713 (1985).

52. Yu. S. Kolesov, *The Problem of the Adequacy of Ecological Equations* [in Russian], Manuscript No. 1901-85 deposited at VINITI, Moscow (1985).

53. Yu. S. Kolesov, "Stability and bifurcation of travelling waves," in: *Nonlinear Oscillations in Problems of Ecology* [in Russian], Yaroslavl (1985), pp. 3–10.

54. Yu. S. Kolesov, "Method of quasinormal forms in the problem on steady-state solutions of parabolic systems with small diffusion," *Ukr. Mat. Zh.*, **39**, No. 1, 28–34 (1987).

55. Yu. S. Kolesov, "On the range of applicability of the method of quasinormal forms," in: *Mathematical Models in Biology and Medicine* [in Russian], No. 3, Vilnius (1989), pp. 99-111.

56. Yu. S. Kolesov, "The null asymptotics of a relaxation cycle of the mathematical model of Belousov's reaction," in: *Dynamics of Biological Populations* [in Russian], Gor'kii (1990), pp. 3–10.

57. Yu. S. Kolesov and A. Yu. Kolesov, "Biological and physical parabolic systems," in: *Mathematical Models in Biology and Medicine* [in Russian], No. 2, Vilnius (1987), pp. 5–23.

58. Yu. S. Kolesov and A. Yu. Kolesov, "Properties of biological and physical parabolic systems," in: *Mathematical Models In Rational Nature Management Problems* [in Russian]. *Theses of the Reports of the Regional 10th School-Seminar*, Rostov-on-Don (1986), pp. 106–107.

59. Yu. S. Kolesov, V. S. Kolesov, and I. I. Fedik, *Self-Oscillations in Systems with Distributed Parameters* [in Russian], Naukova Dumka, Kiev (1979).

60. Yu. S. Kolesov and V. V. Maiorov, "A new method of investigating the stability of solutions of linear differential equations with almost periodic coefficients close to constant," *Differ. Uravn.*, **10**, No. 10, 1778–1788 (1974).

61. Yu. S. Kolesov and V. V. Maiorov, "Substantiation of the algorithm for investigating the stability of solutions of linear almost-periodic equations with aftereffects whose coefficients are close to constant," *Vestn. Yaroslav. Univ.*, No. 10, 70–105 (1974).

62. Yu. S. Kolesov and V. V. Maiorov, "Asymptotically exact estimates of the solutions of nonhomogeneous differential equations with a small parameter bounded throughout the axis," *Differ. Uravn.*, **14**, No. 6, 1013–1017 (1978).

63. Yu. S. Kolesov and V. V. Maiorov, "On a certain estimate of $\|L^{-1}(\varepsilon)\|_{C(R^n) \to C^1(R^n)}$," in: *Investigations of the Stability of the Oscillation Theory* [in Russian], Yaroslavl (1978), pp. 130–131.

64. Yu. S. Kolesov and V. V. Maiorov, "Spatial and temporal self-organization in one-species biocenose," in: *Dynamics of Biological Populations* [in Russian], Gor'kii (1986), pp. 3–18.

65. Yu. S. Kolesov and D. I. Shvitra, *Self-Oscillations in Systems with Time Lag* [in Russian], Mocslas, Vilnius (1979).

66. A. N. Kulikov, "On smooth invariant manifolds of a half-group of nonlinear operators in a Banach space," in: *Investigations of the Stability and Oscillation Theory* [in Russian], Yaroslavl (1976), pp. 114–129.

67. A. N. Kulikov, "On the bifurcations of the origination of invariant tori," in: *Investigations of the Stability and Oscillation Theory* [in Russian], Yaroslavl (1983), pp. 112–117.

68. O. A. Ladyzhenskaya, V. A. Solonnikov, and N. N. Ural'tseva, *Linear and Quasilinear Equations of Parabolic Type*, Amer. Math. Soc., Providence, Rhode Island (1968).

69. Yu. A. Mitropol'skii, *The Averaging Method in Nonlinear Mechanics* [in Russian], Naukova Dumka, Kiev (1971).

70. Yu. A. Mitropol'skii and O. B. Lykova, *Integral Manifolds in Nonlinear Mechanics* [in Russian], Nauka, Moscow (1973).

71. E. F. Mishchenko, "Asymptotic theory of relaxation oscillations in second-order systems," *Mat. Sb.*, **44**, No. 4, 457–480 (1958).

72. E. F. Mishchenko, "Asymptotic methods in the theory of relaxation oscillations," *Uspekhi Mat. Nauk*, **14**, No. 6, 229–236 (1959).

73. E. F. Mishchenko and L. S. Pontryagin, "Almost discontinuous periodic solutions of differential equation systems," *Dokl. Akad. Nauk SSSR*, **102**, No. 5, 889–891 (1955).

74. E. F. Mishchenko and L. S. Pontryagin, "Proof of certain asymptotic formulas for solutions of differential equations containing small parameters," *Dokl. Akad. Nauk SSSR*, **120**, No. 5, 967–968 (1958).

75. E. F. Mishchenko and L. S. Pontryagin, "Derivation of some asymptotic estimates of solutions of differential equations with small parameters multiplying derivatives," *Izv. Akad. Nauk SSSR, Ser. Mat.*, **23**, No. 5, 643–660 (1959).

76. E. F. Mishchenko and N. Kh. Rozov, *Differential Equations with Small Parameter and Relaxation Oscillations*, Plenum Press, New York–London (1980).

77. L. S. Pontryagin, "Asymptotic properties of solutions of systems of differential equations with a small parameter multiplying leading derivatives," *Izv. Akad. Nauk SSSR, Ser. Mat.*, **21**, No. 5, 605–626 (1957).

78. L. S. Pontryagin and E. F. Mishchenko, "Some questions of the theory of differential equations with a small parameter," *Proc. Steklov Inst. Math.*, **169**, No. 4, 103–122 (1986).

79. L. S. Pontryagin and L. V. Podygin, " Approximate solution of a system of ordinary differential equations with a small parameter multiplying derivatives," *Sov. Mat. Dokl.*, **1** (1960).

80. L. S. Pontryagin and L. V. Podygin, "A periodic solution of a system of ordinary differential equation with a small parameter multiplying derivatives," *Sov. Mat. Dokl.*, **1** (1960).

81. N. Kh. Rozov, "Asymptotic calculation of almost discontinuous periodic solutions of second-order differential equation systems," *Dokl. Akad. Nauk SSSR*, **145**, No. 1, 38–40 (1962).

82. N. Kh. Rozov, "Asymptotic theory of relaxation oscillations in systems with one degree of freedom, II. Calculation of a limit-cycle period," *Vestn. Mosk. Univ., Ser. Mat., Mekh.*, No. 3, 56–65 (1964).

83. A. M. Samoilenko, *Elements of the Mathematical Theory of Multifrequency Oscillations* [in Russian], Nauka, Moscow (1987)

84. P. E. Sobolevskii, "Equations of a parabolic type in a Banach space," *Amer. Math. Soc. Transl.*, **49**, No. 2 (1965).

85. V. V. Strygin and V. A. Sobolev, *Division of Motions by the Method of Integral Manifolds* [in Russian], Nauka, Moscow (1988).

86. P. Hartman, *Ordinary Differential Equations*, John Wiley & Sons, New York–London–Sydney (1964).

87. A. N. Sharkovskii, Yu. L. Maistrenko, and E. Yu. Romanenko, *Difference Equations and Their Applications* [in Russian], Naukova Dumka, Kiev (1986).

88. L. P. Shil'nikov, "On a Poincaré–Birkhoff problem," *Mat. USSR Sb.*, **3**, (1967).

89. S. D. Eidel'man, *Parabolic Systems* [in Russian], Nauka, Moscow (1964).

90. C. Bonet, "Singular perturbation of relaxed periodic orbits," *J. Differ. Equ.*, **66**, No. 3, 301–339 (1987).

91. P. Frederickson, J. Kaplan, and J. Yorke, "The Lyapunov dimension of strange attractors," *J. Differ. Equ.*, **49**, 185–207 (1983).

92. J. Grasman, "Asymptotic methods for relaxation oscillations and applications," *Appl. Math. Sciences*, Vol. 63, Springer, New York, (1987).

93. G. E. Hutchinson, "Circular causal systems in ecology," *Ann. N.Y. Acad. Sci.*, **50**, 221–246 (1948).

94. A. Turing, "The chemical basis of morphogenesis," *Philos. Trans. Roy. Soc., London*, **237**, 37–72 (1952).

95. "Dynamics of Bifurcations," *Proc. of Conf. in Luminy, France (1990)*. ed. E. Benoit. *Lect. Notes Math.*, Springer, New York, **1493** (1991).